Bituminous Binders and Mixes

RILEM REPORT 17

BITUMINOUS BINDERS AND MIXES

State of the Art and Interlaboratory Tests on Mechanical Behaviour and Mix Design

Report of RILEM Technical Committee 152-PBM
Performance of Bituminous Materials

Edited by

L. Francken
Belgian Road Research Centre, Brussels, Belgium

CRC Press
Taylor & Francis Group
Boca Raton London New York

CRC Press is an imprint of the
Taylor & Francis Group, an **informa** business
A SPON PRESS BOOK

CRC Press
Taylor & Francis Group
6000 Broken Sound Parkway NW, Suite 300
Boca Raton, FL 33487-2742

First issued in paperback 2019

© 1998 RILEM
CRC Press is an imprint of Taylor & Francis Group, an Informa business

ISBN-13: 978-0-419-22870-7 (hbk)
ISBN-13: 978-0-367-86373-9 (pbk)

British Library Cataloguing in Publication Data
A catalogue record for this book is available
from the British Library

Publisher's Note
This book has been prepared from camera ready copy
provided by the individual contributors.

Visit the Taylor & Francis Web site at
http://www.taylorandfrancis.com

and the CRC Press Web site at
http://www.crcpress.com

Contents

Contributors - RILEM 152 PBM

C. de la Roche   Secrétariat RILEM 152 PBM, L.C.P.C.-Centre de Nantes, Section Matériaux de Chaussées (BP19), Bougenais, France

H. Di Benedetto   Département Génie Civil et Bâtiment, Ecole Nationale des Travaux Publics de l'Etat, Vaulx-en-Vélin Cedex, France

E. Eustacchio   Technische Universität Graz, Universitätsdozent für Mechanische Technologie der Baustoffe, Graz, Streymayrgasse, Austria

A. Fidato   Consortium for the Construction of the Rome–Naples High Speed Train line – Supervision Team 3.1, Rome, Italy

L. Francken   T.C. 152 PBM Chairman, Belgian Road Research Centre, Department of Research, Development and Application, Brussels, Belgium

U. Isacsson   Royal Institute of Technology, Department of Highway Engineering, Stockholm, Sweden

X. Lu   Royal Institute of Technology, Department of Highway Engineering, Stockholm, Sweden

M. Luminari   Autostrade S.P.A., SMA/LAB, Laboratorio Centrale, Fiano Romano (Rome), Italy

I. Pallos   Technical University of Budapest, Budapest, Hungary

M. N. Partl   Swiss Federal Laboratories, EMPA, Dübendorf, Switzerland

Ch. Raab   Swiss Federal Laboratories, EMPA, Dübendorf, Switzerland

J. Verstraeten   Belgian Road Research Centre, Brussels, Belgium

Preface

This report presents the result of activities carried out between the years 1988 and 1995 by two succeeding RILEM Committees: TC 101 BAT " Bitumens and Asphalt Testing" and TC 152 PBM: "Performance of Bituminous Materials".

The general guiding line and philosophy of this work is presented in the introduction. Elements for a new mix design methodology are proposed as a basis for this study. The aim of this RILEM contribution is to evaluate the possibilities of implementing rational concepts and testing procedures for the design and manufacture of bituminous materials in order to cope with the present and future conditions of use in pavement construction. The final purpose of the committee is to recommend significant test procedures for binder evaluation, mix design and performance assessment of bituminous materials.

The three main topics considered in this report are distributed into the following main parts :
- Part 1 : Binder testing
- Part 2 : Mix design
- Part 3 : Mechanical testing of mixtures.

These subjects were not covered under all their possible aspects, a choice was rather made to address some selected issues of great interest at present times such as :
- Characterisation of polymer modified binders.
- Binder rheology.
- Performance based mix design.
- Measurement of complex modulus of coated mixes.
- Measurement and interpretation of fatigue tests.

Some important topics such as, for example, ageing, durability, bitumen emulsions and their use in cold mixtures are not considered at this moment.

Permanent deformation is not considered either although this is a critical problem for bituminous materials.

The present report is a collection of complementary contributions set up by different members or task groups essentially following two alternative approaches :

1) A literature review of existing practices and methods was made for the case of polymer modified binders (Chapter 1), Mixture design (Chapter 3) and mechanical properties of the mixtures (Chapter 5).
2) Interlaboratory tests organised and coordinated by the committee for each of the three main topic: Binders (Chapter 2), mix design (Chapter 4) and testing of asphalt materials (Chapter 6).

The idea of organising large international testing programs on significant testing methods for asphalt pavement materials was raised in 1987 by the RILEM Technical

Committee 101-BAT "Bitumen and Asphalt Testing". Its implementation started after the fourth RILEM International Conference "Mechanical Tests for Bituminous Mixes - Characterisation, Design and Control" organised by this same committee in 1990 at Budapest.

The purposes of the interlaboratory tests were :

1) to promote and develop mix design methodologies and associated significant measuring methods.
2) to compare different test procedures and testing equipments for the measurement of mechanical characteristics and performances of bituminous materials.
3) to release of recommendations for test methods having a high degree of significance and reliability in the determination of fundamental characteristics.

It was considered of first importance to assess whether different equipments and procedures can be used in order to accurately determine fundamental properties of road materials provided that certain recommendations and rules for good practice are followed. Otherwise it might be concluded that a strict standardisation of equipment and testing conditions is compulsory for the purpose of material specification or evaluation in the frame of mix design.

Initially based on measurements and mix design procedures on full mixes, this experimentation was in a later stage (starting in 1994) extended to the case of plain bituminous binders. After much efforts involving many people and laboratories working on a voluntary basis we may now present the full conclusions of three complementary sets of results on respectively binders, mix design and dynamic loading tests.

In a period where deep changes are expected in experimental procedures and specifications for asphalt materials, this work will be a source of information and a reference, not only for research scientists, but also for road engineers and practitioners involved in the design and manufacture of asphalt materials for road construction and maintenance.

The technical and scientific contents of this report are the result of a large effort in collecting information and performing tests. It is impossible here to mention the numerous people who were involved in the RILEM interlaboratory programs, the amount of work they achieved is impressive and had not been possible without their kind and free participation. The discussion and writing of this report was made in close cooperation with the members of the RILEM 152 PBM Technical Committee.

I would like to express my gratitude to all those who contributed to this work, especially to Mrs A. Thomas of the BRRC for taking care of the layout of this report and Mr Jorge Sousa for his valuable input during the final review of the text.

Dr. Sc. Louis Francken
Chairman of RILEM TC 152-PBM
Brussels , August 1997

Introduction

Background of RILEM interlaboratory tests. Basic elements of a testing methodology for bituminous pavement materials and significant features for testing

by M. N. Partl and L. Francken

1. Background of RILEM interlaboratory tests

Due to the physical and structural variety of bituminous pavement materials and the wide range of application in terms of loading and climatic conditions, a great number of different tests methods is now applied worldwide, either based on an empirical, technological, physical or a performance background. Consequently, the selection of suitable testing methods for the design of bituminous pavement materials on performance based requirements is a much more complicated subject than might appear at first sight. However, not only the test procedures are of importance but rather the full methodology beginning with the characterization of the base components, including the mixture design and performance prediction testing procedures and ending with general technical requirements based on full-scale validation and performance testing in the field. Hence, a methodology using a set of homogeneous and complementary operations has to be established leading to clear decisions based on objective criteria and technical assessments relevant to the problem to be solved.

So far, the only methodology which has found a large international audience is the Marshall methodology. Born, in the 40^{ies}, this method has put in the shade several other promising approaches. Its great power and strong points are :

- *Completeness :* It covers all the aspects from manufacture of samples up to follow up on the job sites.
- *Simplicity :* The equipment is rather simple and easy to be handled by personnel with adequate training.
- *Fast :* Results are available in comparatively short time.
- *Low cost :* Due to its simplicity.
- *Coherence :* The different steps are accurately described and supported by criteria.
- *Experience :* Great experience and large data sets are available worldwide.

Bituminous Binders and Mixes, edited by L. Francken. RILEM Report 17. Published in 1998 by E & FN Spon, 11 New Fetter Lane, London EC4P 4EE, UK. ISBN 0 419 22870 5

What makes Marshall methodology more open to criticism are the following points :

- *Unrealistic:* It is unable to reproduce the actual conditions of loading and climate.
- *Not performance based:* There is no methodical link to pavement performance. Performance based test methods are typically conducted under similar conditions as in the road and allow to determine physical properties which are directly related to the performance. This is not the case with the Marshall tests, e.g. there is no direct connection between Marshall stability and fatigue.
- *Inadequate samples:* The sample manufacture procedure associated does not reflect adequately the material as it is on the road.
- *Limited:* Its field of application is restricted to certain mixture compositions and aggregate sizes.
- *Empirical:* It is unable to produce the mechanical characteristics which are essential for the rational evaluation of the pavement structure.

2.   Basic elements of a testing methodology for bituminous pavement materials

In spite of the many useful services provided by the Marshall methodology, the time has come to think of a more accurate and reliable methodology without loosing the advantages of the Marshall methodology. One considerable effort in this direction has been made by the Strategic Highway Research Program (SHRP) ending up recently with a new methodology called SUPERPAVE™ (SUPerior PERforming asphalt PAVEments) which coincides to a large extend with the basic testing methodology and philosophy of RILEM 152 PBM as given in figure 1. This basic methodology is the frame of this state of the art report on bituminous binders and mixtures and consists of three mayor testing parts:

- *Characterization testing* which is referred to within this report in "Part One Bituminous Binders".
- *Mixture design and performance prediction testing* which is discussed in "Part Two Mixture Design" and "Part Three Mechanical Testing of Mixtures".
- *Validation and performance testing* which is dealt with in the Chapter "Final Conclusions and Future Prospects".

The basic testing methodology for bituminous mixture design as illustrated in figure 1 starts with five preparatory steps :

1. *Characterization* of the base components (binder, additives and different aggregate fractions) and of the mixture composition (grading curve, proportions of the constituents including the binder).
    Characterization testing is a type testing procedure which has to be done for identification purposes (finger-prints) and is needed for quality control during production and for development of new products. Hence, it is not necessarily connected to specific in field problems or requirements but it is the basis for a rough general screening of the technical possibilities during definition of the design parameters. To compare different products on a common basis, generally accepted test methods have to be used.
2. *Definition of the design parameters* with respect to requirements (loading, climate, life cycle etc.) and pavement structure (including the position and function of the material in this structure) for a specific design job.
3. *Selection of the type of mixture*, such as asphalt concrete, stone mastic asphalt, open graded asphalt, overlays, etc. which is expected to have the best chance to meet the requirements formulated under step 2.
4. *Selection of test methods as well as type and degree of compaction* suited to assess performance with respect to fatigue, permanent deformation, cracking, etc..
5. *Composition of the mixture for testing*, based either on experience, theoretical considerations or on the results from previous mixture design iterations.

The next three steps concentrate on mixture design and performance prediction testing and consist of:

6. Manufacture of the *samples.*
7. Volumetric and mechanical *testing* including determination of sample composition (binder content, air voids) and testing of modulus, fatigue, permanent deformation and thermal cracking, etc..
8. *Data* processing, collection and analysis with respect to :

    -   volumetric characteristics,
    -   mechanical characteristics,
    -   statistical assessment.

Step 8 will lead to a set of data which can be used in one or both of the following two actions :

9.1 Use in *prediction* methods and models for :

    -   pavement and structural design,
    -   long term pavement performance (LTPP).

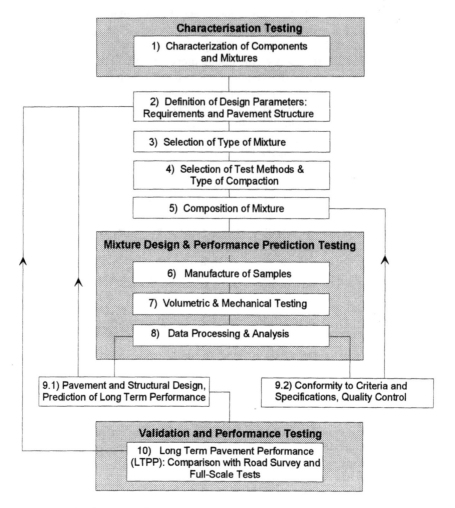

Fig. 1. Basic elements of a methodology for bituminous mixtures (shaded areas are parts where testing has to be done).

9.2 *Conformity to criteria* and specifications as basis for quality control.

The use in prediction models (step 9.1) may finally lead to :

10. *Validation testing* and comparison to long term pavement performance in the field including :

    - results of road survey activities,
    - large scale or full-scale testing.

Steps 9 or 10 may lead to negative conclusions with respect to the material and thus the procedure needs to be repeated again with improvements concerning the selection of the base components and/or the mixture composition, eventually also affecting the selection of type of mixture and the test methods. It can also lead to a review of design parameters and pavement structure, e.g. with respect to loading or life cycle.

3.   Significant features for testing

Main Damages and Distress Mechanisms

The main distress mechanisms to be taken into consideration are those leading to unacceptable levels of :

- safety,
- service life,
- user comfort,
- noise,
- maintenance costs and/or user costs.

The initiation and rate of progress of the main *types of distress* :

- fatigue cracking,
- permanent deformation,
- thermal cracking,
- reflective cracking,
- surface defects.

depend on the following *influence factors* :

- Internal factors :

  - mechanical properties (e.g. material stiffness),
  - aging,
  - material characteristic.

- Pavement structure :

  - flexible,
  - semi-rigid,
  - rigid.

  (For bituminous materials the first two pavement structures are relevant only).

- Manufacturing procedure :

    - mixing and laying temperature,
    - compaction energy and compaction process,
    - climatic conditions.

- External factors :

    - climatic factors (temperature - moisture),
    - loading condition, i.e. axle loads, tire type (single/double tire)
    - traffic intensity, sequence and velocity (stationary - dynamic).

Fatigue cracking appears in the locations of the road structure where the largest tensile stresses are generated under combined action of temperature and traffic loads. It is a long term effect and therefore promoted by aging. The development of fatigue cracking is gradual and can be reduced by :

- the design of the mixture,
- the design of the road structure.

Permanent deformations, such as rutting, are generated by compressive and/or shear stresses. They are enhanced by high temperatures and low traffic speeds. Generally starting in the early stage of pavement life, their development in most cases is gradual but may become critical under high loads and/or summer conditions. Aging reduces the risk of permanent deformation. The affected pavement zone is generally close to the surface and it proves almost impossible to solve the problem by the structural design only. Hence, the main tool to avoid permanent deformations is proper design and manufacturing of the mixture and the improvement of the binder properties as well as the aggregate skeleton.

As compared to fatigue and permanent deformation distress *thermal cracking and reflective cracking* are more specific types of defect as their occurrence and mode of development are characteristic for particular pavement structures and climatic conditions. As for fatigue cracking both effects are promoted by aging.

Climatic factors have a major influence on distress mechanisms. Hence, one of the main requirement to be satisfied by a test procedure is its relevance to temperature conditions at the different locations of the bituminous pavement structures. These temperature conditions are determined by :

- the environmental temperature and it variations (daily and seasonally),
- the solar radiation and its variations,
- the location of the material in the pavement structure.

It has been established that during *summer periods* pavement temperatures can reach values higher than 50°C close to the surface. These conditions are most critical for the problem of permanent deformation. For this reason tests aiming at the assessment of rutting resistance and mechanical stability and stiffness should be carried out in a range of "high temperatures".

In *rainy seasons*, the soil moisture content induces a loss of bearing capacity of the subgrade and hence can give rise to :

- fatigue cracking in flexible road structures,
- reflective cracking in semi-rigid pavements and bituminous overlays on old cement concrete pavements,
- moisture damage and loss of binder-aggregate adhesion.

These conditions are prevailing almost any time of the year and more especially in the intermediate seasons (spring or fall). The recommended temperature range for testing should thus be close to the mean annual temperature.

During the *winter periods* and more specifically in northern countries (for mean monthly air temperatures lower than 0°C) the combination of low temperatures with cyclic variations can promote thermal cracking. Testing procedures for the assessment of this phenomenon should thus be kept lower than 0°C.

Relevant material characteristics and performance laws - Definitions

Material characteristics and performance laws may be used for different purposes :

- To evaluate the quality of the material and its conformity to the requirements.
- As input data for structural design and long term pavement performance prediction based on suitable models.
- For comparisons and ranking of materials.

Material characteristics are required as input data for computing the stress distribution in multilayer models or finite element methods. For the traditional material types and mixture compositions it is possible, as shown in chapter 5, to get estimates of these characteristics by using predictive methods based on correlations between these properties and the mixture composition. But in the case of new types of materials these characteristics should be determined experimentally in any case. In order to facilitate the reading of this report (especially chapters 2, 5 and 6) some of the most important material characteristics are described in the following :

The complex modulus E^*
Building materials such as steel or cement concrete can generally be described by means of a Young's modulus (ratio of an applied stress to the corresponding strain) which is independent on temperature under normal service conditions and nearly independent on loading time. Bituminous materials on the contrary are displaying wide variations with temperature and loading times in all their mechanical properties.

Hence, to characterize the behaviour of such materials (bituminous binders and their combinations with other materials) it is necessary to perform tests over wide ranges of temperatures and frequencies. The most convenient method generally consists in applying sinusoidal loads on test specimens and to retain the measured stress/strain ratio as characteristic value of their mechanical stiffness at the different combinations of temperatures and frequencies covered in the test.

If a sinusoidal stress of frequency Fr and amplitude σ_0 is applied :

$$Stress = \sigma_0 . \sin(2\pi . Fr . t)$$

The resulting strain will also be sinusoidal but out of phase by an angle called the phase shift φ :

$$Strain = \varepsilon_0 (T, Fr) . \sin(2p . Fr . t\text{-}j)$$

The corresponding stress/strain ratio can thus be represented by two components : the stiffness modulus $|E^*|$ (or $|G^*|$ in case of shear) is generally expressed in MPa and the phase shift φ (or δ).

This combination of parameters, called the complex modulus can be split into two components of a complex quantity:

- The real component : $E_1 = |E^*| . \cos(\varphi)$ also called storage modulus.

- The imaginary component : $E_2 = |E^*| . \sin(\varphi)$ also called loss modulus.

When shear stresses are applied the complex shear modulus G^* is determined in the same way as the E^* modulus through formulas (1) and (2). For linear and homogeneous materials the relationship between G^* and E^* is given by :

$$E^* = 2 \ G^* \ (1 + \gamma^*) \ (\gamma = \text{Poisson 's ratio})$$

Note that it is customary to use the symbol δ to define the phase shift of the pure binder and φ for the case of a mixture. Both of them will be used in chapters 2 and 6 of this report.

Basic properties of the complex modulus
The complete characterization of the complex modulus implies the determination of two of its components ($|E^*|$ and φ) or (E_1 and E_2) over a wide range of temperatures and frequencies.

In this report we will use the following different representations currently used to display the relationships between these components and their variations with the physical conditions :

- Black diagram ($|E^*|$ versus φ) : allows the derivation of a single curve which is independent from temperature and frequency.

The purely elastic value E∞ of the complex modulus (also called the glassy modulus reached at very low temperatures or high frequencies) can easily be derived from this type of diagram by extrapolating to 0 values of the phase angle φ.

- The master curve : $|E^*|$ plotted versus a reduced parameter X, where $(X = \log (\alpha_T.Fr))$ gives the temperature frequency dependence of the stiffness modulus for a fixed reference temperature. Such a master curve can be constructed by applying the temperature-frequency equivalency principle.

- Temperature - frequency equivalency principle : The values of the complex modulus are generally given for a combination of temperatures and frequencies. These results can be presented graphically under the form of isotherms of the modulus. Such a master curve can be built up from the modulus values obtained at different temperatures and frequencies by shifting the results along the frequency scale by a factor $\log(\alpha_T)$. One single independent variable $X = \log (\alpha_T.Fr)$ can then be used instead of temperature and frequency.

The advantage of this procedure is that once the master curve is established, it is possible to derive an interpolated value of the stiffness modulus for any combination of temperature T or frequency Fr inside the range covered by the measurements. This gives in addition the possibility to compare the results obtained by two laboratories at different sets of conditions , which avoids the recourse to standardize temperature and frequency conditions.

Two different formulas are generally proposed to compute the shifting factors (in decimal log) :

1) the WLF formula :

$$\log(\alpha_T) = \frac{-C_1 \cdot (T - T_s)}{(T - T_s) + C_2}$$

in which C_1 and C_2 are parameters.

2) the Arrhenius equation :

$$l_n (\alpha_T) = 0.4343 \frac{\delta H}{R} \cdot \left(\frac{1}{T} - \frac{1}{T_s}\right)$$

in which   δH is an apparent activation energy characterizing the material (of the order of 210 KJ/mole).
R is the universal gas constant (8.28 J/mole/°K).
T and Ts are temperatures expressed in K.

This master curve is then a function of a reduced parameter :

$$X = \log (\alpha_T \cdot Fr)$$

The Poisson's Ratio ν

This dimensionless measure is necessary to evaluate the volume changes of the material. Incompressible materials, i.e. materials with no volume change have a Poisson's ratio of ν = 0.5.. The dependency of the modulus and Poisson's ratio on temperature and loading frequency can be expressed by a master curve, applying the temperature - frequency equivalency principle.

To predict long term behavior of a road, *performance laws* have to be used together with the mechanical characteristics mentioned above.

The fatigue law defines the relationship between applied strain or stress and the number of loading cycles at failure. It may be determined either from stress or strain controlled repeated loading tests. The fatigue law is the main information used in most of the structural pavement design methods and is strongly dependent on mixture composition and characteristics of the binder. It depends on the loading history and may be influenced by healing effects as a result of temperature variations and/or rest periods.

The permanent deformation law describes the time dependence of permanent deformation and its relationship with stress conditions. It is generally more complex than the fatigue law. Today, different expressions for creep compliance curves are under discussion. The permanent deformation law can be determined by static creep tests or by dynamic repeated loading tests. However, depending on the type of test, different results are obtained for similar testing conditions (temperatures and magnitude of stress), e.g.

- repeated loading tests will provide higher deformations than static ones,
- triaxial tests may lead to lower deformations than unconfined tests.

Permanent deformation tests are particularly relevant to the high temperature range (e.g. greater than 30°C) where most bituminous pavement materials may show plastic deformation and are no longer linear or elastic. Test methods should aim at the determination of input parameters for the rutting prediction methods. They should also be correlated to rutting simulation tests in the laboratory, such as wheel tracking tests (wheels running back and forth on a sample).

As far as the *range of validity* of the performance laws and properties is concerned, it has to be considered that most models are suited for the particular case where the material is supposed to be linear, elastic, homogeneous and isotropic. These conditions are generally not fulfilled for most of the materials used in practice. Hence, particular attention should be devoted to this when testing and the range of validity of these four assumptions should be determined at least in each case.

PART ONE

BINDER TESTING

1

Testing and appraisal of polymer modified road bitumens

by U. Isacsson and X. Lu

1.1　Introduction

The increase in road traffic during the last two decades in combination with an insufficient degree of maintenance has caused an accelerated deterioration of road structures in many countries. To counteract this process, several types of measures may be effective, e.g. improved design, more optimal use of materials and more effective construction methods. Properties of materials in all layers of the road structure are of great importance for the life of the road. Several factors influence the performance of flexible courses, e.g. the properties of the components (binder, aggregate and additive) as well as the proportion of these components in the mix. Over the years, many different types of materials have been proposed as additives in bituminous mixes. Table 1.1 shows a compilation of groups of such additives, of which only polymers will be discussed in this chapter. Bitumen can also be modified by different types of chemical reactions. Examples of such modifications are also given in Table 1.1.

The purpose of using a special additive in an asphalt pavement is to achieve better road performance in one way or another. For most of the groups listed in Table 1.1, a large number of products are commercially available. The evidence of the effect of a given technical product is not often obvious. There are several reasons for this situation. Laboratory test methods do not always simulate behaviour on the road and field tests are generally too time-consuming and expensive. In this connection, it is also worth mentioning that the mechanisms of action for most of the additives are insufficiently understood.

In recent years, an increased interest in bitumen research has been observed in many European countries. Evidence of this can be seen in the number of papers published at the five Eurobitume conferences held since 1978. There has been an obvious increase in the number of papers published. The corresponding increase in the number of papers describing modified binders or mixes is also evident. Comparison of different types of modifiers shows that interest in polymers as additives is increasing at the expense of other types of modifiers. At the Eurobitume conferences in Madrid 1989 and Stockholm 1993, as many as 87 % of all papers describing modified binders or mixes dealt with polymers (Isacsson and Lu, 1995).

Bituminous Binders and Mixes, edited by L. Francken. RILEM Report 17. Published in 1998 by E & FN Spon, 11 New Fetter Lane, London EC4P 4EE, UK.　ISBN 0 419 22870 5

Table 1. 1 Types of bitumen modification

Types of Modification	Examples
I. Additive (non-polymer) Modification	
1. Fillers	Lime, carbon black, fly ash
2. Anti-stripping additives	Organic amines and amides
3. Extenders	Lignin, sulphur
4. Anti-oxidants	Zinc anti-oxidants, lead anti-oxidants, phenolics, amines
5. Organo-metal compounds	Organo-manganese compounds, organo-cobalt compounds
6. Others	Shale oil, gilsonite, silicone, inorganic fibres
II. Polymer Modification	
1. Plastics	
a. Thermoplastics	Polyethylene (PE), Polypropylene (PP), Polyvinyl chloride (PVC), Polystyrene (PS) Ethylene vinyl acetate (EVA)
b. Thermosets	Epoxy resins
2. Elastomers	
a. Natural rubbers	
b. Synthetic elastomers	Styrene-butadiene copolymer (SBR) Styrene-butadiene-styrene copolymer (SBS) Ethylene-propylene-diene terpolymer (EPDM) Isobutene-isoprene copolymer (IIR)
3. Reclaimed rubbers	
4. Fibres*	Polyester fibres, Polypropylene fibres
III. Chemical Reaction Modification	
	Addition reaction (bitumen + monomer) Vulcanisation (bitumen + sulphur) Nitration reaction (bitumen + nitric acid)

*cf. Table 1.2.

Despite increased interest, modified bitumens will probably never be able to replace conventional bitumens to a great extent due to economic reasons. Reliable statistics showing the present volumes of modified bituminous binders are difficult to obtain. In most countries, only a small proportion of the total bitumen volume used is modified in one way or another, whereas in some countries, e.g. France, Belgium and Hungary, the corresponding figure is higher (10-20 %). The main reason for the low fraction of modified bitumen is most probably the price, which may exceed twice that of conventional bitumen.

In the future, the main use of modified bitumens will probably be for special applications such as :

- surfaces at traffic lights, cross-roads and busy city roads,
- bus stops and parking areas for truck,
- bridges,
- tunnels,
- very busy motorways,
- airports,
- renovation of concrete roads and certain other roads,
- special surfaces, such as thin layers and split-mastic asphalt and slurry seals,
- surface dressing for roads with high traffic,
- stress absorbing membrane interlayers.

Admixture of polymer to binders intended for road application is not a new phenomenon. As early as 1823, an English cork manufacturer was granted a patent for a binder containing natural rubber (Hancock, 1823). After the Second World War, synthetic polymer began to compete with natural rubber as an additive in road bitumen. Based on the number of publications on polymer modified bitumen (PMB) and asphalt (PMA), Zenke (1985) showed that interest in synthetic polymers soon exceeded that in natural polymers. By 1982, as many as 1,100 papers on PMB and PMA had been published (Zenke, 1985). As has already been indicated, there is still growing research interest in this area.

The purpose of this chapter is to provide a compilation of test methods and specifications of polymer modified bitumens. The intention is also to discuss the present situation in this area and the need for further development.

1.2 Polymers used in road applications

Polymer modified bitumen, often abbreviated to polymer-bitumen, is a "modified binder obtained by incorporation of thermoplastic materials, synthetic thermohardening resins, powdered rubber or elastomers in bitumen" (Fritz et al., 1992).

A polymer is a very large molecule comprising hundreds or thousands of atoms formed by successive linking of one or two, occasionally more, types of small molecules into chain or network structures (Hall, 1985). The process of polymerisation may be achieved by one of the two mechanisms (Billmeyer, 1984; Hall, 1985), namely addition and condensation polymerisation, which can be written as follows:

Addition polymerisation : $nM \rightarrow M_n$

$$n/2\ C_2H_4\ \rightarrow\ (-CH_2-)_n$$
Ethylene   Polyethylene (PE)

Condensation polymerisation : $nMN + nPQ \rightarrow (MP)_n + nNQ$

where MN and PQ are monomers, which may be same or different. One example of this type of polymerisation is the formation of polyesters using a diol and a diacid as follows:

$$n\ HO\text{-}CH_2\text{-}CH_2\text{-}OH + nHOOC\text{-}(CH_2)_4\text{-}COOH \rightarrow$$
diol   diacid
$$HO(\text{-}CH_2CH_2\text{-}OCO\text{-}(CH_2)_4COO^-)_nH + (2n\text{-}1)H_2O$$

In addition reactions, the polymer is the only product of the reaction and this type of polymerisation almost invariably occurs through a chain reaction mechanism, including free-radical, and ionic (anionic and cationic) polymerisation reactions (Billmeyer, 1984; Hall, 1985). The term "condensation" is used generally in organic chemistry to refer to reactions in which molecules combine together through the elimination of a small molecule such as water. This means that the polymer itself is not the only product in condensation polymerisation (Billmeyer, 1984; Hall, 1985).

Despite the large number of polymeric products, there are relatively few types which are suitable for bitumen modification. When used as bitumen modifiers, selected polymers, alone or blended with bitumen, should:

- be compatible with bitumen,
- resist degradation at asphalt mixing temperatures,
- improve the temperature susceptibility of the bitumen,
- be capable of being processed by conventional mixing and laying equipment,
- give rise to a coating viscosity at normal application temperatures,
- maintain their premium properties during storage, application and in service when blended with bitumen,
- be cost-effective.

To achieve the goal of improving bitumen properties, a selected polymer should create a secondary network or new balance system within bitumens by molecular interactions or by reacting chemically with the binder. The formation of a functional modified bitumen system is based on the fine dispersion of polymer in bitumen for which the chemical composition of bitumens is important. The degree of modification depends on the polymer property, polymer content and nature of the bitumen.

Polymers can be classified into four broad categories, namely plastics, elastomers, fibres and additives/coatings. Plastics can in turn be subdivided into thermoplastics and thermosets (or thermosetting resins) and elastomers into natural and synthetic rubber as indicated in Table 1.1. In modification of bitumen, plastics, elastomers and fibres have been used, some of which are given in Table 1.1.

Thermoplastics are characterised by softening on heating and hardening on cooling. Some of the principal thermoplastics, such as polyethylene (PE), polypropylene (PP), polyvinyl chloride (PVC), polystyrene (PS) and ethylene vinyl acetate (EVA), have been examined in bitumen modification. These materials, when mixed with bitumen, associate at ambient temperatures and increase the viscosity and stiffness of bitumen at normal service temperatures. Unfortunately, most of them do not significantly increase the elasticity of bitumen and, on heating, they tend to separate, giving rise to coarse dispersions on cooling (Whiteoak, 1991). However, EVA, as statistical copolymers of ethylene and vinyl acetate, are widely used in bitumen modification. This type of polymer is easily dispersed in and has good compatibility with generally available bitumens, as well as being thermally stable at normal mixing and handling temperatures (Whiteoak, 1991).

Thermoset materials (thermosetting resins) are produced by the direct formation of network polymers from monomers, or by crosslinking linear prepolymers (Hall, 1985). The products obtained are insoluble, infusible and do not flow. Heating causes irreversible transformation and reticulation as a result of chemical reactions. Important thermosets include alkyds, amino and phenolic resins, epoxies, unsaturated polyesters and polyurethanes. Thermosets have a strong three-dimensional network structure and cannot be returned to a fluid condition by heating. In fact, these polymers could be used directly in road applications, but the cost would be prohibitive (Downes, 1986).

Elastomers (rubbers) such as natural rubber (NR), poly-butadiene (BR), polyisoprene (IR), isobutene-isoprene copolymer (IIR), polychloroprene (CR), styrene-butadiene copolymer (SBR), and styrene-butadiene-styrene block copolymer (SBS) have been experimented with for a long time in modifying bitumens. The polymers are elastomeric, i.e. their elasticity enables them to totally or partly recover their initial dimension after being subjected to stress or an increase in temperature. The apparent effect of these polymers is an increase in bitumen viscosity rather than elastomeric strengthening (Dinnen, 1985). In some instances, they are used in a vulcanised (cross-linked) state, e.g. reclaimed tyre crumb. However, this type of material is difficult to disperse in bitumen due to the tight molecular structure (Whiteoak, 1991). Among elastomer materials, styrene-butadiene-styrene block copolymers (SBS) have attracted most attention for bitumen modification. These copolymers combine both elastic and thermoplastic properties, and therefore are commonly called thermoplastic rubbers (TR). SBS copolymers consist of styrene-butadiene-styrene triblock chains, and have a two-phase morphology, showing spherical domains formed by the polystyrene blocks within a matrix of polybutadiene. These hard block domains act as physical crosslinks in forming the elastomeric network (Valkering et al., 1992). In addition, they behave as a well-dispersed, fine-particle reinforcing filler in promoting high tensile strength and modulus. The effectiveness of these cross-links diminishes rapidly above the glass transition temperature of polystyrene (about 100°C). On cooling, the polystyrene domains will reform, and the strength and elasticity will be restored (Hall, 1985). The thermoplastic nature of SBS copolymers at elevated temperatures and their ability to provide a continuous network on cooling are the reasons for their great attractiveness as bitumen modifiers. When an SBS copolymer is added to hot bitumen, it absorbs maltenes from the bitumen and swells by up to nine times its initial volume (Valkering et al., 1992). At suitable SBS concentrations (commonly 5 to 6% by mass of SBS) a continuous polymer network can be formed throughout the bitumen. This, in turn, significantly modifies bitumen properties such as viscoelasticity and temperature susceptibility (Valkering et al., 1992).

The defining characteristic of a fibre is its threadlike form, without restriction on composition. Fibres can be separated into natural and synthetic subcategories. Some major fibres are summarised in Table 1.2. In fact, only parts of fibres belong to the scope of this chapter, these being hydrogen-carbon or heterochain polymers, both natural and synthetic. The basic materials of a fibre are often similar to those of earlier mentioned polymers such as polypropylene and polyurethane (Billmeyer, 1984; Hall, 1985). Fibres have traditionally been used as reinforcement and stiffening additives for roofing and industrial waterproofing mastics (Rowlett et al., 1990). Recently, they have been used in experiments in the modification of bitumen and asphalt (Cameleyre et al., 1989; Gordillo et al., 1989; Peltonen, 1989).

Table 1. 2 Some major fibre materials.

Natural Fibres	Synthetic Fibres	
	Organic Type	Inorganic Type
Cellulose	Polypropylene	Carbon
Wool	Polyester	Glass
Asbestos (silicate)	Polyurethane	Steel
	Aromatic polyamides	

1.3   Test methods

Over the years, a large number of test methods have been proposed for characterising polymer modified bitumen (PMB). In most cases, methods standardised for parent bitumen according to ASTM, DIN and other standardisation authorities have been used. Often tests on PMB have been performed according to modified versions of these standard methods with test conditions changed in one way or another (Isacsson and Lu, 1995). Unfortunately, a valid description of improved PMB properties, particularly their tensile and elastic properties, is not obtained using these traditional methods. Therefore, several new test procedures have been developed, some of which are also used in PMB specifications.

The intention of this part of chapter is to describe as fully as possible methods used for testing PMB. As can be seen below, a very large number of test procedures have been described in the literature. In spite of this, the list of methods is probably not complete. Hopefully, no essential test methods have been omitted. In this connection, it is worth mentioning that modifications of a given test procedure are often proposed. However, it is quite possible that not all the modifications proposed in the literature are covered in this chapter.

In this compilation, the different test methods have been divided into two groups, standardised and non-standardised test methods, respectively. The different groups are further divided into subgroups as shown in Tables 1.3 and 1.4. A short description of each non-standardised test method is given in Appendix 1. A corresponding description of the standardised test methods (Table 1.3) is presented in reference (Isacsson and Lu, 1995). It should be emphasised that test methods intended for mixes are not included in this chapter.

Table 1. 3 Standardised test methods.

Type of Test	Standard Conditions	Reference
1. Rheological Tests		
Penetration (dmm)	25 °C, 5 s, 100 g	ASTM D5
Softening Point, R&B (°C)		ASTM D36
Fraass Breaking Point (°C)		IP 80
Kinematic Viscosity (mm²/s)	135°C	ASTM D2170
Dynamic Viscosity (Pas)	60°C	ASTM D2171
Viscosity Using Cone and Plate Viscometer		ASTM D205
Viscosity Using Brookfield Viscometer		ASTM D402
Ductility (cm)	25°C, 5 cm/min	ASTM D113
Tensile test	23°C, 50 cm/min	ASTM D412
Float test	5°C	ASTM D139
2. Ageing Tests		
Thin Film Oven Test (TFOT, %)	163°C, 5 h	ASTM D1754
Rolling Thin Film Oven Test(RTFOT,%)	163°C, 75 min	ASTM D2872
Rotating Flask	165°C, 2.5 h	DIN 52016
3. Solubility Test		ASTM D2042
4. Tests on Flash Point		
Cleveland Open Cup Test (COC Test)		ASTM D92
Pensky-Martens Closed Cup Test (PMCC Test)		ASTM D93
5. Chemical Composition Analysis		
Acc. to ASTM D 4124		
Acc. to IP 143		

Table 1.4 Non-standardised test methods (description in Appendix 1).

---

1. Compatibility/Storage Stability Tests

    1.1 Tube Test (Berenguer et al., 1989; Breuer, 1988c; Fetz et al., 1985; Muncy et al.,
        1987)
    1.2 UV Fluorescent Microscopy (Dong, 1989; Grimm, 1989; Lomi et al., 1989)
    1.3 Crushing Test (Thyrion et al., 1991)

2. Rheological Tests

    2.1 Flow Behaviour Tests
    A. Flow Test (Thompson and Hagman, 1958)
    B. Apparent Viscosity (Denning and Carswell, 1981)
    C. Absolute Dynamic Viscosity Using Coaxial Cylinder (Sybilski, 1993)
    D. Double Ball Softening Point (Texas Transportation Institute, 1983)
    E. Dropping Ball Test (Harders, 1988)

    2.2 Elastic Property Tests
    A. Elastic Recovery Test Using Ductilometer (Breuer, 1988b; Muncy et al.,1987;
        Öster,1989)
    B. Elastic Recovery Test Using Sliding Plate Rheometer (Tosh et al., 1992)
    C. Elastic Recovery Test Using ARRB Elastometer (Oliver et al., 1988)
    D. Elastic Recovery Test Using Controlled Stress Rheometer (Jørgensen, 1988)
    E. Elastic Recovery Test Using Höppler Consistometer (Svetel, 1985)
    F. Torsional Recovery Test (Thompson and Hagman, 1958)

    2.3 Tensile Property Tests
    A. Toughness and Tenacity Test (Benson, 1955)
    B. Toughness and Tenacity Test - Modified Method (Thompson and Hagman, 1958)
    C. Toughness and Tenacity Test acc. to Boussad et al. (1988)
    D. Extraction Test (Woodside and Lynch, 1989)
    E. Force Ductility Test (D-4 P 226; Chaverot, 1989; Verburg and Molenaar, 1991)
    F. Direct Tensile Test acc. To Anderson and Dongre (1993)

    2.4 Static Mechanical Tests
    A. Stiffness Modulus Using the Sliding Plate Rheometer (Tosh et al., 1992)
    B. Stiffness Modulus Using the Bending Beam Rheometer (SHRP Test Method B-
        002, 1993)

    2.5 Dynamic Mechanical Tests
    A. Test Procedure acc. To Jovanovic et al. (1993)
    B. Test Procedure acc. To De Ferrariis et al. (1993)
    C. Test Procedure acc.to Cavaliere et al. (1993)
    D. Test Method Using Mechanical Thermal Analyser (Khalid and Davies, 1993)
    E. Test Method Using Controlled Stress Rheometer (Jørgensen, 1993)
    F. Test Method Based on Balance Rheometer (Kolb, 1985)
    G. Test Method Using Shear Rheometer (SHRP Test Method B-003, 1993)

Table 1.4 Non-standardised test methods, Cont.

3. Adhesion/Cohesion Tests
3.1 Vialit Test (King, Muncy and Prudhomme, 1986)
3.2 Brittleness Temperature (Boussad, Muller and Touzard, 1988)
3.3 Dropping Temperature (Breuer, 1988a)
3.4 Contraction Test (Molenaar, 1991)
3.5 Cohesion Test Using the Vialit Pendulum Ram (Bononi, 1988)
3.6 Boiling Water Stripping Test (Belgian Road Research Centre, 1991)
4. Ageing Test
4.1 Pressure Ageing Vessel Test (SHRP Test Method B-005, 1993)
4.2 Ageing Test acc. to CRR (Choquet and Verhasselt, 1993)
5. Methods of Chemical Analysis
5.1 Spectroscopic Methods
A. Infrared Spectroscopy(Choquet et al.,1991,1992; Fifield et al.,1990, He et al., 1991; Jovanovic, et al., 1991; Little1987; SHRP A-004, 1990; Öster et al., 1989)
B. Nuclear Magnetic Resonance Spectroscopy (Fifield et al., 1990; Giavarini et al., 1989; Goodrich et al., 1986; Santagata et al., 1993)
5.2 Chromatographic Methods
A. Gas Chromatography (Fifield et al., 1990; Ruud, 1989)
B. High Pressure Liquid Chromatography (Fifield et al., 1990; Neubauer, 1988)
C. Thin Layer Chromatography-Flame Ionization Detector (Brule et al., 1988; Sherma et al., 1991; Torres et al., 1993)
D. Gel Permeation Chromatography (Fifield et al., 1990; Johansson et al., 1991; Molenaar, 1991)

1.4 Functional properties and field performance of polymer modified asphalt

In order to relate binder properties to pavement performance, it is necessary to know the fundamental relationships between bitumen properties and mix properties. Mix properties are needed for the pavement response models that provide the necessary input for the pavement performance models (Anderson and Kennedy, 1993).

To minimise the deterioration of a flexible pavement due to influence from traffic and climate, the bituminous layers should

- be stiff enough at elevated service temperatures to avoid permanent deformation (rutting),
- show good load-associated fatigue resistance,
- possess good stripping resistance (low water susceptibility),

- show time-independent properties (good ageing properties),
- have good flexibility at low temperatures (resistance to low temperature cracking),
- be effective against studded tyres (good wear resistance).

All of these performance-related (functional) properties of the mix are influenced to some extent by the binder properties.

A large number of investigations of the relation between binder properties and mix properties have been published. No attempt to summarise all the results obtained in this area has been made in this chapter. Only a few examples will be presented as illustrations. A comprehensive review was published in German by Zenke (1985).

Research (Choyce, 1989; Gschwendt and Sekera, 1993; Khosla and Zahran, 1989; Schüller and Forsten, 1993; Srivastava and Baumgardner, 1993; Terrel and Walter, 1986) has indicated that the addition of polymers, especially SBS and EVA, to bitumens improves resistance to permanent deformation of asphalt mixes. Rutting characteristics may be evaluated using the Wheel Tracking Test. Fig. 1.1 indicates that the resistance to deformation of Hot Rolled Asphalt (HRA) and Asphalt Concrete (AC) wearing course mixes may be markedly improved by the addition of 5 % (by weight) EVA copolymer to bitumen (Choyce, 1989). Fig. 1.1 also indicates that the addition of polymer (in this case EVA) makes the rate of tendency to permanent deformation of the two types of mixes less temperature sensitive. However, excessively far-reaching conclusions drawn from testing only one type of parameter must be avoided.

Load-associated fatigue is often studied using the three-point bending test. An example of the results obtained with this type of test at 0 °C is shown in Fig. 1.2. The tests are carried out by applying a constant alternating strain to the centre of a dense asphaltic concrete beam specimen until "failure" occurs, which is usually when the load required to maintain the level of strain has fallen to half its initial value. Fig. 1.2 indicates that the addition of SBS polymer to a bitumen improves the resistance of dense asphaltic concrete to fatigue cracking (Valkering et al., 1992).

Stripping in asphaltic mixtures, i.e. displacement of the bitumen film from the aggregate surface in the presence of water, often contributes to the deterioration of bituminous layers. To assess debonding effects caused by water, mechanical properties of asphaltic specimens such as tensile strength or Marshall stability are often evaluated before and after exposure to water. The ratio of strength (stability) of "wet" and "dry" specimens is calculated and used as a measure of the mix resistance to stripping. Fig. 1.3, which is based on retained Marshall stability measurements, indicates that PMB (in this case an SBS modified binder) has a positive effect on the water resistance of asphaltic mixtures (Beecken, 1992).

Ageing is induced by chemical and/or physical changes and is usually accompanied by hardening of the bitumen. In road applications, bitumen is exposed to ageing at three different stages: (i) storage, (ii) mixing, transport and laying, as well as (iii) during service life. Ageing is a very complex process in plain bitumens and the complexity increases when polymers are added. The ageing properties of plain bitumen are normally characterised by measuring rheological properties such as viscosity and softening point before and after artificial ageing in the laboratory (Section 3). This procedure is not sufficient in the case of PMB since thermolytic degradation of the polymer may occur during ageing and the fragments formed may contribute to a lowering of the consistency. Therefore, when assessing the ageing properties of PMB, further characteristics, such as elastic recovery and chemical composition (using for example GPC), have to be evaluated (Zenke, 1985). Indications of improved ageing properties by admixture of polymers to the bitumen have been reported in recent publications (Harlin et al., 1989; Mulder et al., 1992; Vonk et al., 1993). However, fundamental studies of the ageing process of polymer modified bitumens are necessary before more definite conclusions can be drawn.

In cold climates, cracking in pavements may be an extensive problem. Low temperature cracking is caused by thermally induced tensile stresses when these exceed the tensile strength of the pavement material. The main factor influencing the degree of cracking at low temperatures is found in the binder properties. Several papers have indicated that the addition of polymers to bitumens may increase resistance to low temperature cracking (Button, 1992; Button et al., 1987; Carpenter and VanDam, 1987). Fig. 1.4, the "theoretical cracking temperature", i.e. the intersection of the cooling and direct tensile curves (King, King and Harders et al., 1993), is given for one type of standard mix containing four straight-run bitumens with or without polymer (styrene-butadiene block copolymer chemically reacted in situ with the asphalt) of three different contents. As can be seen, the softer the base bitumen is, the lower the cracking temperature and, in general, increasing the proportion of polymer further lowers the cracking temperature.

Wear of asphalt pavements is a serious problem in certain low temperature regions where the use of studded tyres is permitted, e.g. in Scandinavia. The most important factor influencing wear from studded tyres is aggregate quality. However, binder properties can also affect wear, as shown in Fig. 1.5, where the results obtained on a vertical wear test track (Blomberg, 1988) are illustrated. The straight-run bitumens used were penetration grade B 120 and of different origin or from different refineries. The figure indicates minimum wear at about 0 °C, and a strong wear increase with decreasing temperature. Compared to mixes containing straight-run bitumen, wear on the PMB mixes (SBS modified) is lower at the minimum point and the effect of decreasing temperature is small.

As described above, there are indications from laboratory investigations of an improvement in asphalt properties due to the addition of polymers. However, validation of laboratory methods by field performance tests is necessary before a more definite opinion on the matter can be given. In such validation studies, the material of the laboratory specimen and the field test tracks should be the same. Careful choice of PMBs and reference bitumen (without polymer) is necessary to make it possible to draw conclusions on the effect of the polymer from comparative studies. Field performance should be characterised by field measurements but not by laboratory testing of samples taken from the road. Control of climate and traffic factors influencing field performance is necessary. Very few papers describing well-designed investigations of the relationships between polymer modified asphalt properties obtained in the laboratory and pavement performance in the field have been found.

To examine performance during construction and service when using different modifiers, a survey of modified asphalt test pavements (Button, 1992) was carried out in 1990. Information was collected on more than 30 test sites where adjacent test pavements were built to evaluate one to five asphalt modifiers. The field performance of about 20 different modifiers was examined. In most cases, no significant difference in performance between test and control sections was established. However, it must be noted that the service life was short (less than 5 years).

A recently published state of the art report conducted in SHRP (Coplantz et al., 1993) describes American studies of the relationships between properties of modified bitumen and pavement performance. Only two (out of 12) field studies involved polymers. One of these studies (Reese, 1989) indicates that the addition of SBS to the bitumen slows down the ageing process as measured by penetration, viscosity and ductility. However, when assessing ageing properties of PMB, further characteristics should also be evaluated as discussed above.

A very positive opinion of the in-service performance of thermoplastic rubber modified bitumens is expressed by Mulder and Whiteoak (1992). The main conclusion of their review is that, in all the cited cases, the thermoplastic rubber modified binder gave improved performance when compared with conventional bitumen.

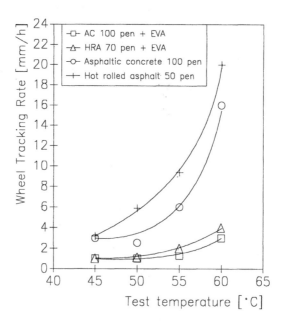

Fig. 1.1 Effect of temperature on wheel tracking rate for HRA and AC mixtures.

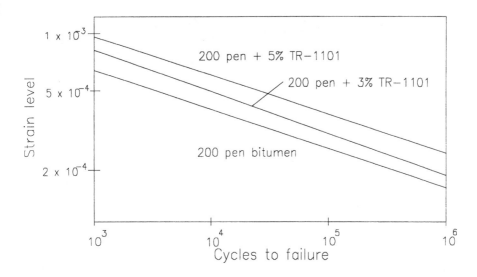

Fig. 1.2 Strain level as a function of cycles to failure.

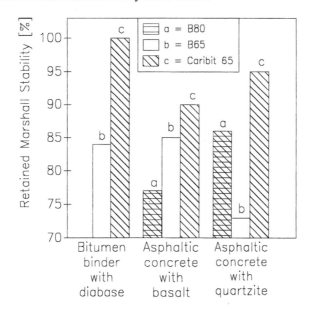

Fig. 1.3 Retained Marshall stability after water storage.

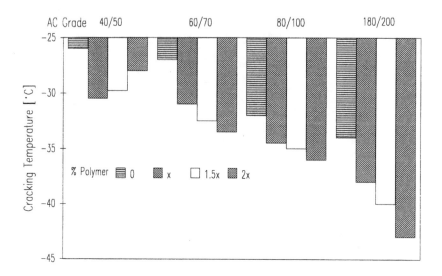

Fig. 1.4 Theoretical cracking temperature for four grades of bitumen and three polymer levels.

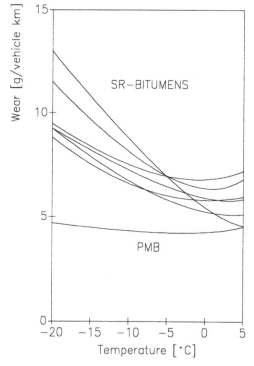

Fig. 1.5 Wear test results obtained from Neste's Test Track.

1.5   Specifications of polymer modified bitumens

In most countries, current bitumen specifications are viscosity or penetration graded and typically based on measurements of viscosity, penetration, ductility and softening point. These measurements are not sufficient to describe properly the linear viscoelastic and failure properties of bitumen that are needed to relate bitumen properties to mixture properties. As has been pointed out by Anderson and Kennedy (1993), these specifications and test methods are not performance-related, because they:

- lack adequate low-temperature measurements,
- do not include fundamental binder properties which may be related to fundamental mixture properties or to pavement performance,
- are not appropriate (capillary viscosity) for measuring consistency at the upper service temperatures, and
- do not consider long-term in-service ageing.

In most case, specification models used or proposed for polymer modified bitumens (see Table 1.5) are derivatives of the specifications of plain bitumens and are supplemented with tests such as tensile strength and elastic recovery. When a polymer is selected as a bitumen modifier, good compatibility between the polymer and the bitumen is a basic requirement because the separation of polymer during storage will result in inconsistent performance (Berenguer, 1988). Therefore, a compatibility test is often incorporated in PMB specifications (Muncy et al., 1987). A typical example of a specification model of polymer modified bitumens (Ruud, 1988) is that proposed by NAMOBIT, a working group of the Nordic Road Association of Asphalt Pavements, in 1986. The specification is based solely on traditional tests and includes the following parameters:

- Softening point.
- Viscosity at 135°C and 180°C.
- Elastic recovery at 25°C using ductilometer .
- Storage stability :

  - penetration (top/bottom),
  - softening point (top/bottom),

- Flash point (PMcc).
- Tests after TFOT :

  - softening point,
  - viscosity at 180°C,
  - elastic recovery.

The specification model is designed to describe PMB performance during production and service life of the pavement. Similar PMB specifications are used or proposed in other European countries, as well as in Australia and Japan, as shown in Table 1.5 (Isacsson and Lu, 1995; Pallotta, 1988; Vercoe, 1990).

Ideally, the specifications should be based on tests which are easily performed, fast and reliable (Valero, 1988). Usually, traditional tests meet the first two requirements, but their reliability is often questioned. This is also the reason for developing new specifications within the Strategic Highway Research Program (SHRP), see Table 1.6 (Petersen et al., 1994). In SHRP, new powerful tools for the evaluation of bituminous binders have been developed.

The SHRP binder specifications are said to be performance related. In this specification, new testing instruments, such as the bending beam rheometer and the direct tension tester, are employed. The instruments are used to measure more fundamental properties such as the inverse of loss compliance, storage modulus, stiffness and strain at failure. The determined parameters have been proposed collectively as being related to the rutting, fatigue and thermal cracking behaviour of bitumens (Anderson and Kennedy, 1993). The specification retains the safety and handling features of current specifications since minimum flash point and maximum high temperature viscosity remain specification items (Table 1.6). Short-term ageing is simulated using Rolling Thin Film Oven Test (RTFOT) and long-term ageing using the Pressure Ageing Vessel (PAV).

It is important to emphasise that proposed SHRP specifications are intended for both plain and polymer modified bitumens and allow selection of a binder based on the climate in which it is expected to perform.

1.6   Discussion

The aim of this chapter is to provide a summary of information found in the current literature regarding test methods and specifications of polymer modified bitumen for road applications.

The main purpose of material testing is to characterise the material in question in such a way that the characteristics measured can be used to predict behaviour in practice. For tests on bituminous binders to be valid, the tests must be sensitive to functional properties of the asphalt pavements, such as resistance to rutting, load-associated fatigue and low temperature cracking. Very little direct evidence of correlation between PMB properties and pavement performance has been found in this study.

In general, when assessing the quality of PMB, traditional methods (Table 1.3) developed for testing plain bitumen, are used. When standard methods are applied to polymer modified bitumens, the test conditions are often modified in one way or another (Isacsson and Lu, 1995).

Table 1.5 Specification tests used or proposed for PMBs*.

Type of Test	Austria	Australia		Belgium		France	Germany		Japan		Poland		Spain	Nordic countries
	E	SBS	EVA	E	P		E	TP	E	TP	E	P		
Penetration @ 25 °C	+	+		+	+		+	+	+	+	+	+	+	
Penetration @ 25°C after overnight cure			+											
Viscosity														
@ 60°C		+	+											
@ 135°C		+	+	+	+									+
@ 180°C														+
Softening Point	+	+	+	+	+		+	+	+	+	+	+	+	+
Plastic Interval						+								
Flash Point	+	+					+	+	+	+	+	+	+	+
Fraass Point	+			+	+	+	+	+			+	+	+	
Flow at 60°C		+												
Ductility														
@ -5°C				+	+									
@ 5°C											+		+	
@ 7°C							+		+					
@ 10°C						+								
@ 13°C	+						+	+						
@ 15°C										+	+	+		
@ 25°C	+						+	+			+	+		
Float Test @ 60°C													+	
Adhesion Test	+													
Vialit Test						+								
Density @ 25°C							+	+			+	+	+	
Toughness &Tenacity									+	+				
Torsional Recovery		+	+											
Elastic Recovery	+	+	+				+				+		+	+
Homogeneity							+	+						
Segregation	+	+	+											
Storage Stability											+	+	+	+
Tracking Test						+								
Con. other polymers and mineral matter		+	+											
Ageing Tests														
Weight change	+						+	+	+	+	+	+	+	
Penetration change	+						+	+	+	+	+	+	+	
Softening point change							+	+			+	+	+	+
Ductility														
@ 5°C											+		+	
@ 7°C							+							
@ 13°C							+	+						
@ 15°C											+			
@ 25°C							+	+			+			
Viscosity @ 180°C														+
Elastic recovery	+						+				+			+

Note: E--Elastomers, P--Plastomers, TP--Thermoplastics

Table 1.6 SHRP binder specification.

Performance Grade	PG-52							PG-58					PG-64						PG-70			
	-10	-16	-22	-28	-34	-40	-46	-16	-22	-28	-34	-40	-10	-16	-22	-28	-34	-40	-10	-16	-22	-28
Average 7-day Maximum Pavement Design Temperature, °C [a]	<52							<58					<64						<70			
Minimum Pavement Design Temperature, °C [a]	>-10	>-16	>-22	>-28	>-34	>-40	>-46	>-16	>-22	>-28	>-34	>-40	>-10	>-16	>-22	>-28	>-34	>-40	>-10	>-16	>-22	>-28
Original Binder																						
Flash Point Temperature AASHTO T48: Min, °C	230																					
Viscosity, ASTM D 4402: [b] Max 3 Pa·s, Test Temp, °C	135																					
Dynamic Shear, SHRP B-003: [c] G*/sinδ, Min, 1.0 kPa Test Temperature @ 10 rad/s, °C	52							58					64						70			
Physical Hardening Index [d], h	Report																					
Rolling Thin Film Oven Test Residue (AASHTO T240, ASTM D 2872)																						
Mass Loss, Max, percent	1.0																					
Dynamic Shear, SHRP B-003: G*/sinδ, Min, 2.2 kPa Test Temperature @ 10 rad/s, °C	52							58					64						70			
Pressure Aging Vessel Residue (SHRP B-005)																						
PAV Aging Temp, °C [e]	90							100					100						100/(110) [e]			
Dynamic Shear, SHRP B-003: G*/sinδ, Max, 5000 kPa Test Temperature @ 10 rad/s, °C	25	22	19	16	13	10	7	25	22	19	16	13	28	25	22	19	16	13	34	31	28	25
Creep Stiffness, SHRP B-002: [f] S, Max, 300,000 kPa, m-value, Min, 0.30, Test Temp, °C	0	-6	-12	-18	-24	-30	-36	-6	-12	-18	-24	-30	0	-6	-12	-18	-24	-30	0	-6	-12	-18
Direct Tension, SHRP B-006: [f] Failure Strain, Min, 1.0% Test Temp @ 1.0 mm/min, °C	0	-6	-12	-18	-24	-30	-36	-6	-12	-18	-24	-30	0	-6	-12	-18	-24	-30	0	-6	-12	-18

[a] Pavement temperatures can be estimated from air temperatures using an algorithm contained in the SUPERPAVE software program or may be provided by the specifying agency.

[b] This requirement may be waived at the discretion of the specifying agency if the supplier warrants the asphalt binder can be adequately pumped and mixed at temperatures that meet all applicable safety standards.

[c] For quality control of unmodified asphalt cement production, measurement of the viscosity of the original asphalt cement may be substituted for dynamic shear measurement of G*/sinδ at test temperatures where the asphalt is a Newtonian fluid (generally above 55°C). Any suitable standard means of viscosity measurement may be used, including capillary or rotational viscometry.

[d] The physical hardening index h accounts for physical hardening and is calculated by $h = (S_{24}/S_1)^{m_1/m_{24}}$ where 1 and 24 indicate 1 and 24 hours of conditioning of the tank asphalt. Conditioning and testing is conducted at the designated test temperature. Values should be calculated and reported. S is the creep stiffness after 60 seconds loading time and m is the slope of the log creep stiffness versus log time curve after 60 seconds loading time.

[e] The PAV aging temperature is 100°C, except in desert climates, where it is 110°C.

[f] If the creep stiffness is blow 300,000 kPa, the direct tension test is not required. If the creep stiffness is between 300,000 and 600,000 kPa the direct tension failure strain requirement can be used in lieu of the creep stiffness requirement. The m-value requirement must be satisfied in both case

Much research (Anderson and Kennedy, 1993; Muncy et al., 1987; Rudd, 1988; Valero, 1988) has demonstrated that the traditional specifications of plain bitumen are not sufficiently adequate for PMBs (cf. Section 1.5). These specifications are in general based on empirical test methods, such as penetration, softening point, ductility and Fraass breaking point, the performance relations of which are not always obvious.

There are several reasons why empirical methods nevertheless are used. Above all, these test methods have been used for a very long time and present knowledge on bituminous binders is for the most part based on results obtained using these methods. When assessing the properties of new binders, a comparison is made with binders which have been characterised with the aid of empirical methods. To make a direct comparison, the same methods are used. Another reason for the popularity of test methods such as penetration and softening point is the fact that these methods are very easy to use and the costs of performing the tests comparatively low. These processes contribute to a "stand still" of research in this area.

To describe the properties of PMBs in an overall situation, the PMB specifications should be supplemented by other test parameters, such as tensile strength, elastic recovery and compatibility. This is also the case in several specifications which are proposed or in use. However, when testing these parameters, the same empirical methods are often used. For example, the elastic recovery of PMB is often measured using the standard ductilometer and the compatibility assessed by measuring penetration and/or softening point before and after storage.

As attested by Ramond et al. (1993), there are three critical working ranges for bituminous binders:

- A range of high temperatures and long loading times during which the binders may flow, entailing a risk of rutting of the mixes.
- A range of low temperatures and long loading times during which the mixes are liable to crack under the effects of thermal stress.
- A range of low temperatures and short loading times during which the binder is brittle and may give rise to mechanical cracking.

As indicated above, the tests currently used for bitumen specifications yield little information on the behaviour of bitumens in these critical ranges. Fortunately, advances in rheometers have made it possible to perform dynamic tests at a wide range of temperatures and frequencies, from which various rheological parameters (e.g. complex modulus and phase angle) in different conditions can be obtained. Such rheometers have been described in Appendix. This type of test equipment has long been considered too expensive and improper to use in routine specification tests. However, recent research (Anderson and Kennedy, 1993; Ramond et al., 1993) has indicated that dynamic parameters are useful for predicting of performance-related properties.

As has already been mentioned (Section 1.5), several new test procedures have been developed in SHRP. A dynamic shear rheometer is used to measure specific rheological properties in the "upper" temperature range before and after artificial ageing according to RTFOT. It is also used to characterise properties of the binder in the "middle" temperature range after PAV ageing (cf. Appendix 1), which is assumed to simulate long-term ageing in service. The testing temperature is chosen according to climate and binder grade. A bending beam rheometer for measuring low temperature stiffness has been developed in SHRP. The beam is loaded at its midpoint with a constant load and the deflection of the midpoint of the beam is measured for 240 s. The stiffness of the binder is calculated from the dimensions of the beam, the applied load and the measured deflection. Tests are performed at the critical temperature, in this case the minimum pavement temperature plus 10 °C. In addition, a novel test procedure based on direct tension test has been developed in SHRP. This test is intended to replace tests such as ductility at 5°C and Fraass breaking point to evaluate the low-temperature failure properties of bitumens. The test is performed at low temperatures where thermal shrinkage cracking occurs, typically the minimum pavement temperature plus 10 °C. All these tests are considered to be related to pavement performance (Anderson and Kennedy, 1993).

As mentioned in Section 1.4, ageing occurs during the production of the pavement and during its service life. The circumstances during different ageing stages vary considerably. Standardised ageing test methods simulate the ageing that occurs during the production of the pavement. To simulate long-term ageing in service, the PAV test (Appendix 1) has been developed in SHRP. The test is performed after RTFOT or TFOT ageing.

The chemistry of bitumen is very complex and is even more complex after the admixture of polymer. No specifications including requirements on the chemical composition of bitumen or PMB have been found in the literature. It is doubtful whether this type of requirement should be included at all in specifications, at least not for plain bitumen. When characterising the ageing properties of PMB, some chemical tests could be suitable for specification purposes, as has already been mentioned in Section 1.4.

Within the European Committee for Standardisation (CEN), a Technical Committee, TC 19, is working on bituminous binders for roads. Today (1994) there is a proposal for a specification model for pure bitumen based on testing traditional parameters such as penetration, softening point and viscosity. No corresponding specification model for PMB has been proposed and such a model cannot be expected before 1996 (Blomberg, 1994) at the earliest.

In conclusion, the development of PMB specifications is a complicated task and it is very difficult to make a single specification fit every product due to the complexity of performance of the various modified bitumens as a function of bitumen nature and the type as well as the content of polymer. A major step towards better understanding of the function of bitumen and PMB has been taken in SHRP. Hopefully, the new SHRP specifications and test methods, which are intended for both plain bitumen and PMB, will be more suitable in predicting binder performance on roads compared to earlier tests. However, many years of experience are needed before a more definite opinion on this matter can be stated.

1.7 References

Anderson D. A. and Dongre R. N. (1993) : Development of the SHRP direct tension test for bitumens, *Proceedings of 5th Eurobitume Congress*, Stockholm, Vol. IA, pp. 250-255.

Anderson D. A. and Kennedy T. W. (1993) : Development of SHRP binder specification, *Proceedings of the Association of Asphalt Paving Technologists*, Vol. 62, pp. 481-507.

Beecken G. (1992) : Ten years' experience with Caribit in Germany, *Shell Bitumen Review 66*, pp. 12-15.

Belgian Road Research Centre Method of Measurement-ME 65/91 (1991) : Boiling water stripping test on coarse aggregates coated with a bituminous binder.

Benson J. R. (1955) : New concepts for rubberized asphalts, *Roads and Streets*, April, pp. 138-142.

Berenguer B. (1988) : Significant methods based on traditional tests or based on modern mechanical or rheological tests, *Proceedings of RILEM Residential Seminar*, Dubrovnik (Yugoslavia), pp. 259-268.

Berenguer B. and Caba, J. S. (1989) : Enhancement of the plasticity range in bitumen-SBS mixtures for roofing purposes, *Proceedings of 4th Eurobitume Symposium*, Madrid, Vol. I, pp. 79-82.

Billmeyer Jr., F. W. (1984) : Textbook of Polymer Science, 3rd edn, John Wiley & Sons.

Blomberg T. (1988) : Better wear resistance with polymer modified bitumen, *Proceedings of RILEM Residental Seminar*,Dubrovnik (Yugoslavie), pp.315-317.

Blomberg T. (1994) : Neste OY, Borgå, Finland, Personal communication.

Bononi A. (1988) : Methodologie de caracterisation des liants modifies, *Proceedings of RILEM Residential Seminar*,Dubrovnik (Yugoslavia),pp.119-133.

Boussad K., Carré D.and Muller, J-M. (1988) : The TTT: Toughness Tenacity Test, *Proceedings of RILEM Residential Seminar*, Dubrovnik (Yugoslavia), pp. 87-90.

Boussad K., Mulle, J-M. and Touzard B. (1988) : Evaluation of brittleness temperature thru plate cohesion test, *Proceedings of RILEM Residential Seminar*, Dubrovnik (Yugoslavia), pp. 225-228.

Breuer, J. U. (1988a) Dropping temperature of chips in the cold (KST) and in the heat (WST), *Proceedings of RILEM Residential Seminar*, Dubrovnik (Yugoslavia), pp. 229-231.

Breuer J. U. (1988b) : Elastic recovery of polymer modified binder, *Proceedings of RILEM Residential Seminar*, Dubrovnik (Yugoslavia), p. 83.

Breuer J. U. (1988c) : Storage stability test-Homogeneity test for hot stored polymer modified binder, *Proceedings of RILEM Residential Seminar*, Dubrovnik (Yugoslavia), p. 85.

Brule B., Brion Y. and Tanguy A. (1988) : Paving asphalt polymer blends: relationships between composition, structure and properties, *Proceedings of the Association of Asphalt Paving Technologists*, Vol. 57, pp. 41-64.

Button J. W. (1992) : Summary of asphalt additive performance at selected sites, *Transportation Research Record 1342*, TRB, National Research Council, Washington, D. C., pp. 67-75.

Button J. W., Little D. A., Kim Y. and Ahmed, J. (1987) : Mechanistic evaluation of selected asphalt additives, *Proceedings of the Association of Asphalt Paving Technologists*, Vol. 56, pp. 62-90.

Cameleyre J., Samanos J. and Tessonneau J. (1989) : In-situ behaviour of new types of fibre-based asphalt, *Proceedings of 4th Eurobitume Symposium*, Madrid, Vol. I, pp. 838-843.

Carpenter S. H. and VanDam T. (1987) : Laboratory performance comparisons of polymer-modified and unmodified asphalt concrete mixtures, *Transportation Research Record 1115*, TRB, National Research Council, Washington, D. C., pp. 62-74.

Cavaliere M. G., Diani E. and Vitalini Sacconi L. (1993) : Polymer modified bitumens for improved road application, *Proceedings of 5th Eurobitume Congress*, Stockholm, Vol. IA, pp. 138-142.

Chaverot P. (1989) : The force ductility test: A simple means of characterization of the polymer modified asphalt, *Proceedings of 4th Eurobitume Symposium*, Madrid, Vol. I, p. 96-100.

Choquet F. S. and Ista E. J. (1992) : The determination of SBS, EVA and APP polymers in modified bitumens, *Polymer Modified Asphalt Binders, ASTM STP 1108*, Kenneth R. Wardlaw and Scott Shuler, Eds., pp. 35-49.

Choquet F. S. and Verhasselt A. F. (1993) : Ageing of bitumen: From the road to the laboratory and vice versa. *International Conference*, The Hague (Netherlands).

Choquet F., Ista E. and CogneauP. (1991) : Determining the composition of roofing membranes manufactured with atactic polypropylene (APP), *Proceedings of Third International Symposium on Roofing Technology*, pp. 436-444.

Choyce P. W. (1989) : Asphalt mixtures containing EVA modified bitumen, *Proceedings of 4th Eurobitume Symposium*, Madrid, Vol. I, pp. 505-508.

Coplantz J. S., Yapp M. T. and Finn F. N. (1993) : Review of relationships between modified asphalt properties and pavement performance, *SHRP-A-631*.

D-4 Proposal P 226 : Proposed test methods for bituminous materials in tension.

De Ferrariis L., Gallino G., Italia P., Mancin, G. and Rebesco E. (1993) : Rheological characteristics of polymer modified bitumens, *Proceedings of 5th Eurobitume Congress*, Stockholm, Vol. IA, pp. 133-137.

Denning J. H. and Carswel, J. (1981) : Improvements in rolled asphalt surfacings by the addition of organic polymer, *TRRL Laboratory Report 989*.

Dinnen A. (1985) : Bitumen-thermoplastic rubber blends in road applications, *Proceedings of 3rd Eurobitume Symposium*, Hague, Vol. I, pp. 646-651.

Dong A. (1989) : Research work into modified bituminous binders with polymer additives, *Proceedings of 4th Eurobitume Symposium*, Madrid, Vol. I, pp. 181-185.

Downes J. W. (1986) : Modified binders to the year 2000, *6th International Asphalt Conference*, Sydney, Australia, pp. 35-39.

Fetz E. and Angst Ch. (1985) : Polymer modified bitumen requirements, *Proceedings of 3rd Eurobitume Symposium*, Hague, Vol. I, pp. 506-511.

Fifield F. W. and Kealey D. (1990) : Principles and Practice of Analytical Chemistry, 3rd edn, Blackie and Son Ltd, London.

Fritz H. W., Huet J. and Urban R. (1992) : English-French-German dictionary of technical terms related to hydrocarbon binders, hydrocarbon composite materials, processes for removing pavement material and for the rehabilitation of asphalt pavements, *Materials and Structures*, Vol. 25, pp. 171-185.

Giavarini C. and Vecch, C. (1989) : Evaluation of bitumen by NMR, *Proceedings of 4th Eurobitume Symposium*, Madrid, Vol. I, pp. 260-263.

Goodrich J. L., Goodrich J. E. and Kari W. J. (1986) : Asphalt composition tests: Their application and relation to field performance, *Transportation Research Record 1096*, TRB, National Research Council, Washington, D. C., pp. 146-167.

Gordillo J., Tomas R. and Gomez F. (1989) : Advantages of the incorporation of fibre into cold micro asphalt concrete, *Proceedings of 4th Eurobitume Symposium*, Madrid, Vol. I, pp. 786-790.

Grimm G. (1989) : Application of microscopic methods in the field of polymer-bitumen binders, *Proceedings of 4th Eurobitume Symposium*, Madrid, Vol. I, pp. 54-59.

Gschwendt I. and Sekera, M (1993) : Bituminous mixes with rubber modified binders, *Proceedings of 5th Eurobitume Congress, Stockholm*, Vol. IB, pp. 516-519.

Hall C. (1985) : Polymer Materials, Higher and Further Education Division, Macmillan Publishers Ltd.

Hancock T (1823) : Brit. Patent No.4768 [1823], From Zenke, G. (1985).

Harders O. (1988) : Dropping of ball-test, *Proceedings of RILEM Residential Seminar*, Dubrovnik (Yugoslavia), p. 91.

Harlin J. P., Jariel J. L. and Blondel J. C. (1989) : Twelve years service life without specific maintenance for the bitumen Cariflex TR-1101 bound wearing course of the Saint-Quentin Bridge, *Proceedings of 4th Eurobitume Symposium*, Madrid, Vol. I, pp. 125-127.

He L. Y. and Button J. W. (1991) : Methods to determine polymer content of modified asphalt, *Transportation Research Record 1317*, TRB, National Research Council, Washington, D. C., pp. 23-31.

Isacsson U. and Lu X. (1995) : A compilation of test methods and specifications of polymer modified bitumens for road applications, *TRITA-IP AR 95-26*, Royal Institute of Technology, Sweden.

Johansson U. and Linde S. (1991) : Polymer modified asphalt binders - Part 1, *SP Report*, No. 2, Swedish National Testing and Research Institute.

Jovanovic, J. A., Djonlagic, J. and Dunjic, B. (1993) A rheological study of behaviour of polymer-bitumen blends, *Proceedings of 5th Eurobitume Congress*, Stockholm, Vol. IA, pp. 256-262.

Jovanovic J., Sokiv M., Smiljanic M., Pap I., Neumann H.-J. and Rahimian I. (1991) : The production of stable bitumen-polymer blends, *Proceedings of International Symposium on Chemistry of Bitumens*, Rome, Vol. II, pp. 906-921.

Jørgensen T. (1988) : Creep testing with the controlled stress rheometer - A versatile tool for measuring elastic recovery of polymer modified bitumens, *Proceedings of RILEM Residential Seminar*, Dubrovnik (Yugoslavia), pp. 93-95.

Jørgensen T. (1993) : Rheological testing of binders with a controlled stress rheometer, *Procedings of 5th Eurobitume Congress*, Stockholm, Vol. IA, pp. 171-174.

Khalid H. and Davies E. (1993) : A dynamic approach to predict the performance of conventional and polymer modified binders and mixes containing them, *Proceedings of 5th Eurobitume Congress*, Stockholm, Vol. IA, pp. 246-249.

Khosla N. P. and ZahranS. Z. (1989) : A mechanistic evaluation of mixes containing conventional and polymer modified (styrelf) asphalts, *Proceedings of the Association of Asphalt Paving Technologists*, Vol. 58, pp. 274-302.

King G. N, Muncy H. W. and Prudhomme J. B. (1986) : Polymer modification: Binders effect on mix properties, *Proceedings of the Association of Asphalt Paving Technologists*, Vol. 55, pp. 519-540.

King G. N., King H. W., Chaverot P., Planche J. P. and Harders O. (1993) : Using European wheel-tracking and restrained tensile tests to validate SHRP performance graded binder specification for polymer modified asphalts, *Proceedings of 5th Eurobitume Congress*, Stockholm, Vol. IA, pp. 51-55.

King G. N., King H. W., Harders O., Arand W., Chaverot P. and Planche J. P. (1993) : Influence of asphalt grade and polymer concentration on the low temperature performance of polymer modified asphalt, *Proceedings of the Association of Asphalt Paving Technologists*, Vol. 62, pp. 1-22.

Kolb K. H. (1985) : Laboratory evaluation of polymer modified bitumen and polymer modified asphalt, *Proceedings of 3rd Eurobitume Symposium*, Hague, Vol. I, pp. 533-536.

Little D. N., Button J. W., White R. M., Ensley E. K., Kim Y. and Ahmed S. J. (1987) : Investigation of asphalt additives, *Report No. FHWA/RD-87/001*, U. S. A. Department of Transportation, Federal Highways Administration, Chapter III.

Lomi C. and Varisco F. (1989) : Compatibility of bitumen with acactic polypropylene, new experimental monitoring procedure, *Proceedings of 4th Eurobitume Symposium*, Madrid, Vol. I, pp. 269-274.

Molenaar J. M. M. (1991) : Polymer modified bitumens - Functional properties and quality aspects, *Proceedings of International Symposium on Chemistry of Bitumens*, Rome, Vol. II, pp. 860-886.

Mulder E. A. and Whiteoak C. D. (1992) : An objective assessment of the in-service performance of thermoplastic rubber modified bitumens for road applications, *16th ARRB Conference*, Perth, Western Australia.

Muncy H. W, King G. N. and Prudhomme J. B. (1987) : Improved rheological properties of polymer-modified asphalts, *Asphalt Rheology-Relationship to Mixture, ASTM STP 941*, Edited by Briscoe,O. E., American Society for Testing and Materials, Philadelphia, pp. 146-165.

Neubauer O. (1988) : Chromatography of SBS in bitumen, *Proceedings of RILEM Residential Seminar*, Dubrovnik (Yugoslavia), p. 155.

Oliver J. W. H., Witt H. P. and Isacsson U. (1988) : Laboratory investigations of polymer modified bitumens using the ARRB elastometer, *Proceedings of RILEM Residential Seminar*, Dubrovnik (Yugoslavia), pp. 105-109.

Pallotta D. S. (1988) : Significant methods based on modern physical and Physicochemical test, *Proceedings of RILEM Residential Seminar*, Dubrovnik (Yugoslavia), pp. 277-284.

Peltonen P. (1989) : Fibres as additives in bitumen, *Proceedings of 4th Eurobitume Symposium*, Madrid, Vol. I, pp. 938-942.

Petersen J. C., Robertson R. E., Branthaver J. F., Harnsberger P. M., Duvall J. J., Kim S. S., Anderson D. A., Christiansen D. W., Bahia H. U., Dongre R., Antle C. E., Sharma M. G., Button J. W. and Glover C. J. (1994) : Binder characterization and evaluation, Vol. 4: Test Methods, *SHRP-A-370*, National Research Council, Washington, D. C..

Ramond G., Pastor M. and Such C. (1993) : Determination of performance of a binder from its complex modulus, *Proceedings of 5th Eurobitume Congress*, Stockholm, Vol. IA, pp. 81-85.

Reese R. (1989) : Letter summary of presentation made by Ron Resse at the 21st Pacific Coast Conference on Asphalt Specifications, California Department of Transportation.

Rowlett R. D., Martinez D. F., Mofor D. A., Romine R. A. and Tahmoressi M. (1990) : Performance of asphalt modifiers: Classification of modifiers and literature review, *Strategic Highway Research Program*, Center for Construction Materials Technology, Southwestern Laboratories, Houston, Texas.

Ruud O. E. (1988) : Polymer modified bitumen for road constrution - A Nordic evaluation, *Proceedings of RILEM Residential Seminar*, Dubrovnik (Yugoslavia), pp. 67-70.

Ruud O. E. (1989) : Characterization of bituminous binders by gas chromatography, *Proceedings of 4th Eurobitume Symposium*, Madrid, Vol. I, pp. 285-289.

Santagata E. and Montepara A. (1993) : An NMR based method for the chemical characterization and rheological selection of modified bitumens, *Proceedings of 5th Eurobitume Congress*, Stockholm, Vol. IA, pp. 143-147.

Schüller S. and Forstén L. (1993) : Effect of binder properties on polymer modified asphalt concrete, *Proceedings of 5th Eurobitume Congress*, Stockholm, Vol. IB, pp. 520-525.

Sherma J. and Fried B. (1991) : Handbook of Thin-Layer Chromatography, Marcel Dekker, Inc., New York, pp. 342-348.

SHRP A-004 Interim Report, (1990) Summary of Phase I Results, Southwestern Laboratories, Texas.

SHRP Test Method B-002 (1993), *Technology Information Sheet*, January.

SHRP Test Method B-003 (1993), *Technology Information Sheet*, February.

SHRP Test Method B-005 (1993), *Technology Information Sheet*, January.

Srivastava A. and Baumgardner G. L. (1993) : Polymer modified bitumen and related properties, *Proceedings of 5th Eurobitume Congress*, Stockholm, Vol. IA, pp. 151-155.

Svetel D. (1985) : Investigation of the elastic behaviour of bitumen, *Proceedings of 3rd Eurobitume Symposium*, Hague, Vol. I, pp. 108-114.

Sybilski D. (1993) : Non-Newtonian viscosity of polymer-modified bitumens, *Materials and Structures*, Vol. 26, pp. 15-23.

Terrel R. L. and Walter J. L. (1986) : Modified asphalt pavement materials - The European experience, *Proceedings of the Association of Asphalt Paving Technologists*, Vol. 55, pp. 482-518.

Texas Transportation Institute (1983) : Asphalt-rubber binder laboratory performance, *Research Report 347-IF*.

Thompson D. C. and Hagman J. F. (1958) : The modification of asphalt with neoprene, *Proceedings of the Association of Asphalt Paving Technologists*, Vol. 27, pp. 494-512.

Thyrion F., Cogneau P. and Zwijsen M. (1991) : Method for the evaluation of bitumens based on the characteristics of bitumen-polymer mixes, *Proceedings of International Symposium on Chemistry of Bitumen*, Rome, Vol. II, pp. 969-999.

Torres J., Gonzalez J. Ma. and Peralta X. (1993) : Correlation between the fractionation of bitumen according to the methods ASTM D 4124 and Iatroscan, *Procedings of 5th Eurobitume Congress*, Stockholm, Vol. IA, pp. 203-208.

Tosh D. J. Taylor M. B. and Robinson H. L. (1992) : Simple but effective, *Highways*, May, pp. 26-28.

Valero L. (1988) : Production control and reception control of polymer-bitumens either for waterproofing or for road construction, *Proceedings of RILEM Residential Seminar*, Dubrovnik (Yugoslavia), pp. 285-286.

Valkering C. P., Vonk W. C. and Whiteoak C. D. (1992) : Improved asphalt properties using SBS modified bitumens, *Shell Bitumen Review 66*, pp. 9-11.

Verburg H. A. and Molenaar J. M. M. (1991) : Force-ductility test for specification of polymer modified bitumen, *Proceedings of Interational Symposium on Chemistry of Bitumen*, Rome, Vol. II, pp. 922-931.

Vercoe J. (1990) : Specification - Overseas experience, *ARRB Research Report 183*, Edited by Oliver, J. W. H., Australian Road Research Board, pp. 65-81.

Whiteoak D. (1991) : The Shell Bitumen Handbook, Shell Bitumen U. K., Chapter 10.

Vonk W. C., Phillips M. C. and Roele M. (1993) : Ageing resistance of bituminous road binders: benefits of SBS modification, *Proceedings of 5th Eurobitume Congress*, Stockholm, Vol. IA, pp. 156-160.

Woodside A. R. and Lynch H. G. (1989) : Assessment of performance of polymer modified binders when used in surface dressing, *Proceedings of 4th Eurobitume Symposium*, Madrid, Vol. I, pp. 685-687.

Zenke G. (1985) : Polymer-modifizierte Straßenbaubitumen im Spiegel von Literaturergebnissen Versuch eines Resümees, *Die Asphaltstrasse*, No. 1, pp. 5-16, No. 4, pp. 170-182 and No. 6, pp. 264-284.

Öste,R. and Sillén B. (1989) : The durability of modified bitumen in the asphalt making process and during one year on the road, *Proceedings of 4th Eurobitume Symposium*, Madrid, Vol. I, pp. 324-328.

2

RILEM interlaboratory test on the rheology of bituminous binders

by L. Francken

2.1   Introduction

Thanks to the research efforts recently invested in many countries, there is a growing interest for the characterisation of road binders on the basis of their rheological properties. In particular the complex modulus measured by means of rheometers is becoming a major indicator of the mechanical behaviour and performance. Hence this will become a basic information in the specification of both plain and modified bitumens. For this reason, the RILEM Technical Committee 152 PBM "Performance of bituminous materials" has initiated in 1994 an interlaboratory test on the rheology of bituminous binders which completes the series of interlaboratory tests carried out from 1989 to 1993 in the field of bituminous mix design and repeated loading tests described in chapters 4 and 6 respectively.

2.2   Working program

The aim of this test program was to compare the experimental methods and devices used by different laboratories to determine the complex modulus of bituminous binders (definitions and general principles are given in the introduction).

The exercise, consisting in measuring the two components of the complex modulus (stiffness modulus $|E^*|$ and phase angle δ) of 4 binders, was based on the following guidelines:

1.   The participating laboratories were invited to use their own equipment, own sample preparation procedures and usual test conditions of temperature and frequencies. In order to avoid additional effects or reasons of discrepancies, it was requested not to make any preliminary ageing test (RTFOT or similar described in chapter 1).

2.   In addition to its currently used testing conditions of temperatures and frequencies, each laboratory was asked to use a set of 15 reference conditions resulting from the combination of :

    -    3 frequencies (Fr = 1,10 and 100Hz),
    -    5 temperatures (T = -25, 0, 25, 50 and 75°C).

Bituminous Binders and Mixes, edited by L. Francken. RILEM Report 17. Published in 1998 by E & FN Spon, 11 New Fetter Lane, London EC4P 4EE, UK. ISBN 0 419 22870 5

3.   Participators were invited to give a comprehensive description of the experimental procedure used (including calibration, specimen preparation, interpretation, etc...). In cases where the shear modulus |G*| was determined, this value was converted to the stiffness modulus |E*| using the relationship : |E*| = 3 |G*| assuming $\nu = 0.5$.

2.3   The binders to be tested

The binders to be tested were selected among those which were used in 1985 to build experimental road sections of the RN5 road between Charleroi and Couvin (Francken, 1990). On this experimental road of the Belgian Road Research Centre (BRRC) eight different binders have been used in road sections of open graded mixtures with the purpose to assess their long term performance.

Samples of all these binders had been taken at the time of construction, and four of them (one straight-run bitumen and three polymer modified bitumens) were selected to be used as material to be distributed among the participants.

Their code numbers and nature are :

*   244, B80 straight run bitumen,
*   245, reclaimed tire Rubber bitumen,
*   248, SBS elastomeric bitumen,
*   249, EVA plastomeric bitumen.

The straight-run bitumen was taken as reference and the three polymer modified bitumens were chosen as examples of currently used polymer modified bitumens. The binders supplied to the participants were identified by their code numbers and the characteristics given in table 2.1, but their nature and composition were not communicated to the participants when starting the test program.

Table 2.1 Characteristics of the binders.

Binder	code	244	245	248	249
Property	Units	80/100	Rubberised	Elastomer	Plastomer
Density	t/m^3	1.028	1.046	1.017	1.027
Pen.(25°C)	mm/10	76	97	96	98
R&B	°C	47	52.5	72.5	52
Fraass	°C	-14.05	-20.3	-13.7	-16.3
Visc.135°C	mPa.s	347	---- *	916	894

*   particles in suspension made this measurement impossible.

2.4　Participating laboratories

The experiment was coordinated and interpreted by the BRRC. It was started in may 1994 with 19 laboratories　from 11 countries applying for a participation on a voluntary basis.

Table 2.2 Participating laboratories.

Lab-Code	Name of the laboratory	Rheometer Type
B1	B.R.R.C, Belgium	Métravib
B2	NYNAS n.v., Belgium	Carri-Med
F1	L.C.P.C., France	Métravib
F2	Ent.　Jean LEFEBVRE, France	Bohlin
F3	ESSO Centre de Recherches, France	Bohlin
F4	Raffinerie BP - Dunkerque, France	Métravib
F5	Centre de Recherches　TOTAL, France	Métravib
F6	Centre de Recherches ELF, France	Rheometrics
NL2	D.W.W.　Rijkswaterstaat, The Netherlands	Contraves
NL6	Koninklijke　SHELL　Lab., The Netherlands	Carri-Med
S1	Royal Institue of Technology, Sweden	Rheometrics
CH1	EMPA., Switzerland	SHRP-DSR
E1	CEDEX , Spain	Métravib
DK1	Ass. Danish Asphalt Industries, Denmark	Own construction
H3	Ministry of Transport, Hungary	Carri-Med
YU1	University of Belgrade, Serbia	Rheometrics
GB1	B.P. OIL, United Kingdom	Carri-Med
GB2	S.W.K.　Pavement Engineering, United Kingdom	Bohlin
I1	Autostrade S.p.A., Italy	Rheologica

2.5　Experimental conditions

2.5.1　TEST ARRANGEMENTS AND SPECIMEN GEOMETRIES

Many different testing conditions were used in the frame of this program. They vary not only with the equipment itself but also, for a given equipment (manufacturer), with the numerous possible combinations of loading modes, specimen geometries and sizes. An overview of all conditions present in this test programme is given Fig. 2.1 and in table 2.2 in which we have used the following classification :

- Equipment type : Seven different equipments (manufacturers) were used, which can be divided into two main types :

  - R :   12 rotational rheometers.
  - L :   6 linear vibrating rheometers.

- Loading mode : These devices were used in conjunction with different loading modes :

  - SH :   shear.
  - TC :   tension-compression.
  - TO :   torsion.

- Specimen geometries :

  - CY :   cylindrical.
  - PR :   prismatic.
  - PP :   parallel plates.
  - AN :   annular.
  - SP :   spherical.

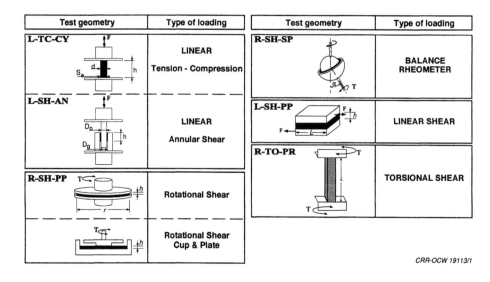

CRR-OCW 19113/1

Fig. 2.1 Test arrangements and specimen geometries.

2.5.2   INFLUENCE OF THE SPECIMEN SIZE AND GEOMETRY

The different combinations of the conditions used in the programme are given in Tables 2.3 and 2.4 for the test arrangements and specimen geometries presented on Fig. 2.1.

It is to be noted that the size and geometry of the specimen determines its mechanical stiffness as well as the magnitude of the applied conditions of stresses and strains.

In many cases different specimen geometries or sizes were adopted over the temperature range in order to obtain a mechanical stiffness in accordance with the accuracy for the measuring equipment used.

For example :

- In the rotational shear mode with parallel plate geometry (RO-SH-PP) this can be achieved by changing the plate diameter d (from 10 to 40mm) and the gap thickness h (from 0.25 to 2mm). Specimen stiffness then increases with increasing plate diameter and decreasing gap width.
- In the linear vibrating mode with annular specimens (L-SH-AN) stiffness is varied by changing the dimensions of the internal diametre Dg and external diametre of Dp of the device (Fig. 2.1.)

2.5.3   RHEOMETER TYPES

The values presented by the different laboratories were generally obtained by different methods or specimen geometries, depending on the temperature/frequency range (see table 2.3 and 2.4) :

2.5.3.1 Linear rheometers (table 2.3)

- **B1, F1, F4 and F5** used a linear vibrating rheometer with two sample geometries: tension compression on cylindric specimens (TC-CY) at temperatures between -25°C and +30°C ; Annular shear (SH-AN) for temperatures higher than or equal to 25°C.
- **E1** used a linear rheometer with the same sample geometry (SH-AN) over the whole range of temperatures.
  In these 5 cases the automatic temperature scanning and frequency sweep did not allow the direct measure at the required reference conditions. The values given in the Tables 2.6 to 2.9 are then generally obtained by interpolation over master curves eventually combining two sample geometries.
- **DK1** used flat parallel rectangular plates in shear mode over the whole range of temperatures. The gap between the plates ranged between 150 and 500mm and the displacements used went up to 200μm in some cases.

Table 2.3 Characteristics of the linear rheometers.

Lab code	Test type	Temperature (°)		Dimensions (mm)					Strain (%)	
				CY Cylinder		AN Annular shear				
		from	to	Dia	Height	Dg	Dp	h	from	to
B1	TC-CY	-25	25	9	18	-	-	-	0.05	0.05
	SH-AN	25	60	-	-	10	8	5	0.5	0.5
E1	SH-AN	-25	50	-	-	10	8	5	0.5	0.5
F1	TC-CY	-25	25	9	18	-	-	-	0.05	0.05
	SH-AN	25	60	-	-	10	8	5	0.5	0.5
F4	TC-CY	-30	20	9	18	-	-	-	0.05	0.05
	SH-AN	20	60	-	-	10	8	5	0.5	0.5
F5	TC-CY	-20	30	10	20	-	-	-	0.05	0.05
	SH-AN	20	60	-	-	10	9	5	1	1
DK1	SH-PP*	-10	60	-	-	-	-	-	6.6	13

*   Dimensions unknown.

2.3.5.2 Rotating rheometers (Table 2.4)

- **B2** carried out two series of measurements with two different temperature control arrangements.
- **S1** gave two sets of results with each of the arrangement R-SH-PP mentioned in Table 2.1 :

    - one with the (8 or 25mm) parallel plate,
    - a second one obtained on a cup-plate system combining plate diameters of 8mm and 25mm in one single specimen arrangement.

- **YU1** used two specimen geometries (respectively noted YUr and YUd in the Tables 2.6 to 2.9) : parallel plates in the high temperature range and prismatic specimens (63.5 x 12.7 x 2-3mm) in torsional shear in the low temperature range.

2.5.4   SPECIMEN PREPARATION

To melt the binders it was recommended to use temperatures 20°C above the softening temperature given in Table 2.1. The binder was then treated by each participant according to his usual procedure. Some specimen geometries (CY, PR) required the use of special moulds. Difficulties for preparing samples of binder 245 were reported by different laboratories. The granular consistency due to rubber particles in suspension made it particularly difficult to prepare homogeneous samples when small gap width were required. Different laboratories mentioned a large scatter on the results for this binder and one of them (B2) was unable in its first serie to make valuable tests for this reason.

Table 2.4 R-PP Rotating rheometers with parallel plates.

Lab	Rheometer		Temperature (°C)		Dimensions (mm)		Strain (%)	
code	Name	Type	From	To	Dia	Gap	From	To
B2	Carri-Med	CSL500	15	50	20	1		<5
CH1	Bohlin	CSR50	20	50	8	2	0.1	<1
F2	Bohlin	DSR50	15	75		1	0.2	1
F3	Bohlin	CS50	5	5	10	2	0.4	4
			25	50	20	1	1	4
F6	Rheometric	RDA II	-25	25	8	1.4 - 1	0.05	25
			50	75	8	1.6 - 2	20	75
GB1	Carri-Med	CSR500	-25	25	10	2	0.025	0.25
			50	75	40	1	2	20
GB2	Bohlin	CSR	25	75	25	1	0.25	25
H1	Carri-Med	CSL500	-25	50	20	2	0.05	0.4
			75	75	10	1.5	3.5	12
I1	Rheologica	Stresstech	0	75	20	2	0.02	370
NL6	Carri-Med	CSL500	0	75	25	1	0.1	20
S1	Rheometric	RDA II	-25	0	8	1.5	0.02	0.08
			25	75	25	1	0.09	42
YU1	Rheometric	RMS600	-25	25	25	2 - 4	0.1	10

2.5.5 RANGE OF TEMPERATURES AND FREQUENCIES

The range of temperatures T and frequencies Fr covered in the frame of the present experience is given in Table 2.5. It varies from one laboratory to another according to the type of equipment used and to the specimen geometry. In most of the equipments, testing temperatures and loading frequencies are computer controlled. Special procedures allow several test temperatures to be run sequentially while frequency sweeps can be generated once the temperature equilibrium of the specimen is reached. In order to compare the ranges covered by the different laboratories, the combinations of temperatures and frequencies used can be reduced to the variation of one single parameter derived from the time-temperature equivalency principle (see definition in the introduction of this report).

If we express the (T, Fr) combinations in terms of the reduced variable $X=\log(\alpha_T.Fr)$ taking the Arrhenius equation as basic expression for the shifting factor defined in chapter 3.2 of the introduction.

Assuming $0.4343 \times \delta H/R \approx 10000$ K as a realistic average value, we obtain in Fig. 2.2 an image of the ranges covered by the different laboratories. Positive values of X correspond to temperatures lower than the reference of 15°C; negative values correspond to higher temperatures.

When comparing these data with the information of Tables 2.3 and 2.4 it can be stated that:

• laboratories which cover a rather large range of high temperatures (negative values) are using rotational shear rheometers (RO-SH) sometimes combined with low frequencies,

- those who are more located in the low temperature range are either of the linear tension compression (LI-TC) type or the rotating torsion type (lab YU1),
- where a large range of temperatures and frequencies can be covered different sample geometries must be used in combination.

2.5.6 GENERAL REMARKS

Some laboratories have compared different specimen configurations (S1, YU1) or have studied the linearity by using different stress levels (GB1).

Several laboratories mentioned that they had little or no experience with their equipment for this type of measurement. The reason being that either the equipment was old and almost not used (NL1), or that it was recently purchased (CH1, H1, I1, S1, E1) or in a development phase (DK1).

2.6   Mode of interpretation

The different reports where delivered to the BRRC in the course of the year 1995. They contain a large amount of information which is hard to report in detail in the frame of this report. The main points of interest to be retained after analysis of this bulk of information will be summarised in the following.

2.6.1   REFERENCE STIFFNESS FOR THE PURE BITUMEN

The analysis of the data relevant to the two components of the complex modulus (stiffness $|E*|$ and phase angle δ) for the 15 imposed conditions of temperature and frequency mentioned here must:

1)   allow a first estimation of the reproducibility of the test,
2)   reveal the outlying results.

Detecting outlying results supposes that somehow a true value is known. In order to be as objective as possible an external set of true values can be proposed as references for the pure bitumen on the basis of Van der Poel's nomogram (1954). Taking the penetration index and softening temperature of the pure bitumen from Table 2.1, this set of values can easily be derived over the full range of temperatures and frequencies covered by the test. These values are given on the first line of table 2.6a under the code VdP. Unfortunately, this procedure does not work for the case of polymer modified bitumen.

Table 2.5 Range of temperatures and frequencies covered by the tests. Values of X are derived from equation (2.1).

N°	Lab code	Temperature °C		Frequency (Hz)		Log (α_T.Fr)	
		From	To	From	To	From	To
1	B1	-25	60	5	250	-3.4177	7.3114
2	B2	15	50	0.01	10	-5.3010	1.0000
3	CH1	20	50	0.1	30	-4.3010	0.9573
4	DK1	-10	60	0.02	64	-5.8157	4.7020
5	E1	-25	50	5	100	-2.6020	6.9135
6	F1	-25	60	7.8	250	-3.2246	7.3114
7	F2	15	75	0.1	10	-6.2524	1.0000
8	F3	5	50	1	40	-3.3010	2.6979
9	F4	-30	60	5	250	-3.4177	8.0394
10	F5	-20	60	7.8	250	-3.2246	6.6123
11	F6	-25	75	1	10	-5.2524	5.9135
12	GB1	-25	75	1	10	-5.2524	5.9135
13	GB2	25	75	0.02	10	-6.9513	-0.0223
14	H1	-25	75	1	40	-5.2524	6.5156
15	I1	0	75	1	100	-5.2524	3.6738
16	NL1	25	75	0.8	20	-5.3493	0.2788
17	NL2	0	75	1	10	-5.2524	2.6738
18	S1	-25	75	0.1	10	-6.2524	5.9135
19	YU1	-25	75	0.16	157	-6.0482	7.1094

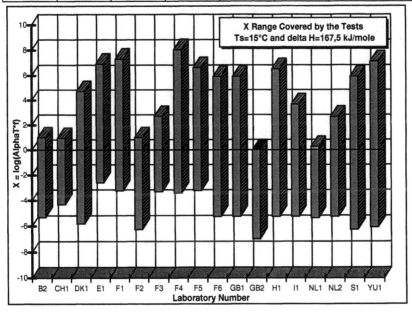

Fig. 2.2 Range of X values covered by the tests.

2.6.2   INTERPOLATED VALUES

Several laboratories gave the results for the reference conditions which were into the range of their equipment.

However values had to be derived by interpolation over the master curve in three typical situations:

1)   when the temperature/frequency pairs did not correspond exactly to the proposed reference conditions (due to the specificity of each equipment for what concerns temperature control, data acquisition or wave generation),

2)   when the amount of results given per binder was so great (360 in some cases) that interpolation had to be carried out in order to derive representative values of $|E^*|$ and δ at the reference conditions,

3)   when enough data were available to do so, a more fundamental interpretation was carried out by deriving a master curve and Black diagrams of the complex modulus from each set of results.

In the two last cases some of the data sets supplied by the participants were analysed on the basis of a procedure for the automatic determination of shifting factors and the derivation of interpolated results from the modulus master curve. This operation was carried out by applying the temperature-frequency equivalency principle.

2.6.3   DERIVATION OF MASTER CURVES AND SHIFTING FACTORS

A special computer procedure (called FIMAST) developed at the Belgian Road Research Centre (Francken and Vanelstraete, 1996) has been applied to determine :

•   the parameters δH of an Arrhenius expression describing the shifting factor (2.1) used to build the master curve,

•   to derive interpolated values of the complex modulus for particular conditions of frequencies and temperatures.

A polynominal power function of the form :

$$F(X) = C_1 + \sum_{i=1}^{i=D} C_{i+1} \cdot X^i \quad (2.2)$$

with $X \equiv \log(\alpha_T \cdot Fr)$

is adopted for adjusting the master curve.

The δH value optimising the scatter of the experimental points around this power function is adopted as the true value. A separate analysis is made on the stiffness modulus $|E^*|$ and on the phase angle δ so that two different values are given for the activation energy δH.

When data are given in the low temperature range (down to -25°C) the computer program also determines the purely elastic modulus E_∞ by extrapolating $|E^*|$ for $\delta = 0$ (on the BLACK diagram).

2.7 Presentation of the results

By combining the results given for the reference conditions with the interpolated results of the laboratories using conditions other than the 15 reference conditions one obtains the overall results presented in Tables 2.6 to 2.9 for the 4 binders.

Tables 2.6a to 2.9a are giving the data corresponding to measured and interpolated values of the stiffness modulus |E*| for the four binders and the different laboratories. Tables 2.6b to 2.9b give the phase angles δ also measured or interpolated for the four binders.

2.8 Analysis of the complex modulus for the reference bitumen 244

The detailed analysis of the data corresponding to the four tested binders led to identical conclusions for what concerns the precision and accuracy of the measurements. The outlying results appeared also to be the same for the four binders. For reasons of concision and clarity the detailed data analysis and relevant conclusions will be illustrated on the basis of the data corresponding to the plain bitumen (number 244). The reader interested in more details of this study on the other binders may use the data given in the corresponding tables.

2.8.1 NORM OF THE COMPLEX MODULUS |E*|

On Fig. 2.3 stiffness modulus results can be compared with the values calculated for bitumen 244 on the basis of Van der Poel's nomogram (first row in table 2.6a noted VDP). Owing to the enormous range covered by the modulus (almost 7 decades) there seems generally to be a good agreement. It is to be noted however that most of the outlying results are to be found in the range of high stiffness values, and that they are systematically too low. In absolute values we see in the data table that strong differences are to be noticed from one laboratory to another.

When Van der Poel's nomogram was published it was supposed to be accurate within a factor of 2. In order to detect the outlying results we have taken the assumption that the values derived from Van der Poel's nomogram are true and that outlying results are those displaying a difference of more than a factor 2 with respect to them. The criterium of acceptance was thus :

$$0.5 < \frac{|E*|}{VdP} < 2 \quad (2.3)$$

The measures rejected in this way are underlined in Table 2.6a.

The number and percentages of acceptable data are given for each set of measurements in the right end columns of Table 2.6a.

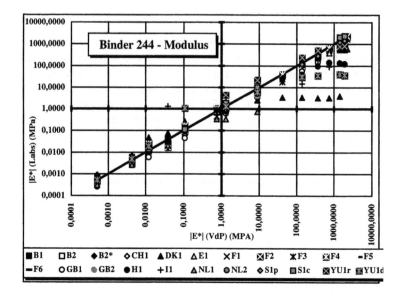

Fig. 2.3 Stiffness modulus |E*| of binder 244; Measured values versus Van der Poel predicted values.

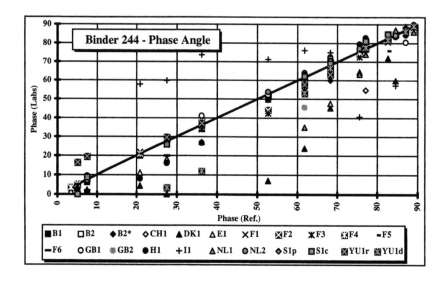

Fig. 2.4 Phase angles - binder 244 Overall measurements versus reference values.

2.8.1.1 Plain bitumen (244)

- 76 percent of the data are accepted on this basis,
- 8 laboratories have a rate of acceptance of 100% of their data (B1, B2, CH1, F1, F4, F5, NL2, S1),
- three test series have a high number of rejected values :

DK1 (27% acceptance) : It was stated that this equipment was in development at the time of the experiment. It generates high strains which might result in non linearity. It is possible also that the mechanical stiffness of the specimen was higher than the device fixing and shaft system for most of the temperature range. Which causes the measured "apparent" modulus to remain constant at a very low value.

GB1 (20% accepted) : Problems with the temperature control system were reported. Subsequently to the work it had been noticed that the calibrated temperatures had changed by between 1.1 to 3.6°C.

I1 (50% acceptance) : The equipment was recently purchased. The results presented a high scatter with some values out of scale by several decades. Unrealistic values obtained at 75°C and 100Hz for the four binders are revealing a specific problem for these particular conditions.

YU1 : (50% acceptance on both equipments).

- In the low temperature range, as already mentioned, it must be noticed that all the outlying results are on the low side. Most of the tests carried out in the shear mode (DK1, E1, F6, GB1, I1 and YU1d) give too low values.
- With the exception of S1 which remains with realistic measurements down to -25°C, it appears that the shear mode is not suitable below a given temperature. This limiting temperature is depending on the test arrangement adopted. For each set of values corresponding to a given laboratory, the difference with respect to Van der Poel's values increases when the temperature decreases.
- Most of the equipments gave results in the intermediate temperature range from 25 to 50°C with a high rate of acceptance.
- At 75°C, the rate of acceptance is still high but here almost all the tests were made in rotating shear mode with parallel plate specimens (R-SH-PP).

2.8.1.2 Polymer modified binders (245, 248 and 249)
The evaluation of the three polymer modified binders considered in this study raises the problem of the reference. It is indeed well known that stiffness moduli derived from Van der Poel's nomogram on the basis of the penetration index and softening temperature are not reliable for Polymer Modified Binders (PMB). Therefore we have taken as reference average values of the measurements obtained by a panel of five laboratories having scored 100% acceptance for the case of binder 244 : B2, CH1, F1, NL2 and S1. These values are given on the first data row labelled REF in Tables 2.7a to 2.9a. The data selection was made with the same criterium as for the pure binder and the results are displayed in the right hand side columns of each table.

The examination of the results of these tests reveals that :

- the outlying results are very similar to those found with binder 244,
- the general figures defining the overall variation and the variation when excluding outlyers are also similar,

2.8.2   PHASE ANGLE δ

The practical importance of phase measurements has been pointed out in the SHRP studies. They have led to the proposal of the first existing performance requirements using this information as a basic piece of information. It is indeed considered in the SHRP-SUPERPAVE method developed in the context of the Strategic Highway Research Program (1994) (see table 1.6) that the basic requirements based on the complex modulus must be :

1)   a minimum value of $|G^*|/\sin(\delta)$ in the high service temperature range to limit permanent deformation,
2)   a maximum value of $|G^*|.\sin(\delta)$ in the medium temperature range to reduce the risk of fatigue cracking.

As there is no reference such as Van der Poel's nomogram to assess the validity of the data, detecting outlying results is more difficult. Reference values were derived from the overall average of the measured phase angles (clearly outlying results being discussed). It can be seen in the data Tables 2.6b to 2.9b that, if a few outlyers are removed, the overall results show less scatter than the stiffness modulus. However we also notice on Fig. 2.4 that there are clearly to categories of results :

1)   Most of the values which are in agreement within 2 to 5° for all conditions.
2)   The other results are not in agreement with the overall average they display very large difference generally greater than 10°.

This second situation appears for example in the case where the shear mode is used at temperatures lower than 25°C (E1, H1, I1 and YU1d). We must also bear in mind that unlike what is the case for the modulus, the absolute value of the error is more important for practice than the relative one because :

1) the range of variation is restricted,
2) the phase angle has more effect on the performance in the medium and high temperature range,
3) a high relative error on the small values (<10°) is of little significance at low temperatures.

The selection of the outlying cases was made on the bases of the overall average phase angles. The analysis carried out reveals that the variation is generally higher for high frequencies combined with low stiffness values, which might reveal some problems of inertial corrections or resonance effects.

When the phase angle lies above 40° the variation is in all cases lower than 20% and drops below 8% in all cases for 82% acceptance of the results.

It can thus clearly be stated that the phase measurements can be made with a good accuracy with most of the existing equipments.

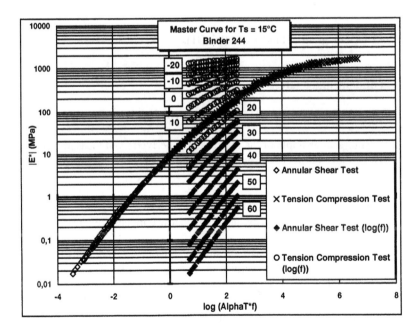

Fig. 2.5 Master curve of bitumen 244.

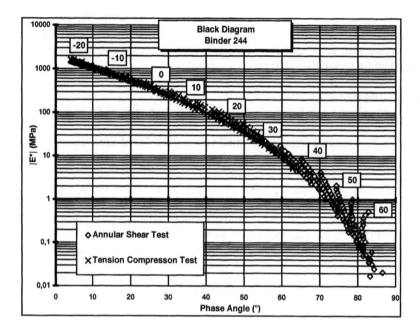

Fig. 2.6 Black diagram of binder 244 (lab B1).

2.9 Fundamental analysis

2.9.1 MASTER CURVES AND SHIFTING FACTORS

Fig. 2.5 displays an example of the master curve derived from measurements carried out at temperatures between -30 and +60°C on a linear vibrating rheometer with two test modes :

- Annular shear for temperatures higher than 25°C.
- Tension compression on cylinders in the low temperature range.

This curve was built on the basis of shifting factors derived from the Arrhenius equation by determining the activation energy optimising the scatter of the experimental points.

This master curve can then be used to determined particular values of moduli for any combination of temperatures or frequencies located in the experimental range.

Several laboratories (between 12 and 13 depending on the binder) gave a sufficient number of data over the full temperature range to allow the determination master curves and shifting factors on the basis of the method proposed by Francken and Vanelstraete (1995).

The activation energies derived in this way are given in the last column of the Tables 2.6a to 2.9a for the $|E*|$ modulus and in the Tables 2.6b to 2.9b for the phase angle δ.

The analysis of the data leads to the following statements :

2.9.1.1 Stiffness modulus $|E*|$

The activation energy defining the shifting factors for the modulus master curve can be determined with a variation factor of less then 10%.

Its value is almost the same for the four binders tested :

$$\delta H_{|E*|} = 158 \text{ kJ/mole}$$

2.9.1.2 Phase angle δ

The activation energy defining the shifting factors of the phase angle are determined with a variation factor of 15% for the PMB binders and 12% for the reference bitumen 244.

Its value is almost the same for the four binders tested :

$$\delta H_\delta = 226 \text{ kJ/mole}$$

We come thus to the surprising conclusion that there is a large difference in the shifting factors to be applied to each component of the complex modulus.

This conclusion is clearly established on the basis of all the results presented in the frame of this experiment.

2.9.2   BLACK DIAGRAMS

Fig. 2.6 represents the $|E*|$ versus δ relationship also called Black diagram for the same set of results. This curve allows the derivation of the purely elastic modulus E_∞ by extrapolating the curve to zero phase angles. This operation yields values of the glassy modulus E_∞ which are very close to one another for the four tested binders.

2.10   Conclusions

The results presented by the RILEM interlaboratory test programme on binder rheology indicate that :

1.   In the present state of development of the equipments and procedures the interlaboratory variation (standard deviation divided by average value) is depending on the temperature. The interlaboratory test shows that when using different selected equipments (outlyers being removed) the variation on the |E*| modulus is of the order of 20% at medium temperatures and rises to 40% at high (>50°C) and low (<0°C) temperature.

2.   The precision achieved on the measurement of the phase angle is much better (1 to 12 % depending on the magnitude of δ). It is in the most interesting range of temperatures (20°C and above according to *Kennedy T.W, Huber G.A, Harrigan E.T, Cominsky R.J, Hughes C., Von Quintus H., and J.S. Moulthrop*) that the precision is the best.

3.   The statements based on the analysis of data relevant to the pure bitumen 244 taken as reference remain unchanged for the three polymer modified binders used in the test. This means that the different methods used will work with the same accuracy and precision on these binders.

4.   Owing to the precision required for practical purposes this situation is in progress with respect to the factor 2 generally accepted as realistic in the past (*Van der Poel, 1954*). It represents also a substantial progress with respect to a former RILEM interlaboratory test carried out in 1984 (*Verstraeten, 1983*).

5.   The AMRL analysis made in 1994 in the frame of the SHRP SUPERPAVE Program (*Haleem Tahir, 1994*) reveals interlaboratory variation on the stiffness of the order of 13 to 15 % for different laboratories using the same type of equipment and a standardised testing procedure (AASHTO MP1). This indicates that further improvement is possible and that practical recommendations have to be set up in order reach this goal.

6.   The variability observed in this experiment can to a large extend be attributed to a lack of accuracy and control of the test temperature. This variation is moreover of the same order as the variation in bitumen stiffness resulting from a difference in temperature of 1°C. This means that further improvements are already to be expected if this parameter is better calibrated and controlled.

7.   The detailed analysis of the master curves and Black curves (log |E*| vs δ) are showing that the purely elastic moduli of the different binders is almost the same and that the shifting factors to be used for the norm |E*| of the modulus and for the phase angle δ are not the same.

Table 2.6a Norm of the complex modulus |E*| of binder 244

N°	Code	T=-25°C 1 Hz	10 Hz	100 Hz	T=0°C 1 Hz	10 Hz	100 Hz	T=25°C 1 Hz	10 Hz	100 Hz	T=50°C 1 Hz	10 Hz	100 Hz	T=75°C 1 Hz	10 Hz	100 Hz	Total	Acc.	%	Dh
	VdP	1416,0	1920,0	2223,0	140,0	367,0	737,0	1,32	8,95	41,66	0,012	0,107	0,749	0,0005	0,0043	0,038	12	12	100	40,00
1	B1	1162,0	1553,0		88,7	260,7	593,8	1,58	8,32	36,24	0,022	0,152	0,966	0,0007	0,0052	0,024	6	6	100	35,00
2	B2							0,93			0,021	0,147		0,0009	0,0068	0,038	5	5	100	
2*	B2*							1,66			0,019	0,153					4	4	100	
3	CH1							1,79	9,15		0,020	0,135					4	4	100	
4	DK1	4,0			3,5	3,3	3,3	1,06	2,52	3,42	0,046	0,252	1,071			0,075	11	3	27	35,00
5	E1	547,0	546,0	606,0	97,7	225,0	392,0	2,09	8,88	31,96		0,144	0,853	0,0005			11	8	73	45,00
6	F1	1024,0	1426,0	1679,0	75,5	223,3	516,1	1,45	7,52	32,34	0,014	0,153	0,940	0,0005	0,0030	0,017	11	11	100	39,50
7	F2							1,09	6,54			0,094	0,737				8	6	75	38,50
8	F3				108,3			1,21	4,31	18,32	0,011	0,110	0,445				7	5	71	50,00
9	F4	1369,0	1886,0	2217,0	100,0	291,9	668,7	1,78	9,20	39,51	0,024	0,166	1,038			0,025	13	13	100	40,50
10	F5	1096,0	1418,0		90,8	259,9	569,9	1,50	7,82	34,22		0,146	0,842				10	10	100	41,00
11	F6	518,0	481,0		116,9	245,0		1,23	7,08		0,014	0,114		0,0005	0,0048		10	8	80	
12	GB1	592,0	722,0		56,1	143,3		0,60	3,51		0,006	0,044		0,0003	0,0026		10	2	20	
13	GB2							0,96	4,06		0,011	0,086		0,0004	0,0038		6	5	83	
14	H1	130,0	119,0		39,0	96,9	136,6	0,83	5,15	24,61	0,020	0,128	0,843	0,0006	0,0053	0,035	14	9	64	35,00
15	I1				14,8	178,8	90,9	0,60	9,60	18,66	0,013	0,126	0,510	0,0006	0,0069	1,281	12	6	50	
16	NL1							0,33	0,74		0,012	0,076	0,330	0,0007	0,0026	0,021	8	5	63	35,00
17	NL2				157,0	357,0		1,28	7,32		0,011	0,092		0,0006	0,0052		8	8	100	38,50
18	S1p	1190,0	1394,0		120,8	279,4		0,92	5,46		0,013	0,103		0,0005	0,0045		10	10	100	
18*	S1c	1995,0	2329,0		205,5	477,0		1,87	9,93		0,014	0,110		0,0006	0,0057		10	10	100	
19	YU1r	744,0	820,0		233,9	507,3		4,50	21,91								6	3	50	
19*	YU1d	41,0	37,0		30,0	35,8		1,29	6,12		0,019	1,052		0,0005	0,0049		10	5	50	
Average values for all results																				
AVG		801	1061	1501	96,2	239	371,4	1,39	7,26	26,59	0,017	0,171	0,779	0,0005	0,0047	0,1894				39,4
STD		556	684	670	61,6	136,2	241,8	0,81	4,16	10,81	0,008	0,201	0,236	0,0002	0,0013	0,4129				4,35
n		13	12	3	16	15	8	22	20	9	18	21	11	13	13	8	202	202	100	12
% Var		69	64	45	64	57	65	58	57	41	49	118	30	30	28	218				11
Average after removal of outlyers with 76,24% acceptance																				
AVG		1226	1668	1948	127	313	548	1,36	7,72	33,15	0,016	0,124	0,824	0,0005	0,0047	0,0362	202	154	76	
STD		360	340	2	49	97	92	0,36	1,48	4,59	0,004	0,026	0,198	0,0002	0,0013	0,0182				
n		7	6	2	11	10	5	18	14	6	16	18	10	12	13	6				
% Var		29	20		38	31	17	27	19	14	27	21	24	30	28	50				

Table 2.6b Phase angle δ of binder 244.

Binder N° 244 - Phase Angle

N°	Code	T=-25 °C			T=0°C			T=25°C			T=50°C			T=75°C			Dh
		1 Hz	10 Hz	100 Hz	1 Hz	10 Hz	100 Hz	1 Hz	10 Hz	100 Hz	1 Hz	10 Hz	100 Hz	1 Hz	10 Hz	100 Hz	
	Mean	7,17	4,45	2,77	35,81	24,67	20,63	66,44	57,21	47,99	82,54	78,10	71,08	88,73	86,43	77,34	
1	B1	6,20	4,60		35,90	27,60	20,00	67,40	60,40	52,50	83,20	79,60	75,20				58,50
2	B2							69,90			83,10	79,30	74,80		87,30	84,70	51,50
2*	B2*							60,10			82,70	77,40		88,50	87,70		
3	CH1							69,90	62,70		83,10	54,80					
4	DK1				27,00	0,10	4,50	45,40	23,90	7,10	72,00	76,80	64,60			59,90	51,50
5	E1	2,20	2,30	1,50		18,20	11,10	66,50	58,90	49,90		79,00	75,50				64,50
6	F1	7,60	5,10	3,40	35,70	28,40	21,80	66,70	60,00	52,50	80,70	78,80	75,10				62,00
7	F2							68,40	59,20	44,30	83,60	79,70	74,40	87,60	86,60	84,90	51,00
8	F3				33,90			66,60	54,30	42,50	84,00	79,70	72,70				45,50
9	F4	8,50	5,20	3,40	38,00	29,00	20,90	68,90	60,60	51,50	84,90	80,90	75,20				51,00
10	F5	6,60	5,20		34,70	27,50	20,90	67,30	60,40	52,90	76,00	77,60	75,70				64,50
11	F6	2,00	5,00		33,00	20,00		71,00	63,00		84,00	80,00		89,00	86,00		
12	GB1	9,40	2,60		41,10	29,50		72,50	62,30		84,90	78,00		88,60	80,50		
13	GB2							65,50	45,60		84,30	79,90		89,10	87,90		
14	H1	1,30	2,50		26,60	16,10	8,00	71,00	61,30	49,60	83,60	82,80	79,50	89,80	84,40	83,50	57,50
15	I1				73,70	59,80	57,80	75,00	76,10	71,30	84,80	82,20	40,70	89,70	88,10	57,30	57,30
16	NL1							47,70	34,90		82,50	74,10	63,40	86,00	88,90	86,90	57,50
17	NL2				35,70	26,70		72,00	64,10	53,80	83,00	80,70	77,30	89,10	86,40	84,20	52,50
18	S1p	7,00	0,00		37,10	28,10		70,80	63,30		83,90	80,20		89,20	87,60		
18*	S1c	8,70	0,00		35,90	25,80		69,10	62,30		84,70	81,10		89,10	85,90		
19	YU1r	19,40	16,40		37,00	29,70		63,30	57,80								
19*	YU1d				11,90	3,50		66,70	53,10		81,80	77,60		89,10	86,30		

Table 2.7a Norm of the complex modulus |E*| of binder 245.

N°	Code	T=-25°C 1 Hz	10 Hz	100 Hz	T=0°C 1 Hz	10 Hz	100 Hz	T=25°C 1 Hz	10 Hz	100 Hz	T=50°C 1 Hz	10 Hz	100 Hz	T=75°C 1 Hz	10 Hz	100 Hz	Total	Acc.	%	Dh
	Ref.	785,0	1248,0	1562,0	32,7	109,2	298,2	0,62	2,72	13,06	0,028	0,116	0,512	0,0031	0,0090	0,043	12	12	100	37,50
1	B1	683,0	1195,0		25,5	94,1	284,6	0,65	2,82	12,11	0,034	0,131	0,536	0,0038		0,043			100	37,50
2	B2																6	5	83	
2*	B2*							0,69	2,82		0,029	0,134	0,218	0,0038	0,0020		4	3	75	
3	CH1							0,36	1,70		0,017	0,057				0,071	10	3	30	43,00
4	DK1				5,1	8,7	9,7	0,33	0,89	2,43	0,072	0,107					11	8	73	41,00
5	E1	363,0	453,3	564,5	23,5	74,0	181,3	0,70	2,44	8,92		0,134	0,464				12	12	100	38,50
6	F1	834,0	1386,0	1768,0	28,6	107,4	329,6	0,69	2,88	12,29		0,141	0,532			0,046	8	8	100	35,00
7	F2							0,34	1,79		0,018	0,071	0,342	0,0023	0,0080	0,027	8	8	100	41,50
8	F3				43,2	88,3		0,66	2,69	13,43	0,024	0,130	0,425				13	13	100	37,50
9	F4	658,0	1040,0	1357,0	23,5	93,2	288,9	0,58	2,40	10,74	0,040	0,133	0,478	0,0020	0,0100	0,049	10	10	100	37,50
10	F5	672,0	1067,0		27,1	98,3	289,9	0,79	3,11	12,52	0,020	0,164	0,636				10	9	90	38,00
11	F6	490,0	518,7		47,3	132,0		0,62	2,92		0,009	0,044		0,0011	0,0060		10	9	90	
12	GB1	397,0	583,0		16,2	50,9		0,26	1,30								10	3	30	
13	GB2							0,96	4,06		0,011	0,086		0,0004	0,0040		6	3	50	
14	H1	150,0	142,1		23,2	70,9	130,9	0,55	2,59	12,16	0,040	0,126	0,488	0,0030	0,0160	0,051	14	11	79	36,50
15	I1	20,5			20,5	36,6	81,9	0,43	2,14	7,71	0,022	0,097	1,035	0,0023	0,0170	1,296	12	8	67	
16	NL1							0,18	0,52		0,017	0,058	0,178	0,0016	0,0070	0,025	8	3	38	35,00
17	NL2	38,7			38,7	118,1		0,46	2,84	17,62	0,028	0,085	0,379	0,0033	0,0130	0,034	11	11	100	37,00
18	S1p	689,0	961,5	1230	36,2	99,3		0,40	1,86		0,020	0,091		0,0027	0,0130		10	10	100	
18*	S1c	1077,0	1550,0	1562	52,7	144,4		0,71	3,18		0,019	0,086		0,0023	0,0130		10	10	100	
19	YU1r	734,0	1089,0		43,6	116,1		0,82	3,74								6	6	100	
19*	YU1d	50,0	47,9		11,3	24,0		0,57	2,24		0,018	0,078		0,0022	0,0120		10	6	60	
	Average values for all results																			
AVG		566	836,2	1230	29,13	84,76	199,6	0,56	2,424	10,99	0,026	0,103	0,476	0,0022	0,01	0,182				38,2
STD		278	459,1	499,5	12,95	37,25	109,1	0,196	0,846	3,809	0,014	0,032	0,21	0,0009	0,005	0,394				2,48
n		12	12	3	16	16	8	21	21	10	17	20	12	12	12	9	201	162		11
% Var		49	54,91	40,62	44,44	43,95	54,65	35,09	34,9	34,65	56,35	31,19	44,16	39,97	45,57	216,2				6,49
	Average after removal of outlyers with 80,59% acceptance																			
AVG		693	1184	1562	33,34	103	274,9	0,595	2,677	11,95	0,025	0,111	0,428	0,0027	0,012	0,043				
STD		182	195,5	205,7	10,35	20,97	49,52	0,17	0,608	2,657	0,008	0,026	0,163	0,0006	0,004	0,014				
n		9	7	2	13	12	5	19	18	9	14	17	10	9	10	8			81	
% Var		2	17	13	31	20	18	29	23	22	32	23	38	21	31	32				

Table 2.7b Phase angle δ of binder 245.

Binder N° 245 - Phase Angle

N°	Code	T=-25 °C			T=0°C			T=25°C			T=50°C			T=75°C			Dh
		1 Hz	10 Hz	100 Hz	1 Hz	10 Hz	100 Hz	1 Hz	10 Hz	100 Hz	1 Hz	10 Hz	100 Hz	1 Hz	10 Hz	100 Hz	
	Mean	7,17	4,45	2,77	35,81	24,67	20,63	66,44	57,21	47,99	82,54	78,10	71,08	88,73	86,43	77,34	
1	B1	6,20	4,60		35,90	27,60	20,00	67,40	60,40	52,50	83,20	79,60	75,20		87,30	84,70	40,00
2	B2							69,90			83,10	79,30	74,80				35,00
2*	B2*							60,10			82,70	77,40		88,50	87,70		
3	CH1							69,90	62,70		83,10	54,80					
4	DK1					0,10	4,50	45,40	23,90	7,10	72,00	76,80	64,60			59,90	35,00
5	E1	2,20	2,30	1,50	27,00	18,20	11,10	66,50	58,90	49,90		79,00	75,50				45,00
6	F1	7,60	5,10	3,40	35,70	28,40	21,80	66,70	60,00	52,50	80,70	78,80	75,10				39,50
7	F2							68,40	59,20	44,30	83,60	79,70	74,40	87,60	86,60	84,90	38,50
8	F3				33,90			66,60	54,30	42,50	84,00	79,70	72,70				50,00
9	F4	8,50	5,20	3,40	38,00	29,00	20,90	68,90	60,60	51,50	84,90	80,90	75,20				40,50
10	F5	6,60	5,20		34,70	27,50	20,90	67,30	60,40	52,90	76,00	77,60	75,70				41,00
11	F6	2,00	5,00		33,00	20,00		71,00	63,00		84,00	80,00		89,00	86,00		
12	GB1	9,40	2,60		41,10	29,50		72,50	62,30		84,90	78,00		88,60	80,50		
13	GB2							65,50	45,60		84,30	79,90		89,10	87,90		
14	H1	1,30	2,50		26,60	16,10	8,00	71,00	61,30	49,60	83,60	82,80	79,50	89,80	84,40	83,50	35,00
15	I1				73,70	59,80	57,80	75,00	76,10	71,30	84,80	82,20	40,70	89,70	88,10	57,30	
16	NL1							47,70	34,90		82,50	74,10	63,40	86,00	88,90	86,90	35,00
17	NL2				35,70	26,70		72,00	64,10	53,80	83,00	80,70	77,30	89,10	86,40	84,20	38,50
18	S1p	7,00	0,00		37,10	28,10		70,80	63,30		83,90	80,20		89,20	87,60		
18*	S1c	8,70	0,00		35,90	25,80		69,10	62,30		84,70	81,10		89,10	85,90		
19	YU1r	19,40	16,40		37,00	29,70		63,30	57,80								
19*	YU1d				11,90	3,50		66,70	53,10		81,80	77,60		89,10	86,30		

Table 2.8a Norm of the complex modulus |E*| of binder 248.

N°	Code	Binder N° 248 - |E*| T=-25°C 1 Hz	10 Hz	100 Hz	T=0°C 1 Hz	10 Hz	100 Hz	T=25°C 1 Hz	10 Hz	100 Hz	T=50°C 1 Hz	10 Hz	100 Hz	T=75°C 1 Hz	10 Hz	100 Hz	Total	Acc.	%	Dh
	VdP	833	1236	1619	48,89	146,5	337,3	0,885	4,347	19,91	0,035	0,153	0,635	0,0026	0,015	0,047	13	13	100	37
1	B1	636	1073	1492	33,07	110,3	298,5	0,924	4,132	17,14	0,037	0,171	0,784	0,0031	0,015	0,05	6	6	100	35
2	B2							0,712	4,038		0,043	0,181		0,0031		0,066	6	6	100	
2*	B2*							0,983			0,038	0,179		0,0032	0,018		6	6	100	35
3	CH1							1,209	5,76		0,043	0,209					4	4	100	
4	DK1	3			3,725	3,822	3,41	0,805	2,076	3,435	0,066	0,227	0,81		0,041	0,091	12	5	41,7	35
5	E1	416	494,9	565,9	39,69	114,8	249,9	1,121	4,695	18,19		0,18	0,81				11	8	72,7	39
6	F1	784	1278	1690	37,6	127,4	349,2	0,94	4,189	17,63	0,017	0,164	0,733			0,047	12	12	100	38
7	F2							0,496	2,791		0,024	0,08	0,453	0,0017	0,006	0,024	8	6	75	36
8	F3				43,18	88,3		0,662	2,688	13,43	0,024	0,13	0,425				8	8	100	41,5
9	F4	767	1147	1675	67,02	171,1	342,6	0,782	4,191	18,85	0,028	0,067	0,302			0,026	13	11	84,6	45
10	F5	779	1137		41,73	137,2	359	1,02	4,399	18,1	0,028	0,171	0,736	0,0018	0,013		10	10	100	39
11	F6	433	458,5		62,94	158,5		0,83	4,32		0,011	0,14		0,0011	0,007		10	9	90	
12	GB1	407	553,7		19,8	63,93		0,387	2,138		0,011	0,064			0,013		6	0	0	
13	GB2							0,72	2,942		0,026	0,13		0,002	0,013		6	6	100	
14	H1	98	117,8		54,9	111	139,3	0,441	2,338	11,23	0,012	0,033	0,126	0,0057	0,006	0,01	14	4	28,6	49
15	I1				17,28	20,37	72,3	0,312	1,752	6,06	0,03	0,012	0,981	0,0285	0,023	1,275	12	3	25	35
16	NL1							0,3	0,645		0,022	0,095	0,302	0,0013	0,007	0,035	8	3	37,5	35
17	NL2	721	901,5		51,08	154,2		0,612	4,064	27,81	0,032	0,123	0,618	0,0023	0,013	0,047	11	11	100	
18	S1p	1199			46,44	121,5		0,689	3,522		0,031	0,155		0,002	0,013		10	10	100	
18*	S1c		1547		62,85	178,7		0,784	3,999		0,024	0,115		0,0018	0,013		10	10	100	
19	YU1r	873	1186		53,82	143,6		0,999	4,749								6	6	100	
19*	YU1d	52	53,55		2,795	5,817		0,497	1,037		0,024	0,126	0,706	0,0017	0,011	0,008	10	5	50	
	Average values for all results																			
	AVG	551	828,9	1356	39,87	106,9	226,8	0,737	3,356	15,19	0,03	0,131	0,59	0,0043	0,014	0,167	210	210	100	38,7
	STD	342	459,7	462,7	19,44	54,59	128,9	0,252	1,297	6,635	0,012	0,055	0,251	0,0071	0,009	0,37				4,27
	n	13	12	4	16	16	8	22	21	10	18	21	12	13	14	10				12
	% Var	62	55	34	49	51	57	34	39	44	42	42	43	164	62	221				11
	Average after removal of outlyers with 74,29% acceptance																			
	AVG	774	1181	1619	49,53	134,7	319,8	0,821	3,926	17,8	0,033	0,152	0,706	0,0022	0,015	0,048	210	156	74	
	STD	203	183,6	90,07	10,54	26,18	40,64	0,195	0,855	4,531	0,011	0,038	0,169	0,0005	0,003	0,021				
	n	8	7	3	12	12	5	18	16	8	15	17	9	9	9	8				
	% Var	26	16	6	21	19	13	24	22	25	33	25	24	25	23	43				

Table 2.8b Phase angle δ of binder 248.

Binder N° 248 - Phase Angle

N°	Code	T=-25 °C			T=0°C			T=25°C			T=50°C			T=75°C			Dh
		1 Hz	10 Hz	100 Hz	1 Hz	10 Hz	100 Hz	1 Hz	10 Hz	100 Hz	1 Hz	10 Hz	100 Hz	1 Hz	10 Hz	100 Hz	
	Mean	9,75	5,81	5,98	40,25	32,28	24,93	61,17	55,93	52,58	64,45	63,17	61,33	75,43	69,17	60,84	
1	B1	12,20	9,20	7,70	39,70	33,70	27,90	61,90	58,00	53,10	62,10	65,10	65,50				64,50
2	B2							63,00			63,60	60,90	60,70		72,50	65,90	51,50
2*	B2*							58,70	44,70		63,80	62,30		72,00	67,00		
3	CH1							64,80	60,70		64,90	48,80					
4	DK1	6,00	3,40	2,80	0,30	-0,90	3,80	51,50	32,30	13,10	55,90	63,90	59,50			53,00	45,50
5	E1	12,50	9,60	7,60	39,80	29,70	20,20	64,40	60,30	53,40	63,10	64,30	65,60				51,50
6	F1				40,20	33,70	27,60	64,10	60,10	54,80	66,50	66,50	67,30				64,50
7	F2							62,60	62,00	55,90	64,00	64,60	62,90	62,30	65,70	67,10	57,50
8	F3				44,50			61,30	55,60	49,30	50,70	62,60	63,00				45,50
9	F4	14,30	8,90	5,80	45,20	36,00	27,10	65,60	61,40	54,40	71,60	62,40	66,10		46,80	46,80	45,50
10	F5	10,20	7,80		40,90	33,60	26,40	63,80	60,60	56,10	68,00	68,50	66,90				64,50
11	F6	3,60	2,60		42,00	27,00		65,00	62,00		69,10	65,00		82,00	72,00		
12	GB1	15,50	10,10		49,80	36,90		64,50	58,80		64,80	62,60		65,80	66,50		
13	GB2							59,60	48,10		64,60	64,10		77,40	69,00		
14	H1	1,50	1,40		34,40	22,80	12,60	62,80	56,90	48,00	65,40	65,60	65,00	74,90	65,90	64,30	48,00
15	I1				52,60	49,40	53,80	52,60	64,30	81,80	65,50	61,80	43,40	77,40	60,60	55,20	
16	NL1							40,30	27,20		65,30	57,10	47,10	80,20	74,10	67,10	45,50
17	NL2				44,30	34,80		64,00	62,30	58,50	65,10	64,40	64,30	77,30	71,50	67,30	58,50
18	S1p	9,80	4,10		41,60	33,20		63,70	60,40		68,70	64,60		79,50	71,30		
18*	S1c	10,80	0,00		44,70	35,00		64,00	62,70		66,30	64,90		79,40	72,90		
19	YU1r	15,20	11,30		42,80	35,50		64,00	59,10			66,50					
19*	YU1d	5,40	1,30		41,20	43,80		63,50	57,10					76,90	70,20		

Table 2.9a Norm of the complex modulus |E*| of binder 249.

N°	Code	T=-25°C 1 Hz	10 Hz	100 Hz	T=0°C 1 Hz	10 Hz	100 Hz	T=25°C 1 Hz	10 Hz	100 Hz	T=50°C 1 Hz	10 Hz	100 Hz	T=75°C 1 Hz	10 Hz	100 Hz	Total	Acc.	%	Dh
	Ref.	1122,0	1507,0	1758,0	93,9	251,8	390,0	1,31	7,12	34,50	0,038	0,197	0,886	0,0017	0,0090	0,051	13	13	100	39,50
1	B1	1134,0	1539,0	1684,0	66,9	216,4	545,1	1,36	6,39	27,51	0,038	0,203	0,915	0,0015	0,0100	0,042	6	6	100	35,00
2	B2							0,98			0,037	0,250		0,0023	0,0020	0,061	6	5	83	
2*	B2*							1,23	4,57		0,042	0,309					4	4	100	
3	CH1							2,59	10,89		0,057	0,360					4	4	100	
4	DK1				4,2	4,1	3,9	1,37	2,94	4,08	0,080		1,262				10	3	30	36,50
5	E1	470,0	477,0	544,0	79,6	188,4	336,5	2,35	8,95	30,37		0,259	1,239			0,111	11	8	73	43,00
6	F1	1089,0	1556,0	1831,0	64,4	207,6	524,2	1,52	6,71	27,95	0,021	0,237	1,088	0,0010	0,0050	0,054	12	12	100	39,00
7	F2							1,07	5,45			0,129	0,839			0,028	8	8	100	37,50
8	F3				78,9			0,62	2,30	12,63	0,019	0,146	0,332				7	2	29	48,00
9	F4	1006,0	1457,0		102,4	235,7	442,7	1,47	7,10	28,21	0,033	0,103	0,483	0,0015	0,0130	0,045	11	11	100	47,00
10	F5	903,0	1280,0		53,1	171,1	438,0	1,32	5,91	24,29		0,201	0,981	0,0006	0,0060		11	11	100	38,50
11	F6	578,0	571,0		127,7	279,8		1,35	8,04		0,028	0,173		0,0015	0,1250	0,112	10	9	90	
12	GB1	584,0	727,0		36,7	104,0		0,40	2,15		0,010	0,068					10	2	20	
13	GB2							1,30	4,32		0,030	0,193					6	5	83	
14	H1	190,0	177,0		47,3	111,0	174,0	1,87	7,83	28,40	0,078	0,375	1,665	0,0018	0,0190	1,260	14	7	50	36,50
15	I1				7,8	28,7	86,4	0,38	2,00	6,36	0,022	0,160	0,888	0,0023	0,0160	0,038	12	5	42	
16	NL1							0,38	0,75		0,023	0,113	0,377	0,0009	0,0060		8	5	63	35,00
17	NL2				129,9	344,0		1,01	8,01	64,54	0,035	0,162	0,964	0,0017	0,0110	0,052	11	11	100	35,50
18	S1p	1069,0	1226,0		126,3	263,7		1,19	5,96		0,034	0,210		0,0015	0,0130		10	10	100	
18*	S1c	1476,0	1701,0		146,6	336,0		1,62	7,38		0,024	0,117		0,0015	0,0130		10	10	100	
19	YU1r	1046,0	1293,0		81,6	195,6		1,63	7,35								6	6	100	
19*	YU1d	35,0	33,0		15,4	24,3		0,69	3,32		0,023	0,130		0,0012	0,0100	0,046	10	5	50	
	Average values for all results																			
	AVG	798	1003	1353	73,06	180,7	318,9	1,259	5,634	25,43	0,035	0,194	0,919	0,0015	0,019	0,18				39,3
	STD	408	552	575	43,52	104	192,8	0,562	2,589	16,02	0,019	0,08	0,37	0,0005	0,031	0,361				4,28
	n	12	12	3	16	15	8	22	21	10	18	21	12	13	13	10				12
	% Var	51	55	43	60	58	60	45	46	63	53	41	40	31	32	200				10,9
	Average after removal of outliers with 76,7% acceptance																			
	AVG	987	1436	1758	92,06	243,8	457,3	1,439	6,99	33,04	0,032	0,201	1,032	0,0016	0,01	0,046				39,3
	STD	262	162	74	32,05	57,47	73,98	0,452	1,639	12,97	0,01	0,076	0,297	0,0004	0,003	0,01				4,28
	n	9	7	2	12	15	5	18	15	7	14	20	10	12	10	7	206	158	77	12
	% Var	r 27	11	4	35	24	16	31	23	39	30	38	29	26	32	22				10,9

Table 2.9b Phase angle δ of binder 249.

Binder N° 249 - Phase Angle

N°	Code	T=-25 °C			T=0°C			T=25°C			T=50°C			T=75°C			Dh
		1 Hz	10 Hz	100 Hz	1 Hz	10 Hz	100 Hz	1 Hz	10 Hz	100 Hz	1 Hz	10 Hz	100 Hz	1 Hz	10 Hz	100 Hz	
	Mean	6,01	3,37	3,00	34,63	25,77	22,81	60,00	53,02	49,31	74,12	70,17	62,90	86,33	80,51	71,07	
1	B1	5,80	3,80	3,00	35,60	28,20	21,30	62,40	57,80	52,10	73,30	70,50	67,90		83,80	78,20	64,50
2	B2							61,60			75,50	70,50	65,70				51,50
2*	B2*							53,90	37,70		72,90	64,90		85,70	79,40		
3	CH1							60,00	55,50		73,90	83,60					
4	DK1				0,10	0,50	4,10	38,50	21,40	7,60	66,30	66,60	56,10			58,60	54,00
5	E1	2,60	2,50	1,60	26,80	18,70	11,80	60,30	54,50	47,10		69,20	66,70				64,50
6	F1	8,30	5,30	4,40	38,20	30,60	23,20	63,20	57,70	51,40	74,20	71,70	68,40				56,50
7	F2							61,30	57,90	51,00	75,80	71,20	66,80	84,60	82,60	80,00	64,00
8	F3				38,30			61,60	56,40	48,00	76,20	66,60	63,30				45,50
9	F4	10,20	4,70		38,80	31,60	24,60	65,10	58,00	50,20	61,20	69,70	69,60				51,50
10	F5	7,70	5,40		37,00	28,80	21,40	64,90	60,90	55,20	77,40	71,60	68,80				64,50
11	F6	1,50	1,00		34,00	19,00		64,00	55,00		78,00	71,00		87,00	80,00		
12	GB1	3,50	9,80		45,10	34,00		66,00	57,70		80,40	70,20		87,20	77,90		
13	GB2							56,40	39,70		75,60	70,00		86,10	80,40		
14	H1	0,10	0,90		29,10	17,70	8,20	62,50	54,40	43,80	71,80	70,00	66,70	87,70	76,50	72,50	48,00
15	I1				59,30	56,30	67,90	64,60	67,60	87,50	78,60	71,20	40,30	86,40	78,40	56,50	
16	NL1							44,30	33,70		73,10	62,70	51,10	85,20	82,30	75,10	45,50
17	NL2				36,30	26,80		63,60	57,50	48,50	75,30	70,40	66,30	86,60	82,80	76,60	45,50
18	S1p	8,40	0,00		31,90	25,10		63,10	58,90		73,90	70,10		86,60	81,40		
18*	S1c	7,40	0,00		35,40	26,40		60,40	58,20		74,80	71,90		87,00	81,00		
19	YU1r	10,60	6,90		38,90	30,80		62,20	57,00								
19*	YU1d		0,10		29,20	12,00		60,00	56,00		74,10	70,00		85,80	80,10		

2.11   References

Francken L. : Belgian Experience with modified binders Highway research : Sharing the benefits London October 1990.

Francken L and Vanelstraete A : Complex moduli of bituminous binders and mixtures. Interpretation and evaluation. Paper 4.047 EURASPHALT & EUROBITUME Congress Strasbourg 1996.

Van der Poel C. : A general system describing the viscoelastic properties of bitumen. Journal of applied Chemistry, vol 4, part 5 1954.

Verstraeten J. : Essais comparatifs de mesures de modules d'un bitume; Proceedings of the third RILEM International Symposium devoted to Testing of Hydrocarbon binders and materials. Beograd September 1983.

Haleem Tahir A. : An update from AASHTO, State Department of Transportation and Superpave. FOCUS september 1994 pp 4-5.

Kennedy T.W, Huber G.A, .Harrigan E.T, Cominsky R.J, Hughes C., Von Quintus H., and J.S.Moulthrop: Superior Performing Asphalt Pavements (SUPERPAVE) : The product of the SHRP Asphalt Research Program.

Strategic Highway Research Program. National Research Council. Washington DC, 1994.

PART TWO

MIX DESIGN

3

State of the Art Report on Mix Design

by M. Luminari and A. Fidato

3.1 From traditional to modern mix design

During the past 20 years there has been an increase in the aggressive action of traffic on pavements due to the rise in axle loads and number of passages. There has been also a growing concern for safety, driver comfort, maintenance needs and environmental considerations, which has led to the demand for more durable and stronger pavements. In this context greater attention has been devoted to bituminous mixes, given their widespread use in flexible and semi-rigid pavements.

The need for more suitable testing methods aimed at better forecasting the performance of the materials, not only during laying, but over the entire working life of the pavement, has been recognized. Alongside the traditional "empirical" type tests, we have seen the development of two other categories of tests, defined by J. Bonnot at the third RILEM symposium (Belgrade, 1983) as "fundamental" and "simulation." However, none of these latter mechanical tests can in themselves provide complete indications of the qualities required of a bituminous mix, and, more specifically, none of them are sufficient to completely characterize mix durability. More details on these types of tests can be found in the "Inventory of bituminous mix design methods" in Appendix 2, where the different mix design methods of the countries studied will be analyzed.

The tests defined above as empirical permit only a generic assessment of the performance of a mix, and do not permit the determination of the intrinsic properties of the materials. Their manner of stressing specimens is very different than the ways in which the mix is stressed *in situ*, and hence in certain cases they have even been termed "arbitrary".

The fundamental tests serve to determine the intrinsic characteristics of the materials, independent of the test conditions. The intrinsic mix properties will be introduced in the algorithms used to predict pavement performance. While these tests lack the simplicity of the traditional mechanical tests, they permit better interpretation of the distribution of stress within the specimens, and a more precise correlation with the actual load conditions to which a pavement is subjected.

The simulation tests serve to simulate and reproduce in the laboratory those processes involved in mix production and laying (i.e. gyratory shear compactor), or during the pavement working life (i.e. large scale wheel tracking test).

While new types of test methods were being developed especially for wearing courses, there was a growing tendency to adopt special bituminous mix formulations based on the specific pavement problem to be resolved ("special purpose mix design") and also depending on the layer of the pavement to be constructed.

Bituminous Binders and Mixes, edited by L. Francken. RILEM Report 17. Published in 1998 by E & FN Spon, 11 New Fetter Lane, London EC4P 4EE, UK. ISBN 0 419 22870 5

Each layer performs a very distinct function within the pavement as a whole, and consequently the qualities of the raw materials employed therein will vary accordingly. Examples of these special mixes include Stone Mastic Asphalt, Gussasphalt, Hot Rolled Asphalt, Porous Asphalt, Thin and Ultra-Thin Asphalt and Very Soft Asphalt, subjects of the current European standards harmonization process (by the WG1 of the CEN TC 227). These mixes, together with Asphalt Concrete for wearing courses, basecourses and roadbases, cover nearly the entire range of materials used in Europe for construction and maintenance of bituminous pavements.

The difficulties encountered in designing bituminous mixtures derive from the fact that required properties are, sometimes, very difficult to achieve simultaneously, and therefore it is often necessary to settle for a compromise when selecting the most suitable bituminous mixture to be laid. The results of experimental studies carried out in the past years on bituminous mix design have shown the influence of some mix composition factors on mix properties, thereby helping the formulator decide which approach is most appropriate.

The mix design process consists of a series of procedures and/or tests followed by or executed to select aggregates, aggregate gradation, binder type and content, additive/modifier type and content to produce a bitumen-aggregate mix that satisfies specific requirements.

It should be noted that the purpose of mix design is to formulate a mix of aggregates and binder which is both the most economical and at the same time has:

- sufficient binder to ensure durability;
- sufficient percentage of voids in the mineral aggregate, so as to minimize post-compaction by traffic, without giving rise to bleeding or flashing, loss of stability or harmful effects due to the action of air and water;
- sufficient workability to permit laying the mix without risk of segregation;
- sufficient performance characteristics over the service life of the pavement.

The concept underlying modern mix design is to determine the proportions of each mix component, using available materials, such that one achieves optimum in-service performance, commensurate to the traffic and environmental conditions and structural factors to which the pavement will be exposed. The performance of a mix will be judged on whether it fulfills its intended function, i.e., whether all the requirements for the mix are actually met.

In the last few years performance-related and performance-based mix design methods have been developed. In these new formulation approaches, the simulation and/or fundamental tests used allow us to have the best correlation between the results measured in the laboratory and the *in situ* behavior, measuring material response to various states of stress.

Performance-based mix design methods should allow, from the materials properties of the mix and the performance prediction models, estimations of the development of pavement distresses expected over a selected service life, for any pavement structure, in any traffic and environmental condition. The efficiency of these new performance-based mix design methods has still to be proven. However, they represent the basis of the future mix design system.

The new mix design methods are no longer based on the criterion of seeking the optimum mix composition to guarantee maximum value mechanical properties, but rather on that of ensuring definite minimum performance levels. Examples of this concept are presented by the American SHRP Superpave level 2 and 3 and SHRP-A-698 methods, the Great Britain approach from Nottingham University and the French approach, the Australian NARC guide level 2 and 3, the Finnish ASTO/Asphalt Specifications, and the Dutch CROW guideline.

These mix design methods (SHRP, ASTO, NARC) contain distinct levels. With more tests, and more time to complete the mix design process, higher reliability of the design is expected. This distinction permits the selection of a design process that is appropriate for the traffic loads and volumes expected for a given paving project so as to design for optimal service and economic performance, rather than maximum achievable performance.

Later in this chapter a mix design methodology will be proposed which attempts to combine all the positive, important and interesting features of the existing mix design systems, also taking into account the conclusions reached by the first RILEM Interlaboratory Test Program (described in Chapter 4).

3.2   Classification of mix design methods

The international scientific-technical literature was consulted regarding: a) existing standardized and non-standardized mix design methods, b) new mix design methods which are continually being perfected and c) methods which are still being experimented on and analyzed.

Bituminous mix design methods, with particular reference to asphalt concrete, are divided into the following six categories: recipe, empirical, analytical, volumetric, performance-related and performance-based. The definition of these six categories, followed by a short reference to the national methods, is described below.

The various mix design methods used and under development in different countries (Australia, Belgium, Finland, France, Germany, Italy, Switzerland, The Netherlands, United Kingdom and USA), are described and analyzed in the "Inventory of bituminous mix design methods" in Appendix 2.

Nine criteria were taken into consideration in classifying the national methods studied. These criteria were chosen on the basis of the definitions given for the six mix design categories. Depending on which of the nine criterion is taken into consideration, the mix design methods will fall into at least one of the above-mentioned six categories (see Tables 3.1 and Fig. 3.1).

Table 3.1 Mix Design Methods, Criteria and Categories.

Mix Design Method	Based on Experience of Mixes of Known Composit.	Specimens Fabricat. and Compact.	Volum. Criteria	Volum. Analysis/ Compos.	Specimens Compact. which reprod. the in-situ Compaction process	Emp. Tests	Simul. Tests	Fundam. Tests	Fundamental Properties with Proven Relation to Performance Used in Perform. Prediction Model	Mix Design Category
Australia - NARC guide '96 level 1	yes	yes	yes	yes (-)	yes	no	yes	no	no	Recipe\ Volumetric
Australia - NARC guide '96 level II/III	yes	yes	yes	yes (-)	yes	no	yes	yes	no	Recipe\ Volumetric\ Performance Related
Belgium - CRR R61/87	no (°)	no (^)	yes (^)	yes	no	yes(^)	no	no (°)	no	Analytical\ Empirical
Belgium CRR 1996 (draft)	no (°)	no (^)	yes (^)	yes	no (§)	yes(^)	yes (^)	no (^)	no	Analytical\ Empirical\ Performance Related
Finland ASTO / PANK '95	yes	yes	yes	yes (-)	yes	no	yes	yes	no	Recipe\ Volumetric\ Performance Related
France AFNOR	yes	yes	yes	yes (-)	yes	yes	yes	yes (#)	no	Recipe\ Volumetric\ Performance Related
Germany - DIN	yes	yes	yes	no	no	yes	no (°)	no (°)	no	Recipe\ Empirical
Italy - CNR, ANAS & AUTOSTRADE	yes	yes	yes	no	no (°)	yes	no	no (^)	no	Recipe\ Empirical
Switzerland-SN 640431a	yes	yes	yes	no	no	yes	no	no	no	Recipe\ Empirical
The Netherlands RAW Standards	yes	yes	yes	no	no	yes	no	no	no	Recipe\ Empirical
The Netherlands CROW (draft)	no (°)	yes	yes	yes (-)	no	yes	yes	yes	no	Volumetric\ Performance Related
United Kingdom BS 594 / BS 4987	yes	no	no	no	no	yes	no	no	no	Recipe
UK - BS 598	yes	yes	no	no	no	yes	no	no	no	Recipe\ Empirical
UK Nottingham Univ.	no (°)	yes	yes	yes (-)	yes	no	yes	yes	no	Volumetric\ Performance Related
USA - Asph. Inst. '84/91	no (°)	yes	yes	no	no	yes	no	no	no	Empirical
USA SHRP Superpave Level I	no (°)	yes	yes	yes (-)	yes	no	yes	no	no	Volumetric
USA SHRP Superpave Level II	no (°)	yes	yes	yes (-)	yes (=)	no	yes	yes	yes (+)	Volumetric\ Performance Based
USA SHRP Superpave Level III	no (°)	yes	yes	yes (-)	yes (=)	no	yes	yes	yes	Volumetric\ Performance Based
USA SHRP-A-698	no (°)	yes	yes (/)	yes (-)	yes	no	no	yes	yes	Performance Based

(°) Yes, only for some criteria used.
(^) Yes, only for the verification of the base composition.
(§) Yes, only for simulation and fundamental tests.
(°°) Compaction and/or Test procedure used only for special design studies, and not for routine mix design.
(#) Test carried out only for study of a completely new formula with non traditional materials or materials of unknown performance.
(+) No reference was found with a comparison between actual in situ performance and predicted laboratory performance.
(-) Only volumetric analysis.
(/) It should only be noted that at the expected in-situ volumetric characteristics, the mix still has the desired performance.
(=) Some researchers agree that the gyratory compactor may not produce specimens suitable for performance based analysis.

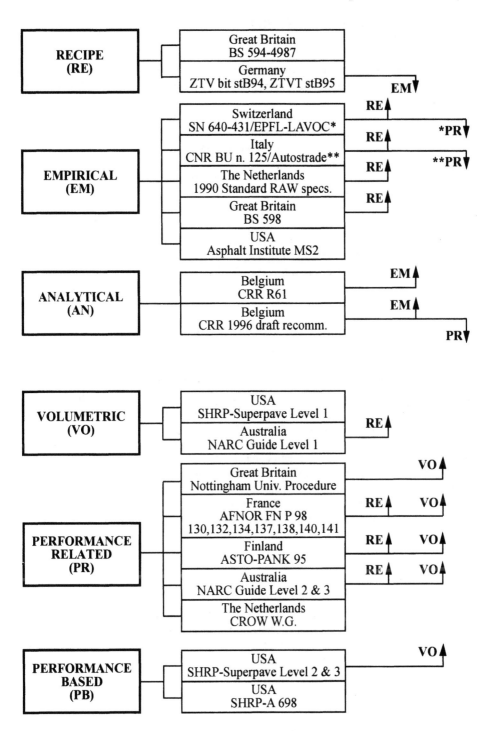

Fig. 3.1 State of the Art Report: Flow Chart of the Mix Design Methods studied.

3.2.1 MIX DESIGN METHODS BASED ON THE USE OF STANDARD MIXES (RECIPES)

The recipe method is based on the experience of traditional mixes of known composition which, over long time periods and under given site, traffic and weather conditions, have performed successfully.

A given recipe defines the bituminous mixture in terms of the aggregate gradation, binder penetration grade, mix composition, layer thickness and mix characteristics during in-plant manufacture, laying and compaction. This method does not entail the preparation of any specimens.

In Great Britain bituminous mixtures are designed trusting in the use of the so-called "recipe" method, with the exception of a version of the Marshall method of the Asphalt Institute used as an alternative to the recipe method of design for hot rolled asphalt wearing course mixes.

In Germany bituminous mixtures for base courses and wearing courses are formulated with the recipe method and the Marshall test is used only for the voids analysis.

In Finland the recipe method, is still being maintained under the new specifications for light traffic volumes (mix design class C and D) and for gussasphalt and porous asphalt.

3.2.2 EMPIRICAL MIX DESIGN METHODS BASED ON EMPIRICAL TESTS, ESPECIALLY THE MARSHALL TEST

The basic concept in the empirical mix design method is the selection of the binder content based on the optimization of several variables, taking into account the limits set thereon based upon prior experience, including those determined by void analysis. These variables are not a direct measure of performance.

In these methods specimens are compacted using a test procedure which does not reproduce, in the laboratory, the in-situ compaction process.

The specimens are then tested using a mechanical empirical test method that does not measure the fundamental (intrinsic) properties of the materials.

The mix design method based on the empirical Marshall test, standardized by the American Society for Testing and Materials (ASTM D 1559), is used in the United States by numerous State agencies.

The British Marshall mix design method is used for constructing wearing courses of Hot Rolled Asphalt. In order to take into account the gap in the aggregate grading of the mix, this method first calculates the optimal bitumen percentage in a manner similar to that used in the ASTM method. Secondly it applies a corrective factor to this latter in function of the percent of large aggregate, so as to obtain the "design binder content."

The mix design method used, for example, in Germany for roadbases, and in Italy, the Netherlands and Switzerland, starts with a "recipe" which dictates the percentage of aggregates, filler and the type and percentage of bitumen, bitumen penetration grade. If further determines the optimal composition of the mix in conformity with Marshall requirements regarding stability, flow and residual voids (in Italy and the Netherlands the Marshall quotient is also required).

3.2.3 ANALYTICAL MIX DESIGN METHODS

The analytical mix design method permits the calculation in volume terms of the mix composition (gradation of the aggregate classes, bitumen and void percentages). This method does not consider the preparation of any specimen. Composition is determined exclusively with analytical computations that can be computerized.

The Belgian analytical method (as presented in the "Code de bonne pratique" and in the recently developed version still under discussion) employs analytical computations to determine the base composition of bituminous mixtures in volume terms. This method considers it equally important to verify the base composition by means of testing, which are not used to determine the optimal bitumen content.

In the 1987 version of the "Code de bonne pratique" the basic composition is verified by applying the Marshall empirical test. According to the latest version of the Code, the choice of test to be used to verify the base composition is based on the mineral aggregates skeleton of the mix subjected to mix design. The Marshall mechanical test is still recommended for sand skeleton mixes, and the wheel tracking test for mixes rich in mastic such as SMA, while the Cantabro test is used for mixes poor in mastic such as porous asphalt.

3.2.4 VOLUMETRIC MIX DESIGN METHODS

In the Volumetric mix design method, the choice of design binder content and aggregate gradation is obtained by analyzing the proportional volumes of air voids, binder and aggregate for mixtures which have been compacted using a test procedure that is assumed to reproduce, in the laboratory, the in-situ compaction process.

The mechanical properties of specimens are not tested. The volumetric proportions of the laboratory-compacted specimens, which are equivalent to those obtained for the mix in-situ, influence the mechanical performance of the mixes.

Level 1 of the Superpave mix design system, developed recently in the United States within SHRP, is defined as volumetric, and it can be used to obtain a mix which would perform satisfactorily *in situ* under low traffic conditions, without having recourse to any mechanical tests. In this method, the final choice of the optimum bitumen content is made by considering the measurements of three parameters namely, the percentage of residual voids, the percentage of voids in the mineral aggregates and the percentage of voids filled with bitumen.

On the other hand, this method, in which the specimens are compacted with the SHRP Gyratory Shear Compactor, also utilizes certain performance-based bitumen specifications and empirical performance-related specifications for the aggregates.

In the Australian NARC level 1 mix design method, which can also be defined as volumetric, the aim is to achieve a composition with correct volumetric proportions starting from a fixed grading curve provided by the Australian Standard, and preparing three mixes with bitumen contents close to the trial one. The trial bitumen content can be selected either on the basis of experience with mixes of similar composition which have shown good performance, or it can be calculated from a mathematical formula in such a way that a minimum bitumen film thickness is achieved. Specimens are then compacted in the Australian Gyratory Shear Compactor (Gyropac), each with a different number of gyrations, and their densities and other volumetric parameters are calculated. The design bitumen content is the bitumen percentage that corresponds to a compacted mix with a voids percentage value equivalent to the design value, depending on the traffic level.

In Finland, when the raw materials are well known, level one of the ASTO mix design system is used. This system, which can also be defined as volumetric, consists of determining the optimal mix based on measurements of the volumetric parameters (voids in the mineral aggregate (VMA), voids filled with bitumen (VFB) and void content (VC) of the mix). These values are then compared with the design volumetric criteria depending on the type of mix.

The volumetric and compaction properties are determined using the Finnish Gyratory Shear Compactor known as the Intensive Compactor Tester (ICT). The suitable binder content is considered to be that which is equivalent to 85% of the voids in the mineral aggregate, when the VMA falls to its minimum value. The optimal binder percentage can be determined for 1, 2 or 3 different aggregate gradings, and one can then choose that which performs best.

3.2.5   PERFORMANCE-RELATED MIX DESIGN METHODS

In the performance-related mix design methods, mixes that meet established volumetric criteria are compacted and tested with simulation and/or fundamental tests to estimate and/or measure their properties which are related to the pavement performance and to determine the optimal mixture on the basis of these additional criteria.

The French mix design is based on the utilization of simulation tests, including the Gyratory Shear Compactor - G.S.C. (Presse à Cisaillement Gyratoire - P.C.G.) and the Wheel Tracking Test (Orniéreur). The P.C.G. workability and compacting test must be performed on the type of mix selected, varying certain parameters regarding the materials and the aggregate grading; the formulas selected are then subjected to the Duriez test to access their sensitivity to water. The mechanical properties specified in the standards regarding bituminous mixes are:

- Resistance to permanent deformations (specified as the maximum rut depth resulting from the wheel tracking test).
- Stiffness modulus (specified by the complex modulus or by a secant modulus measured in a direct tensile test) and

- Fatigue strength (specified by a minimum deformation value for 10^6 load applications in a constant deformation fatigue bending test).

The wheel tracking test is only obligatory for roads with high freight traffic volumes, whereas the tests for the mechanical characterization of the mix behavior, because they are very demanding in terms of cost and time, are not obligatory for all cases. The tests to determine the modulus and the fatigue strength must be performed only in cases of pavements differing from those presented in the French Pavement Catalogue. It should be stressed that the French bituminous mix design method also introduces references to raw materials, with an extensive definition of the composition itself. Once its limits are set, the mix is optimized taking into consideration the mechanical property specifications.

The mix design method recently developed at the University of Nottingham in Great Britain performs a first selection of optimal mixes by setting certain limits on the following volumetric parameters: the VC, the VMA and the VFB. The material properties are then determined for those mixtures selected on the basis of these volumetric proportion limits. The Nottingham Asphalt Tester (NAT), is used to measure the stiffness modulus and fatigue resistance, with the repeated-load indirect tensile test and to measure resistance to permanent deformation with the repeated load test and the axial static creep test.

Level 2 of the Australian National Research Council (NARC) mix design method, can also be defined as performance related. This design method uses a test device manufactured in Australia called the Materials Testing Apparatus (MATTA). This device is similar to the British NAT for mechanical characterization of mixes. It is configured in such a way as to be able to determine the stiffness modulus (measured using the repeated loading indirect tensile test), the resistance to permanent deformation (using the dynamic creep test) and the fatigue strength (measured with a repeated load bending beam test). The cylindrical test specimens utilized for the first two tests are compacted using the Australian Gyratory Shear Compactor (Gyropac) which is first used for the volumetric analysis of the mix in level one of the design procedure. Mixes for roads with heavy traffic volumes (NARC level 3), where greater resistance to permanent deformation is required, are also subjected to the wheel tracking test and refusal density check.

Level 2 of the Finnish ASTO mix design method measures some performance properties of the mix selected on the basis of volumetric criteria. The wear resistance is considered to be the most important characteristic of the mix *in situ* and is determined using the SRK test. Thanks to this test, the rutting depth, due to wear of the surface layer of the pavement, can be predicted in laboratory. The other performance properties required at in the level 2 analysis are the evaluation of the resistance to permanent deformation and to the water sensitivity. The first is normally measured using the repeated load creep test or the wheel tracking test; the second can be measured using the indirect tensile test.

Level 3 of the Finnish mix design entails the determination of mix resistance to low temperature cracking by means of the indirect tensile test. This level considers also the stiffness and the fatigue resistance, which are measured only for purposes of determining the pavement design input parameters. In the Finnish specifications the mixes used for roads of high and medium traffic (mix design class A and B) are formulated exclusively with the performance related mix design method, which is similar to the above-described ASTO level 2 and 3.

The mix design method currently being defined in the Netherlands (proposal of the CROW Working Group "Asphalt Technology") which should replace the current mix design method based on the Marshall test, might also be considered as performance-related. Depending on the type of mix and the pavement layer in which it will be employed, a series of formulation criteria are selected, and the properties associated with these are defined by means of suitable tests (this is done with the same aggregate grading and for different bitumen contents). The mean value corresponding to that interval in which all the chosen criteria are respected indicates the bitumen content for the "job mix formula". This method uses the dynamic creep test to determine the resistance to permanent deformation of specimens compacted with the gyratory shear compactor.

3.2.6 PERFORMANCE-BASED MIX DESIGN METHODS

The mix, initially designed with any of the above mentioned methods, is then subjected to performance-based tests and to an integrated assessment system in order to determine how the mix will behave over a period of time so as to accept or reject it.

Specimens are compacted and tested in order to determine the fundamental properties that have proven to be related to performance and that will be used as input in the material property model.

Starting from the project data (pavement structure, traffic and climate), different models regarding material properties, environmental effects, pavement response and distress are applied to predict pavement performance, supplying realistic estimates of the evolution of different kinds of distress over the working life of the pavement.

The conversion of the observed laboratory performance into expected field performance is obtained by means of shift factors calibrated according to previous experience.

One of the initial proposals for a performance-based mix design methodology was presented in the USA within the SHRP context and corresponded to levels 2 and 3 of the Superpave mix design system for intermediate and high traffic levels. Although no longer proposed for implementation by FHWA (considered as unfeasable and due to errors encountered in the prediction models) (ref. Witczac et al 1997) it is worth while presenting here the concepts in which this methods were based. The initial choice of the design bitumen percentage was determined using the volumetric method. Following this, the proposed mixes were prepared by adopting different bitumen contents, greater, lesser and equal to the design bitumen percentage. The mixes were then subjected to a series of tests based on the performance levels selected for the mix design routine. The Superpave software used the results of these tests to predict pavement performance with respect to permanent deformations, fatigue and low temperature cracking. The optimal bitumen content was determined as a function of these performance forecasts.

Besides the use of the gyratory shear compactor, specified also for level 1, Superpave levels 2 and 3 also required the use of the Shear Test Device and the Indirect Tensile test. Level 3 mix design required a considerably longer testing time and a greater number of tests than the level 2, but was expected to provide more reliable estimates of pavement performance.

The SHRP A-698 mix design method proposed by some researchers who took part in the SHRP project, can also be considered performance-based. This design concept is based on measurements of physical properties in the mix that can be linked to its performance in the various distress modes. The objective of this performance-based mix design method is to select the optimum bitumen content in a mix that will simultaneously satisfy the resistance to permanent deformations and fatigue cracking requirements for a given set of conditions. To evaluate rut and fatigue cracking resistance of the mixes, this method employs the fundamental tests developed during the SHRP A-003A project: the repetitive simple shear test at constant height (RSST-CH) and the flexural bending beam test.

3.3   Advantages and disadvantages of mix design methods and test procedures

Following the analysis of the main bituminous mix design it is useful to summarize here both the advantages and the disadvantages of each of the six categories into which the methods have been classified.

Mix Design Method Based on Use of Standard Mixes (Recipes)

In numerous national road laboratories, bituminous mixtures are still designed by choosing among a series of standard formulas defined simply on the basis of the positive practical experience with said mixes under different weather and traffic conditions. As mentioned earlier, this approach requires careful selection of mixture raw materials, according to standard recipes calculated in terms of mass. This type of mix design is consistent with the approach defined in the indications given in the descriptive tender specifications in which precise checks are required on all the variables of the process of preparation, transport and laying of the mix. Listed below are the different advantages and disadvantages of the so-called "recipe" mix design method.

Advantages
- Mix formulas derived from the recipes have been shown to ensure optimal performance *in situ*.
- They are simple to use and can be applied to any type of mix.
- The mix designer can easily specify the materials required in the pavement layers.

- It is easier for the supplier of such mixes, to fulfill the requirements of a given recipe, than to have to change mix composition because the quality of the raw materials has altered.
- It is relatively easy to control that every material and mix property is in compliance with the specifications.

Disadvantages
- Traffic and weather conditions to which the final mix will be subjected may differ from those on which the recipe-derived mix is based.
- The performance of a bituminous mixture for surface layers subjected to heavy traffic may depend not just on the composition of the mix, but also on the production and laying processes, and it is difficult to specify these latter operations with the recipe method.
- If the tests show that the supplied raw material does not conform to the specifications, even though the differences are only minor ones, there is no way to evaluate the effects of these differences on the "recipe" mix, and thus determine its *in situ* performance.
- Mix design methods based on recipe-type specifications frequently do not permit the use of aggregates possibly available locally, thus entailing unnecessary costs.
- Given the great number of recipe-type specifications available, it is possible that an inexperienced engineer might choose an unsuitable one.
- Whereas it is in the interest of quality control to follow the specifications to the letter, it is incorrect to think that a mix that does not conform to the specification is in some way necessarily defective.
- By designing a mix in terms of the mechanical characteristics required, rather than a specific recipe-type composition, it is easier to introduce innovations.
- Even when aggregates that conform to requirements are used, mixes prepared with the same recipe but with aggregates of different origins may have very different mechanical properties, and consequently perform quite differently *in situ*. In fact, the shape, texture, porosity and mineralogical composition of the aggregates have a marked influence on the properties of the bituminous mix.

The recipe-type method has the undoubted advantage of meeting the expectations of various protagonists of the road construction process. However, the impossibility of being able to prescribe every characteristic of the materials or to evaluate performance variations *in situ* made it practically necessary to have recourse to mix design methods based on simple, effective and low-cost mechanical tests.

Mix Design Method Based on Empirical Tests, especially the Marshall

In order to move beyond traditional recipe-based bituminous mix design methods, designers introduced simple and effective test methods which served to avoid the difficulties connected with the particular visco-elastic nature of the binder, and test procedures suited to define bituminous mixture stability in standardized manner.
  Of the many procedures proposed, mention should be made of the:

* Unconfined compression tests (for example, the Duriez method used by the LCPC, ASTM D1074-83, DIN 1996, Part 4, 1984 and Part 12, 1985),
* Indentation tests (among those utilizing the cone penetrometer in accordance with DIN 1996, Part 13, 1984 and certain BS for the mastics), and
* Flow tests [for example, the Hveem stabilometer (ASTM D1560-81a), Hubbard-Field (ASTM-D1138-52 deleted 1980), Marshall (ASTM D1559-60T)].

Of all these types of tests, it was the Marshall test which proved most practical, and was thus introduced not only in the bituminous mix design method of the same name, but also in various European, as well as American, tender specifications. More recent testimony of this emerged from two recent international co-operative studies. The first was the survey of standard mixture design methods conducted by the P.I.A.R.C. Technical Committee on Road Materials Testing for the XIX World Road Congress held in 1991. The second was the part devoted to mix design methods of the first interlaboratory test program organized by the RILEM Technical Committee BAT 101 in the 1991-1993 period. Below are listed the advantages and disadvantages of the design method identified with the traditional Marshall test.

Advantages
* This mix design method is based on a simple and inexpensive test.
* It is the most widely used in the world for formulating bituminous mixes.
* It requires no particular training or qualified staff.
* Vast amounts of information and results are available to permit defining quality criteria for a wide range of specific uses.

Disadvantages
* This method, which uses an empirical test, is not based on the fundamental properties of a bituminous mix.
* It is not suited to new traffic conditions. For this reason, in recent years we have seen a steady increase in rutting problems with mixes prepared using the Marshall design method.
* In the manufacture of specimens, the test method does not simulate the compaction that occurs in the pavement. The *in situ* compacting action of compactors or vehicle passages (post compaction) is different and more effective than that of the Marshall hammer.

- This method is unsuited for roadbase mix design. In these mixes the aggregate nominal size may reach 37.5mm, with the result that specimens compacted in small molds (diameter of ~101 mm) are not representative of the performance *in situ*.
- This method is unsuitable for gap-graded mix design. Originally, in the United States, the Marshall method was intended for continuously graded bituminous mixes.
- This method cannot be used *in situ*ations that exceed circumstances for which this mix design procedure was developed more than fifty years ago. In fact the basic reason for the problems with the usage of the Marshall method was precisely the introduction of new raw materials and in particular the adoption of modified binders.
- In this method, the application of simple, more or less defined stresses makes it difficult or impossible to interpret the test results correctly in relation to real conditions.

Analytical Mix Design Method

The need to extend the working life of pavements subjected to increasingly heavy traffic, in terms of both loads and frequency, has increased the demand for ever more stable and durable bituminous mixes. The materials are pushed to their physical limits in terms of their behavior under extreme temperatures (cracking and rutting), of their compactibility (degree of densification) and of their aging. The added demands placed on traditional materials has led to the use of different types of tests in mix characterization, design and control, with a subsequent rise not just in the number of tests, but also in the complexity and costs of the laboratory study stage. As we shall see this is has been accompanied by the demand for more perfect mechanical characterization of the performance of the materials to be used as input in rational pavement design procedures.

The subsequent research into the relations existing between the raw materials and the mixes created therefrom, and the possibility of limiting the costs of such studies thanks to the application of mathematical relations, has led to the analytical type approach to mix design. The related advantages and disadvantages of this approach are as follows:

Advantages

- The analytical mix design method offers the possibility of forecasting the intrinsic properties of bituminous mixes, such as the visco-plastic behavior and the fatigue strength, thanks to the volumetric approach in determining mix composition.
- Using the volumetric approach linked with this method, one can determine the correct mastic content (filler plus binder) necessary for the bituminous mix, thus avoiding excessive deformation and bleeding (due to the extra mastic content) or fatigue cracking and mix segregation (insufficient mastic content).
- The analytical mix design approach correlated with a mechanical test is of great practical interest, insofar as it reduces the preliminary studies necessary to define mix composition parameters, speeding up the whole design process.
- In cases where the requirements of the mechanical test correlated to the analytic design method are not satisfied, the latter enables one to determine the causes much more rapidly.

Disadvantages
- This mix design method requires excessively laborious calculations, which preclude practical use without the support of special interactive software.
- Recourse solely to the analytical method does not guarantee the good mechanical performance of a bituminous mix, and for correct forecasting it is necessary to introduce at least one test.
- The verification of the base composition (calculated in terms of volume) is a necessary step to guarantee that the mix is correctly formulated.

Volumetric Mix Design Methods

The mix design method based on the traditional Marshall test has been combined with certain criteria related to the volumetric characteristics of the bituminous mixture compacted in laboratory. These characteristics are thus linked to the volumetric acceptance criteria of the mixes *in situ* established as a function of their suitability as judged on the basis of experience.

Studies based on fundamental testing of the mixes have served to clarify the correlations between the mechanical behavior aspects, mix performance *in situ*, and the principal volumetric characteristics after compaction in laboratory.

New laboratory compaction testing attempts to more closely simulate *in situ* compaction (manner and intensity of compaction, as well as the dimensions of the specimens in relation to the maximum aggregate dimensions). Within this framework the development of the gyratory shear compactors of differing and increasingly versatile conception permitted the development of new mix design methods based on the volumetric approach.

Advantages
- It can be stated that the volumetric mix design method can be applied with relative ease, as it requires no particularly demanding or costly innovations as far as equipment or procedures are concerned, thanks to more recent developments in the gyratory shear compactor.
- With respect to traditional empirical methods, such as the Marshall one, these types of methods ought to provide greater reliability. It has been shown, in fact, that a mix with good volumetric characteristics normally meets the requirements for Marshall stability and quotient, and has good probability of performing suitably *in situ*. On the other hand, it is less probable that mixes having sufficient stability but unsuitable volumetric characteristics will perform well.
- Most of the volumetric mix design methods examined include mixture compaction using a Gyratory Shear Compactor which monitors the increase in bulk density as the compactive effort increases and permits laboratory simulation of the *in situ* compaction process.
- The Gyratory Shear Compactor is used also to predict the evolution *in situ* of the percentage of residual voids in a bituminous mix.

Disadvantages
- Volumetric mix design methods do not directly measure mix mechanical properties.
- These mix design methods do not directly predict mechanical performance.
- For very heavy traffic recourse only to this mix design method does not guarantee the good mechanical performance of a bituminous mix, and for correct forecasting it is necessary to introduce a series of tests or at least one mechanical test.
- The Gyratory Shear Compactor analysis only provides indications of the mix workability and its tendency to rutting. It must necessarily be run in tandem with a test to evaluate permanent deformation resistance in extremely severe conditions such as those encountered under a high degree of slow, channeled freight traffic under high temperatures.
- When using a Gyratory Shear Compactor, the compaction data (air voids content in function of number of gyrations) is a function of the angle of gyration, axial pressure and number of gyrations per minute. However a small change in any of these parameters can have a significant impact on the compaction data and criteria.
- The Gyratory Shear Compactor test permits the designing of bituminous mixtures only through the study of the influence of aggregate grading on workability. The test is in fact conducted at a temperature that does not permit assessment of the effect of the binder on workability.

When using specimens compacted with the Gyratory Shear Compactor for determining a fundamental property, one should keep in mind the fact that bulk densities vary throughout the specimen, resulting in a lack of homogeneity. Generally one can say that bulk density in the middle of the specimen is the highest (lowest void content). Laboratory investigations have shown that even when specimens compacted in the laboratory and *in situ* have the same average bulk density, their mechanical characteristics can still differ if the void distribution is not homogeneous. To be able to determine the mechanical properties of a bituminous mixture as fundamental material properties, it is essential that the stress-strain condition of the test specimen be reasonably uniform.

With reference to Superpave level 1, the indication of the "restricted zone," which tends to limit the use of natural sands, is somewhat innovative. However the adoption of this criterion as a rigid restriction could block the use of economical and easily obtainable materials which in the past have already demonstrated their suitability (a similar problem exists with the recipe method.)

As mix design requires the definition of numerous parameters, it may be illusory to think that one can ensure the good mechanical behavior of these materials simply by having recourse to the volumetric method, no matter what type of equipment is used in producing the specimens. Indeed, knowledge of the volumetric parameters is a necessary but insufficient condition to ensure good mixture quality control. True control of these parameters can be reached only with experimental methods based on several different laboratory tests, possibly correlated with or based on performance.

Performance-Related Mix Design Methods

The development of rational type pavement design methods which conceive the pavement as a series of superimposed elastic or visco-elastic layers requires the study of the states of tensions and stress produced by external actions. This has led to the development of increasingly sophisticated tests to measure the intrinsic mechanical properties of the mixes used in constructing the different layers. Some of these tests, defined earlier as fundamental tests, not only provide inputs for the design calculations, but also constitute an integral part of the mix design method. In fact, the inability of traditional methods to guarantee suitable performance of materials in the pavement has led designers to devise new methods involving integration of this new type of test with those defined as simulation tests. These latter permit classifying different mixes and comparing expressly in terms of the performance parameters measured.

Advantages
- The tests used in the performance-related mix design methods measure the material response to various states of stress.
- These methods show good correlation between the results measured in the laboratory and the *in situ* behavior.
- These methods encourage innovation and use of new materials.

Performance-Based Mix Design Methods

The mix design methods falling in this category grew out of American studies conducted with the context of SHRP, and combine both fundamental and simulation tests with computer programs to predict the main types of pavement deterioration.

Advantages
- Performance-based mix design methods encourage innovation and use of new materials, just as do the performance-related type.
- These methods allow one to estimate, in the laboratory, from the materials properties of the bituminous mixture, the development of pavement distresses expected over a selected service life.
- Using the performance-based approach there could be no need to set explicit limits on volumetric parameters, as long as the mix shows the desired performance at the expected in-situ volumetric characteristics.
- The selection of binder content is based on the desired design level of performance for each mode of distress.

Disadvantages
- The efficacy of the tests used in these mix design methods has to be proved by *in situ* tests or full-scale tests.
- Due to the considerable complexity of the application and the delicate equipment required by these method, it seems that it would be difficult to adopt these in

working situations as part of routine design and production control methods within mix production plants.

• Only in particularly appropriate and justified cases would it be feasible to adopt the complex procedures required by this method. In the case of low traffic volumes the incidence of the overall costs of the design method would be considerable.

While these methods appear quite attractive, the critical analysis is limited by lack of knowledge regarding the validation of the calculation methods used in the performance-related methods, the limited use of the related equipment up to now, and the relatively few roads built up to now using these methods.

3.4   From new mix design concepts to a proposal for the general outline of mix design method

3.4.1   INTRODUCTION

This part of the chapter sets out the general criteria proposed for the design of hot bituminous mixtures so as to provide a guideline for operative practice that incorporates the new concepts recently introduced in the field. This proposal combines methodological aspects from various sources referred to in the related literature, including semi-theoretical predictive hypotheses verified experimentally in the laboratory and *in situ*, as well as practical syntheses drawn from experience on the road, which have up to now served as valid guides to mix designers. The following is thus derived also from the analysis of the 19 methods that provided the basis for the classification of the majority of current mixture design methods in 6 different categories.

The proposal described below is also intended to contribute to the discussion of mix design methods, and particularly within the context of the RILEM working group on bituminous mixture design (under TC 152 PBM). It constitutes a preparatory document for the drafting of a recommendation for a mix design method.

Fig. 3.2 shows a breakdown of the main stages of the related logical process, which is conditioned, among other things, by both the pavement design and by the maintenance strategy, insofar as both of these influence the choice of bituminous mix to be used and the related design procedure.

3.4.2   DESIGN PARAMETERS

The general criteria proposed below have as their objective the formulation of a bituminous mix to reach desired levels of performance. The end result is to obtain, from the available materials, a composition that is not only technically suited to the proposed intervention, but is also economically advantageous.

As an initial approximation, the logical process of the design can be essentially reduced to the basic aim of optimising the aggregate grading blend and the binder dosage.

Before commencing the design process, it is necessary to define the particular problem to be resolved with the mixture under design, so as to have some useful guidance in choosing the raw materials. These materials, once their availability has been determined, must then be characterised.

In fact, proper mix design must always be preceded by consideration of the problems which might arise during the working life of a pavement, and which the mixture to be formulated will be called upon to resolve. These problems are essentially linked to the main distress mechanisms which, in the case of pavements, can lead to unacceptable levels of safety, shortened working life and excessive costs to users and for maintenance. Consequently, prior to the design process, it is necessary to select the distress mechanisms that can influence the choice of component materials and the design method itself.

The main types of pavement distress are fatigue cracking, permanent deformation due to repeated load application and thermal cracking. Surface layers may require other performance characteristics such as skid resistance, durability, hydraulic conductivity and tire noise abatement.

The information necessary to define the problem is:

- Pavement type (i.e., flexible, semi-rigid, composite),
- Structure (i.e. layer thickness),
- Importance and location of the road,
- Traffic levels (i.e. intensity, speed, design ESALs, etc.),
- Climatic and environmental conditions (temperature and precipitation, in particular),
- Function of the layer within the pavement.

Analysis of the problem to be resolved, also in relation to the intended use of the mixture, will provide indications as to the materials, the tests, and the conditions for carrying out the tests. Moreover, the analysis of the problem to be resolved must enable one to establish a hierarchy among the different mechanical properties which the mixture must possess, and possibly permit the determination of other details, also in function of the type of intervention.

3.4.3 MATERIAL CARACTERIZATION

The complete characterisation of the available materials, involves subjecting the coarse and fine aggregates and the bituminous binder to a series of physical and chemical tests with the goal of giving useful design indications. These materials should be identified and selected on the basis of the technical-economic criteria derived also from the foregoing analysis.

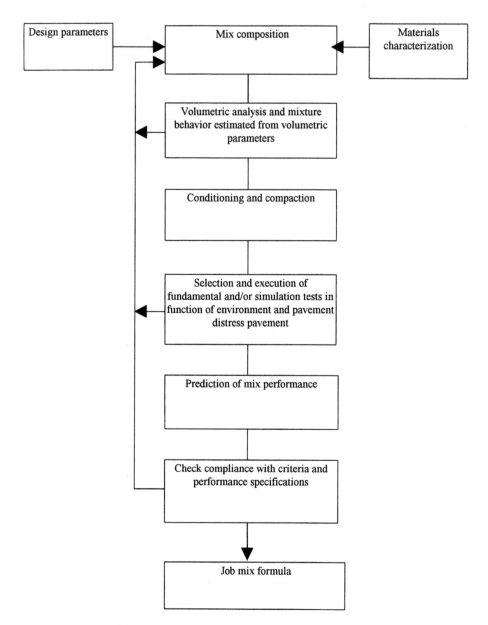

Fig. 3.2 Proposal for the General Outline for a Bituminous Mixture Design Method.

Particular care must be taken to ensure that the specimens are truly representative of the materials to be supplied, and that what is characterised and formulated with it corresponds to what is actually utilised *in situ*.

Within this proposed mix the current traditional tests for aggregate characterization are considered such as:

- the grading analysis,
- the apparent and real density,
- the shape, flat and elongated coefficients,
- the angularity,
- sand equivalent,
- clay content,
- toughness and polished stone value, chosen in function of the maximum aggregate dimension and the intended use of the pavement.

Also, the aggregate should be analysed in terms of the selected binder, to determine the aggregate-binder compatibility, including the ability of the aggregate to absorb the binder.

The bituminous binder will be subjected to tests to determine its consistency at various temperatures and its thermal susceptibility. These will include both traditional tests, and several developed more recently.

The former include determination of the apparent density, penetration, and softening point, using the Ring and Ball method, the Fraass breaking temperature or Fraass brittle point, flash point, with possible exceptions in the case of modified bitumens.

The newly developed tests include the various types of rheological tests, which should be performed, after suitable ageing, at different temperatures (including, for example, determining the stiffness modulus). These latter tests have been introduced especially to provide certain preliminary data useful at a subsequent stage of the mix design, and are related with the effects of the binder on the various pavement distress mechanisms.

3.4.4 MIXTURE COMPOSITION

The first step in the proposed bituminous mixture design process is a selection of the alternative grading envelopes relative to the type of mixture selected (strongly discontinuous, slightly discontinuous, closed, semi-open, open, etc.). This step lays the groundwork for the experimental choice of the design grading performed at the next stage.

The percentage of voids in a grading blend can be predicted by various analytic methods such as those developed by Nijboer, by Goode and Lufsey, by Hudson and Davis, by Huber and Shuler, and by Francken and Vanelstraeten. The value of these methods lies in their ability to reduce the number of alternative blends and the ease with which they can be automated and integrated into the design process.

The choice of grading blend corresponding most closely to the performance objectives of the mixture to be designed emerges, in the case of the lowest-cost operational choice, from the comparative testing of the alternative combinations using identical binder. The initial binder content value for each grading combination is determined using analytical formulas (based, among other things, on the specific surface area of the aggregate, on the percentage of binder absorbed by the aggregate, on the specific gravity and on the minimum binder thickness). It can also be obtained from recipes (the fruit of applied experience in the past with similar mixtures and reported in standards or tender specifications).

In summary, the main steps in the preliminary study of the aggregate grading of a mixture are the selection of the trial aggregate gradings and the calculation or selection of an estimated binder content.

These steps ensure the technical-economic balance of the design process, by limiting the number of blends to be tested in the subsequent stages of mixture design.

3.4.5 VOLUMETRIC ANALYSIS

This step of the proposed method entails the selection of the design aggregate grading curve using, for example, the gyratory shear compactor (GSC). This is achieved through the manufacture and possible conditioning of mixtures consisting of several different grading blends mixed with the same initial binder content and the analysis and related measurement of the their volumetric parameters by subjecting them to a known compactive effort.

The selection is made by analysing the densification curve upon which are represented the maximum theoretical specific gravity values of the alternative mixtures as a function of the compactive effort. Several parameters are monitored such as residual voids value and voids in mineral aggregate value. Workability and compactibility are complementary, but sometimes contradictory, characteristics of a bituminous mixture, which the trial volumetric approach using the gyratory shear compactor permits to consider.

It is appropriate to insert several observations here regarding the manufacture of the specimens produced by combining each of the alternative grading blends with the estimated binder content identified in the preceding section. The laboratory procedures used to prepare the mixtures and the specimens must be carefully defined. Depending on the quantities of mixture to be prepared in the laboratory, suitable thermostat mixers have to be provided to ensure homogeneous manufacture, in terms of both temperature and spatial distribution of the mineral aggregate fractions and the quantities of binder.

As far as the analysis of the volumetric parameters is concerned, the purpose of these parameters is to provide a preliminary prediction of the performance of the mixture, both during the construction stage and during operation. These predictions are based on the correlations between the relative volumetric proportions and the corresponding physical and mechanical characteristics. Such predictions, which are useful in limiting the number of specimens required for the subsequent trials, must nevertheless be confirmed by subsequent mixture tests selected in conformity with the evaluation of the performance characteristics of the mixture under design.

It should also be noted that the percentage of residual voids of the compacted layer to be produced using the bituminous mixture under design must already have been defined before commencing the actual design stages.

The analysis of the mixture compaction is based on its air void content, which depends on the voids in the mineral aggregate (indicated as VMA, and equivalent to the sum of the volumes of the voids Vv and of the effective binder Veb). The VMA parameter represents one of the main volumetric parameters, insofar as it correlates with the measurements of several of the main mixture properties, such as permanent deformations and the elastic component of the complex modulus. Of a certain interest is the effective percentage of voids filled with binder (VFB) as it is realted with certain mixture properties (for example, the elastic component of the complex modulus). Also of interest is the ratio between the volume occupied by the aggregate (Va) and the volume occupied by the binder (Vb), especially because of its effect on mixture fatigue strength.

After having chosen the appropriate test methods, the next step is the statistical characterisation of the measurements performed, possibly in duplicate or in triplicate, to obtain the volumetric values, keeping in mind that precision and accuracy even in the preliminary stages can have significant effect on the final result.

The final result of this stage is the choice of the grading blend, which must then be followed by the determination of the optimum design binder content. This is obtained through experimental measurement of the properties already identified in the analysis of the problems foreseen and to be resolved by the mixture under design.

3.4.6 CONDITIONING AND COMPACTION

The third stage of the proposed method (indicated as "conditioning and compaction" in Fig. 3.2) comprises the manufacture of the specimens for the selection of the optimal design binder content. The specimens, consisting of the aggregate grading blend selected in the previous stage and a binder content varying in lower or higher regular percentage intervals, will be produced in numbers and with physical characteristics (shape, dimensions and mass) determined as a function of the requirements of the performance-related tests.

The operations involved include the manufacture of the mixture, the conditioning, the compaction and taking of the specimens. The manner in which this latter operation is conducted can influence the behaviour of the mixture during the tests.

Conditioning (the simultaneous or separate action of water, freeze-thaw, ageing, traffic loads) is applied either to the binder or to the mixture in the loose state, or to the specimens prior to testing. The choice of the manner of conditioning the mixture or the specimens (times, temperatures, oxidation pressure, and moisture) will depend on the climatic factors one intends to simulate for specific defined pavement working lives. In particular, the procedure for accelerated evaluation in laboratory must take into consideration the ageing of the mixture, both in plant and *in situ*, entailing several stages of conditioning, and simulating differentiated levels.

In developing a mix design method, one of the main problems is to identify and employ a method of laboratory specimen compaction that simulates the *in situ* conditions. True comparability between the properties of the mixture taken from the laboratory specimens and those obtained by testing specimens taken from the pavement is necessary.

According to the literature, the laboratory compaction which best simulates the *in situ* conditions is that obtained with the rolling wheel compactor. Obtaining specimens from slabs presents slight difficulties due to the possible effects of the slicing or core extraction operations. It is also important to check that the apparent specific gravity of the sets of specimens to be used in the tests is of the same order of magnitude.

3.4.7 SELECTION AND EXECUTION OF FUNDAMENTAL/SIMULATION TESTS

The stage of selecting and executing fundamental and/or simulation tests as a function of the type of pavement distress is devoted to identifying and executing several performance-oriented tests so as to determine the optimal design binder content of the mixture to be formulated. The tests must be conducted on specimens consisting of the grading blend selected as a function of the preceding volumetric tests, and with binder content ranging in constant intervals from more to less than the estimated content defined earlier.

The tests methods, which should be applied to specimens obtained using the aforesaid manufacturing, conditioning and compaction procedures similar to actual operational ones, must have the following characteristics:

- able to be performed with identical experimental test apparatus and operational procedures, not requiring highly specialised operational techniques of long execution times;
- able to reproduce the same conditions to which the mixture will be subjected *in situ*;
- sensitivity of the results to variations in several basic components of the mixture (for example, the binder type and content, aggregate type and grading);
- able to yield measurements of the performance and intrinsic properties without inordinately high costs or long execution times;
- able to produce results which are not only apt for insertion in the performance prediction models, but also applicable practically.

All this will further facilitate their routine utilisation in a design procedure based on the general criteria set out thus far.

The main functional properties of the mixtures, including the stiffness and the resistance to the accumulation of permanent deformations, fatigue and low-temperature cracking, can be analysed by subjecting the specimens to various load stresses under standardised condition. It should be noted that in the case of stiffness and fatigue strength, the R.I.L.E.M. Technical Committee No. 101 BAT conducted an interlaboratory testing programme in 1991-1993; a description and critique in the light of more recent practical and interpretative developments is provided in chapter 6 of this report.

Stiffness (see chapter 5)

One of the general criteria proposed as the basis for the development of a bituminous mixture design method is the use of fundamental type testing to measure stiffness.

The stiffness can be determined by measuring the unit deformation of the mixture subjected to the application of static or dynamic loading. Measurements can be executed with or without considerations of inertia forces or lateral confinement and under controlled stress conditions (homogeneous stress field), temperature and level of unit deformation (strain <50 μm/m). With this latter condition, the characteristic measured can be considered independent of the level and direction of load applied. Because of the non-linear and visco-elastic behaviour of bituminous mixtures, the complex modulus is determined at various load application frequencies (so as to obtain the master curve and the stage curves on the Black plane) and various temperatures (so as to define its susceptibility to temperature). The effect of the mixture air voids content (or the degree of compaction) on this parameter which enhances the elastic or viscous components under given conditions should also be considered.

The stiffness can be measured from the application of dynamic, sinusoidal or repeated loads at various test conditions and specimen geometries considered in chapter 6. In the case of static loading, compression test or direct tensile test (on cylindrical specimens) are often used. Today bituminous mixture designers are increasingly using fundamental type tests to determine this intrinsic parameter, because of the possibilities of performing rapid, repeatable, non-destructive measurements, easily processed and interpreted, and usable also at the quality control stage.

Permanent Deformation

The definition of the optimal design binder content also includes the experimental evaluation of the permanent deformation resistance. Such deformations are produced by the combined action of shear and compression stresses. The results of the related testing provide the inputs for the computerised procedures employed in modern mixture design methods for predicting the development of permanent deformations in the pavement, utilising various models. In cases where the models on which the predictions are based are not available or not adequately calibrated, it is necessary to set certain thresholds or intervals on the test parameter(s). These are selected as indicator(s) of potential permanent deformations as a function of the temperatures and the traffic. For example, in static tests the maximum percentage deformation in the loading stage or the ratio of permanent deformations to the amount of stress applied after a predetermined load recovery time, and, in the case of dynamic tests, the minimum slope value of the curve of the unit load-deformation cycles.

This parameter is measured using fundamental and simulation type tests. In the former case the loads applied are normally either static compression type (cylindrical specimens) applied uniaxially in so-called static creep tests, or dynamic, repeated compression type (cylindrical specimens) applied uni-, bi- and triaxially.

The fundamental tests most widely used initially, because of their ease of execution and due to their comparable results on full-scale test sections, were the uniaxial static tests. The tests are conducted at one or more temperatures and measure, as a function of time, the unit vertical deformations obtained by applying known levels of load and after fixed load recovery times. These tests proved inadequate for use in mixture design, however, due to the fact that they do not bring into play, in a realistic manner, the resistance component attributable to attrition. The use of these tests, which do not permit discrimination between two mixtures in which the resistance to permanent deformations is due respectively to the attrition contributed by the aggregate grading, and to the consistency of the binder, is nowadays limited just to the quality control stage. The repeated, dynamic type tests have lately become predominant because they better simulate the loads passing over the pavement and, in the case of triaxial application, the lateral confinement of the layer. These tests are highly sensitive, however, to variations in mixture composition, and even in cases of identical grading blends, can discriminate the behaviour of mixtures with different binder content or with binders of differing rheological behaviour. More recently an attempt has been made to include this type of test in mixture design methods, applying repeated shear stresses on cylindrical specimens kept at constant height.

Difficulties in the past in introducing these fundamental type tests in the design procedure led to the adoption of simulation tests to measure resistance to permanent deformations. The equipment used to perform these tests ranges from extremely large full-scale pavement test facilities to the smaller laboratory-scale wheel tracking test. The latter measures the evolution of the rut depth as a function of the number of passages of a rubber-tyred wheel or the tyre imprint gradient expressed in mm/h, under particular wheel load conditions and temperatures. This test is still widely used today, thanks to the development of laboratory equipment (wheel tracking) belonging to the category of tests which permit application of repeated type loads simulating the passage of traffic over beam specimens, and show good correlation with the onset of plastic deformations on road sections.

Fatigue (see chapter 5)

The stage of optimising the design binder content of a bituminous mixture should also provide for the experimental verification of the resistance to fatigue cracking. Recent developments in testing and data interpretation, including the so-called energetic approach based on accumulation of energy dissipated by each cycle, are set out in chapter 5. On the other hand, the complexity and high costs of the related test equipment available up to now, as well as the excessive time required for the fatigue tests, have considerably limited routine use.

The comparison of the mixtures in terms of fatigue life at similar levels of deformation can contribute to the selection of the final composition. It must be kept in mind, nevertheless, not only that the fatigue performance of the mixture depends on the type of load application selected, but also that the fatigue life obtained represents a lower limit. Such factors as rest periods, traffic wander, and periodic healing which occur during operations, may improve the fatigue performance of the mix *in situ*.

Low Temperature Cracking

Although low temperature cracking resistance is more closely correlated with the properties of the binder the difficulty of identifying suitable test methods has tended to delay the development of direct test criteria to evaluate the behaviour of mixtures at low temperatures.

Test methods have been developed recently to define mixture resistance to thermal tensions and cracking. In the so-called cooling down test, a bound specimen (i.e. with contraction blocked) is subjected to cooling at a constant thermal gradient, or to repeated thermal expansion-contraction cycles. The temperature value at which the specimen cracks is combined together with the tensile strength of the mixture, obtained (under isothermal conditions and for values around zero) by applying direct or indirect type monotonous load or applying direct or indirect constant tensile load (static creep).

These tests, conducted at the selected binder interval, can be included at the stage of selecting the optimal design value with other tests related to similar damage phenomena (cracking caused by thermal cycles).

Durability

Before arriving at the definition of the final mixture formula, it is still necessary to check the durability of the optimal formula. The resistance of the mixture to an accelerated action simulating the main types of aggression expected to act on the pavement over time due to weather conditions (moisture and freeze-thaw cycles) and other external agents or combinations of these needs to be checked. Since this resistance is an increasing function of the binder content, it follows that this check should take place as the final optimisation of the design value previously obtained by balancing the test results with the primary properties.

At this stage of mixture design it is proposed to insert a suitably reliable test of mixture sensitivity to water action or investigation of possible loss of binder-aggregate adhesion in the presence of moisture.

Other Characteristics

Other tests measuring fundamental mixture properties such as, for example, those that contribute to surface skid resistance, permeability and rolling noise, over and above the mechanical properties, can be selected. This will contribute not only to more complete characterisation of the mixtures, but also to the definition of the main design objectives.

3.4.8 PREDICTION OF MIX PERFORMANCE

Once the above-mentioned tests have been performed, and depending on the primary performance characteristics established at the time of defining the problems foreseen and to be resolved with the procedure in question, the subsequent stage should facilitate the prediction of the in-situ behaviour of the mixture under design. This should be done using suitable prediction models based on both the behavioural laws of the mixtures, and on the evolution over time of the performance characteristics as determined from processing the long-term historical data regarding the main condition indicators of existing road pavements, whether experimental or not. This analysis should also be integrated with the preliminary considerations regarding the pavement design.

The test results should be input into the appropriate mixture functional behaviour models (non linear elastic, linear and non-linear visco-elastic, plastic), and the resulting outputs introduced, in terms of fundamental properties in the pavement perfromance models.This analysis should be executed with respect to the main modes analysed, such as permanent deformations and fatigue cracking due to load applications and low temperatures, for the purpose of predicting performance over time.

The use of these models, if appropriately available and calibrated, together with the general criteria previously introduced, serve to update the design procedure, such that it conforms more closely, on an experimental and theoretical-predictive basis, to the performance characteristics of the mixture as compared to what is traditionally done.

3.4.9 CHECK COMPLIANCE

It is an intincic part of the design procedure to check that the mixture performance evaluated experimentally and predicted with the computer models complies with the criteria and performance specifications.

3.4.10 JOB MIX FORMULA

If all criteria are met in the compliance check stage the job mix formula has then been achieved.

3.5   Implementation and Verification of Mixture Design

No less important from the quality assurance standpoint, in light of possible problems of application or faulty performance forecasts, are the stages following the design process, including the definition of the working formula and its quality control, both at the execution stage and during pavement management. Since several of the procedural steps, besides being intimately connected with specific contractual aspects, are still subject to harmonisation at European level, where the implementation stages will be defined, a few observations should be made based on general considerations of a practical nature.

Once the design process has been completed and the final mixture composition obtained, it is necessary to prepare the job mix formula at the mixing plant level, by converting the volumetric parameters into mass values. The mixture produced in plant may differ from the above-described process essentially in the variability either of the raw materials (commercial aggregate sizes may have "tails", with consequent excess of filler) or of the composition parameters affecting especially the aggregate coating ability of the binder. These variations must nevertheless fall within the permitted tolerances and must be provided for also in the recommendations, standards and specifications relating to mix design.

At this stage reference must be made to the tests employed during the entire design process, but especially those connected with the definition of the mixture components (aggregate grading and binder dosage). The tests must be quick and easy to perform, but must also be sufficiently sensitive to detect even slight variations in the component materials, thus safeguarding the statistical reliability of the results.

It should also be noted that the more strictly defined and precise the correlations are between mixture composition and related performance required *in situ*, the easier it will be to predict accurately the possible effects variations (introduced in plant) will have on situ performance as compared to the design mixture.

In fact, both initial quality control and surveillance and monitoring over time can be done using high-performance equipment capable of continuous and high-speed measurement of certain selected pavement condition indicators, both superficial and structural.

The correct management of these latter involves the analysis of these measurements over time, a situation that permits one not only to calculate residual working life, but also to determine the local and global behaviour of the mixture comprising the layers in which the distress mechanisms are at work.

Finally, it can be stated that the mixture design method proposed here, even allowing for the advantages of the empirical methods, is geared toward the selection of test methods which:

1) provide knowledge not only of the main performance characteristics but also of the intrinsic characteristics of the material, not only at the moment of mixture manufacture, but also over time;
2) permit estimation and evaluation of new types of materials; and
3) provide data regarding the behaviour of mixtures, ensuring the greater reliability of modern prediction methods and reducing the margin of error of decisions undertaken.

The implementation of this mix design system will depend on the capability of the technical community to demonstrate its superior performance prediction capabilities. Full acceptance of this mix design system is only likely after demonstrate monitored performance in studies such as the RILEM Pavement Performance Prediction Evaluation studies.

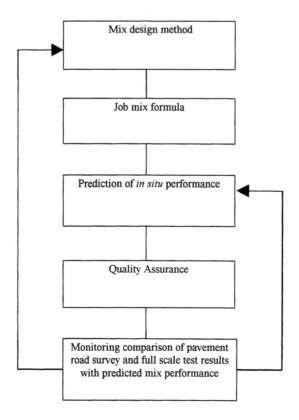

Fig. 3.3 Implementation and Validation of Mix Design Method.

3.6   References

Zakar P. - Huet J. : "Tests on bitumens and bituminous materials", proceedings of the 2nd International R.I.L.E.M. Symposium, Budapest, 9-12 September 1975.

Huet J. - Ajour A.M. : "Testing of hydrocarbon binders and materials", proceedings of the 3rd International R.I.L.E.M. Symposium, Belgrade, 12-16 September 1983.

Bonnot J. : "Rapport général sur les essais mécaniques pratiques de formulation et de controle des enrobés bitumineux", proceedings, vol.II, of the 3rd International R.I.L.E.M. Symposium, Belgrade, 12-16 September 1983.

Huet J. - Ajour A.M. : "Mix design and quality control for bituminous mixes", proceedings of the International R.I.L.E.M. Residential Seminar, Olivet, 24-26 September 1986.

Fritz H. W. : "Méthodes de formulation pour la fabrication des enrobés basées sur les essais mécaniques traditionnels", proceedings of the International R.I.L.E.M. Residential Seminar on the "Mix design and quality control for bituminous mixes", Olivet, 24-26 September 1986.

Dumont A. G. : "Méthodes basées sur les essais mécaniques modernes", proceedings of the International R.I.L.E.M. Residential Seminar on the "Mix design and quality control for bituminous mixes", Olivet, 24-26 September 1986.

Francken L. : "Méthodes analytiques (à partir des characteristiques des constituants) avec ou sans essais mecaniques consecutifs", proceedings of the International R.I.L.E.M. Residential Seminar on the "Mix design and quality control for bituminous mixes", Olivet, 24-26 September 1986.

Fritz H. W. - Eustacchio E. : "The role of mechanical tests for the characterization, design and quality control of bituminous mixes", proceedings of the 4th International R.I.L.E.M. Symposium, Budapest, 24-26 October 1990.

M. Luminari :"Rapport général sur les essais à charge unique", 2nd session, proceedings, vol.II, of the 4th International R.I.L.E.M. Symposium, Budapest, 24-26 October 1990.

Proceedings of the 1st Eurobitume Seminar: "The challange of the future for asphalt road", London, England, 14-15 November 1978.

Proceedings of the 2nd Eurobitume Symposium: "Bitumen, the solution for today and tomorrow", Cannes, France, 7-9 October 1981.

Proceedings of the 3rd Eurobitume Symposium: "Bitumen, flexible and durable", The Hague, The Netherlands, 11-13 September 1985.

Proceedings of the 4th Eurobitume Symposium, Madrid, Spain, 4-6 October 1989.

Proceedings of the 5th Eurobitume Symposium, Stockolm, Sweden, 5-7 June 1993.

Proceedings of the Euroasphalt& Eurobitume Congress, Strasbourg, France, 7-10 May 1996.

Proceedings of the International Conference: "The United States Strategic Highway Research Program (SHRP): Sharing the benefits", London, England, 29-31 October 1990.

Proceedings of the International Conference: "Strategic Highway Research Program (SHRP) and Traffic Safety on Two Continents", The Hague, The Netherland, 22-24 September 1993.

Proceedings of the International Conference: "Road Safety in Europe and Strategic Highway Research Program (SHRP)", Lille, France, 26-28 September 1994.

The Asphalt Institute: "Principles of Construction of Hot-Mix and Asphalt Pavements", Manual Series N. 22 (MS-22), College Park, Maryland, January 1983.

The Asphalt Institute: "Mix design methods for asphalt concrete and other hot mix types", Manual Series N. 2 (MS-2), College Park, Maryland, May 1984.

Santucci L. E. et alii: "Asphalt Mix Design", A.A.P.T. Symposium, Proceedings Association of Asphalt Paving Technologist, vol. 54, San Antonio, Texas, February 11-13, 1985.

Pigois M. L. et alii: "Methods of specimens preparation for mechanical tests", paragraph 4 of the 2nd chapter on "tests on bituminous materials" of the report of the Technical Committee on Testing of Road Materials, chaired by prof. Moraldi, XVIIIth P.I.A.R.C. World Road Congress, Bruxelles, 13-19 September 1987.

Celauro B. - Giuffre O. : "Criteri di formulazione dei conglomerati bituminosi per pavimentazioni stradali", Quaderno N. 9, Istituto di Costruzioni Stradali, Università degli Studi di Palermo, 1987.

Whiteoak D. : "The Shell bitumen handbook", Published by Shell Bitumen U.K., Surrey,1990.

Valero L. et alii: "Survey of standard methods for the design of bituminous concrete", paragraph 3 of the 2nd chapter on "tests on bitumen and bituminous materials" of the report of the Technical Committee on Testing of Road Materials, chaired by prof. Moraldi, XIXth P.I.A.R.C. World Road Congress, Marrakesh, 22-28 September 1991.

AA.VV. :" Heavy Duty Pavements", Status Report by the Technical Committee of the European Asphalt Pavement Association, EAPA, Breukelen, The Netherland, September 1991.

AASHTO - FHWA - NAPA - SHRP - TAI - TRB: "European Asphalt Study Tour", Report published by AASHTO after the visit of an american experts group in Denmark, France, Germany, Italy, Sweden, United Kingdom, June 1991.

Mahoney J. , Bell C. : "Pavement Design and Materials, Past, Present and Future", Proceedings of the 7th International Conference on Asphalt Pavements, Nottingham, August 16-20, 1992.

Eustacchio E. , Gubler R., Pallos I. : "Part 1 mix design of the Interlaboratory Test Program", proceedings of the final meeting, Dubendorf, may 1993.

O' Flaherty C.A.: " Highway engineering", Vol. 2, Edward Arnold, 1993.

Bell C.A. : "Introduction to Mix Design", Lecture Note K of the Residential Course on Bituminous Pavements, Materials, Design and Evaluation of the University of Nottingham, Department of Civil Engineering, Nottingham, 11-15 April 1994.

Cooper K. E. : "Developments in Mix Design", Lecture Note L of the Residential Course on Bituminous Pavements, Materials, Design and Evaluation of the University of Nottingham, Department of Civil Engineering, Nottingham, 11-15 April 1994.

Celauro B. : "Conglomerati bituminosi chiusi", parte II del quaderno AIPCR sui bitumi modificati per usi stradali e loro impiego nei conglomerati e nei trattamenti superficiali, Giugno 1994.

Eustacchio E. , Francken L. , Fritz H. W. : "Mix design and repeated loading tests: an account of a R.I.L.E.M. Interlaboratory Test Program", Proceedings of the International Conference: "Road Safety in Europe and Strategic Highway Research Program (SHRP)", Lille, France, 26-28 September 1994.

Fidato A. :" Bituminous mixture design methods", Vol. I e II, Tesi di Laurea (Relatore Prof. Ing. M. Cupo Pagano, Correlatori Ingg. M. Luminari e A. D'Andrea), Dipartimento Idraulica Trasporti e Strade, Università degli Studi di Roma "La Sapienza", Anno Accademico 1993-1994.

AA.VV. :" Heavy Duty Pavements: Asphalt - meeting the performance demands of Heavy Duty Pavements", Status Report by the Technical Committee of the European Asphalt Pavement Association, EAPA, Breukelen, The Netherland, 1995;

The Asphalt Institute: "Superpave Level 1 Mix design", Superpave Series No.2 (SP-2), Research Park Drive, Lexington, KY, July 1995.

Verstraeten J. et alii: "Enrobés bitumineux à haute resistance a l'ornierage par fluage", Publication 08.03.B du Comité Technique A.I.P.C.R. des Routes Souples CT8, Paris, 1995.

Eckmann B. : "Rapport general sur les propriétés fonctionnelles des enrobés", 4th session, Proceedings of the Euroasphalt& Eurobitume Congress, Strasbourg, France, 7-10 May 1996.

AA.VV.: "Application made of Gyratory Shear Compactor for mix design and control of asphalt mixes works", Proceedings of the International Workshop on he Use of the Gyratory Shear Compactor (topic 2), Nantes, France, December 12-13, 1996.

AA.VV.: "Topic III on Mix Design: Sessions III.A on Mix versus Component Properties and Session III.B on Mix Design Procedures", Proceedings of the 5th International RILEM Symposium on Mechanical Tests for Bituminous Materials: Recent Improuvements and Future Prospects carried out at Lyon on May 14-16, 1997, A.A. Balkema Publisher, Rotterdam, 1997.

Fidato A. : "Bituminous mixtures: mechanical tests and mix design", Lecture Note of the Course of "Costruzione di strade II" of the University of Rome "La Sapienza", Facoltà di Ingegneria, Dipartimento di Idraulica Trasporti e Strade, Edizioni Dedalo,may 1997.

Luminari M. : "Mix Design: State of the Art and RILEM Interlaboratory Test Program", presentation of the activity of a RILEM TG 4 on Mix Design of a TC 152 PBM and TG3 on Mechanical Tests of a TC 101 BAT, Proceedings of the 5th International RILEM Symposium on Mechanical Tests for Bituminous Materials: Recent Improuvements and Future Prospects carried out at Lyon on May 14-16, 1997, A.A. Balkema Publisher, Rotterdam, 1997.

Luminari M., Fidato A. :"State of the Art of Bituminous Mixtures Design: Inventory of the Existing Methods", from a Book "Bituminous Binders and Mixtures: State of the Art and Interlaboratory Tests on Mechanical Behaviour and Mix Design" by Francken et alii, Chapman & Hall, 1997.

Witczak M.W., Von Quitus H. and Schwartz C.A. : "SUPERPAVE support and Performance Models Management : Evaluation of the SHRP Performance Models System. Presentation made at the 8th International Conference on Asphalt Pavements, Seattle, 1997.

4

Interlaboratory tests on mix design

by E. Eustacchio, I. Pallos and Ch. Raab

4.1   Introduction

The idea of organizing an international testing program on significant testing methods for asphalt pavements was raised in 1987 by the RILEM Technical Committee TC 101-BAT "Bitumen and Asphalt Testing" and accomplished by RILEM TC 152-PBM "Performanced Based Materials". Its implementation started after the fourth RILEM International Conference "Mechanical Tests for Bituminous Mixes - Characterisation, design and control" organised by this same committee in 1990 at Budapest.

The aim of the interlaboratory test on "mix design" was to find the best composition, i.e. "the optimum mix", for a mixture using given raw materials with the purpose to fullfil certain boundary conditions (traffic, temperature etc...). The same raw materials were used as for the First OECD Road Common Experiment (FORCE) - project, OECD-report, 1991.

4.2   Working program

Each of the participating laboratories was provided with as much material as was needed to produce about 250 kg of bituminous mix, considering the composition of the bituminous mix at the OECD-FORCE project carried out on the LCPC's circular test track in Nantes, OECD-report, 1991. The laboratories were further asked for a complete report containing all steps carried out and all measurements as well as the final results and possibly some additional comments. After having designed the optimal composition, each laboratory had to manufacture a given quantity of its "optimum" mixture and send it to EMPA for further analysis and complementary tests. Table 4.1 shows the participating laboratories.

4.2.1   METHODOLOGY

Every participating laboratory had the possibility to carry out the tests according to its own method or following the national standard. To be able to realise a complete evaluation of the results and a correct comparison between the adopted design and the test methods the following information had to be given by each laboratory :

Bituminous Binders and Mixes, edited by L. Francken. RILEM Report 17. Published in 1998 by E & FN Spon, 11 New Fetter Lane, London EC4P 4EE, UK. ISBN 0 419 22870 5

1. Principles of the method.
2. Preparation of aggregates (masses, temperature, fractions).
3. Preparation of bitumen (masses, temperature, procedure).
4. Procedure of mixing (temperature, time, method, masses).
5. Test methods and specimens (testing arrangement, type, masses, measures, compaction, volumetric and mechanical characterstics).
6. Test results.
7. Evaluation of the results (procedure, statistical methods).
8. Result : **"The optimum mixture".**

Table 4.1 Participating laboratories.

Lab.-Code	Name of the Laboratory	Part 1	Mat. to EMPA
A1	Amt der Kärntner Landesregierung - Abtg. Bautechnik BT-17 Klagenfurt/Austria	+	+
B1	Centre de Recherches Routières - Dpt. Recherches et Dévelopment Bruxelles/Belgium	+	+
CH1	Eidgenössische Materialprüfungs- und Forschungsanstalt - EMPA Dübendorf/Switzerland	+	+
CH2	École polytechnique fédérale de Lausanne - Laboratoire des voies de circulation Lausanne/Switzerland	+	+
CRO	Institute gradevinarstve Hrvatske Zagreb/Croatia	+	+
D2	Bundesanstalt für Straßenwesen Bergisch Gladbach/Germany	+	+
F1	Central Laboratories for Roads and Bridges Bouguenais/France	+	+
H1	Technical University Budapest - Chair for Road Construction Budapest/Hungary	+	+
H2	Scientific Institute for Transport of Hungary Budapest/Hungary	+	+
I1	Autostrade S.p.A. - Laboratorio Centrale Fiano Romano/Italy	+	-
N1	Veidekke Asphalt Laboratories Division As/Norway	+	+
NL2	Ministerie van Verkeer en Waterstaat - Directoraat-Generaal Rijkswaterstaat-Delft/Netherlands	+	
NL3	Zuidelijk Wegenbouw Laboratorium Vught/Netherlands	+	+
S1	Lulea University of TechnologyRoad Laboratory Lulea/Sweden	+	+
SLO	IRMA d.o.o - Institut za raziskavo materalov in aplikacije Ljubljana/Slovenia	+	+
US1	Federal Highway Administration - FHWA McLean/USA	+	-

4.2.2   GENERAL BOUNDARY CONDITIONS

Mix design was based on the following assumptions and boundary conditions

- *Load* : 20.10^6 18-kip equivalent single-axle loads (ESAL) for the performance period of 20 years. This corresponds to a Traffic Factor TF of about 2600 18-kip equivalent single-axle loads per day and lane.
- *Temperature* : Annual average air temperature of 10 °C, with the monthly average of -1/-2°C in January/February and 21/22°C in July/August. Daily air temperature : minimum in winter -20°C, maximum in summer + 42°C.
- *Road structure :* Total thickness of the asphalt concrete structure of 300 mm was considered to avoid frost damage. Thickness of the surface layer : approx. 50mm and approx. 250mm for the base course. Subbase consisting of a suitable granular layer extremely compacted with heavy rollers.

4.3   Sample division

The EMPA Dübendorf was assigned to extract paving sections from the circuit in Nantes and to collect the corresponding mineral aggregates, fillers and binders. This was done in December 1989. All mineral aggregates were filled separately according to fraction into hot-galvanised steel tanks (dimensions 600 x 700 x 1200 mm) and stored in a closed hall at the EMPA Dübendorf. The filler was filled directly into 10l steel buckets with lids, the bitumen into 5l and 20l steel buckets with lids and stored at EMPA. Sample division on mineral aggregates sand and gravel was carried out in dry, sunny weather on the EMPA site in August 1990.

The mineral aggregates were further divided up with the EMPA sample divider according to the desired packed weight as follows :

- Gravel 6/10   60 sacks à 50 kg.
- Gravel 2/6   20 sacks à 60 kg.
- Sand   0/3   80 sacks à 20 kg.

4.3.1   CHARACTERISATION TESTS

After sample division, a gradation analysis according to Swiss Standard SN 670 710d with square-mesh sieves was carried out on the mineral aggregates. This yielded the results in % by mass, given in Table 4.2 and Fig. 4.1.

The binder was tested according to Swiss Standards SN 671 740a and SN 671 743a for penetration at 25 °C and for ring & ball softening point (Table 4.3).

Table 4.2 Sieve analysis of the aggregates (passing % by mass).

Sieve [mm]	Gravel 6/10	Gravel 2/6	Sand 0/3
16.0	100.00		
11.2	99.00	100.00	
8.0	53.70	99.90	
5.6	3.70	78.60	100.00
4.0	1.40	39.40	99.90
2.8	0.70	11.30	94.70
2.0	0.40	4.20	74.80
1.0	0.20	2.00	42.40
0.5	-	-	26.90
0.25	-	-	18.00
0.125	-	-	12.50
0.09	-	-	10.40

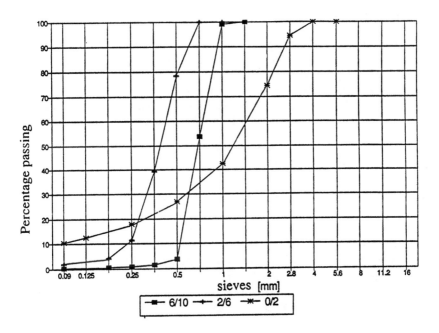

Fig. 4.1 Gradation of the aggregates determined by EMPA.

Table 4.3 Binder properties.

Binder property	Mean value
Penetration at 25 °C [0.1 mm]	53.6
Softening point ring & ball [°C]	53.1

4.3.2 PACKING, SHIPPING

The sacks and steel buckets were packed in plywood crates on pallets and marked for transport. This work was carried out by a specialised packing company. The following quantities were packed and shipped in October 1991 to 18 laboratories :

Gravel 6/103 x 50 kg	=	150 kg
Gravel 2/6 1 x 60 kg	=	60 kg
Sand 0/2 4 x 35 kg	=	140 kg
Filler 1 x 12 kg	=	12 kg
Bitumen B 60/70 17 + 4 kg	=	21 kg
Total		383 kg

4.4   Design methods

Design methods of the participating laboratories are based on their national standards/specifications which rely almost exclusively on the Marshall method. Another type of method (LCPC Design Method) significantly different from the Marshall method was used only by one laboratory (F1).

The majority of laboratories participating in the interlaboratory test program applied a traditional design method using some additional mechanical and informative tests.

4.4.1 TESTING OF COMPONENTS

Mineral aggregates (shipped for designing and making the asphalt mixture) and especially filler were tested completely by several laboratories (CH2, CRO, F1, SLO). Other laboratories considered it sufficient to analyse only grading and density (A1, B1, CH1, D2, H1, H2, I1, NL2, NL3, S1, US1).

When testing bitumen, most laboratories only defined significant parameters - like density, softening point, penetration. In some cases viscosity measurements were also added. Three laboratories (CH1, SLO, US1) did detailed bitumen testing. One of the participating laboratories (CRO) extracted and tested bitumen after two months and demonstrated the changes of parameters.

4.4.2 DESIGN OF MINERAL AGGREGATE MIXTURE

According to national specifications and thus by maximum grain size, three types of asphalt wearing courses can be distinguished :

D_{max} = 10 mm   B1, F1, US1
D_{max} = 11 mm   A1, CRO, CH1, CH2, D2, N, NL3, SLO
D_{max} = 12 mm   H1, H2, I1, S1

- one aggregate gradation mixture was made by CRO, D2, H1, S1, US1 :
  CRO with exponent value m = 0.45 according to Talbot function.
  D2   designing the average value of lower and upper standard grades at sieve size of 2 mm.
  H1   according to results of previous tests.
  S1   with grade close to lower limit grade of national specification.
  US1 with grade close to Talbot function with exponent 0.45.
  I1
- two aggregate gradation mixtures were made by B1, CH2 :
  B1   one mixtures for grading according Talbot.
          one mixtures for stone mastic asphalt.
  CH2 with continuous and discontinuous grades.
- three aggregate gradation mixtures were made by A1, H2 :
  A1   with exponent values m = 0,45, m = 0,55, m = 0, 65 according to Talbot function.
  H2   with fine quantity changes of mineral aggregate components, close to lower limit grade of national specification.
- more than three aggregate gradation mixtures were made by CH1, F1, NL3,CH1 with all practical variants fulfilling the national grade specifications.
  F1   four mixtures "Semi grenu" varying with binder content, sand and filler content tested with the gyratory.
  N1   with considerable variation of quantities of mineral aggregate components.
  NL3 according to so called "Basic matrix test" method. Quantity of 2/6 mm and 6/10 mm fractions was constant, while "filler/sand" rate was changed.
- special remarks concerning adding of the delivered filler :
  no filler has been added by : A1, CH2, CRO, S1,
  0 to 3,0 % by mass filler   : CH1, D2, H1, H2, I1, US1,
  > 3,0 % by mass filler   : B1, F1, SLO,
  filler (% by mass) has been rationed in the wide spread of :
  - 0 to 7.0 % by mass : N1
  - 1.5 to 7.0 % by mass : NL3.

4.4.3 BINDER CONTENT

Bitumen content and composition steps showed significant differences (see Table 4.4).

Table 4.4 Variation of bitumen content.

Code of lab	Bitumen content (related to the mass of asphalt mixture)
A1	with three different mineral aggregate mixtures : - in two cases : 4.5 - 5.0 - 5.5 - 6.0 - 6.5 - 7.0 - in third case with lowest sand content : 5.0 - 5.5 - 6.0 - 6.5 - 7.0
B1	with two different mineral aggregate mixtures : - 5.8 - 6.1 - 6.4 and 6.3 - 6.5 - 6.8
CH1	with ten different mineral aggregate mixtures - in seven cases : 5.4 - 5.55 - 5.7 - 5.85 - 6.0 - 6.15 - 6.3 - in three cases : 5.7 - 6.0 - 6.3
CH2	with two mixtures : - first case : continuous grade close to average of lower and higher grade specification : 5.6 - 6.0 - second case : semicontinuous grade close to lower grade specification : 5.5 - 5.8 - 6.3
CRO	for one mixture : 6.3 - 6.6 - 6.9 - 7.2 - 7.5 with B = 6.9 further two mixtures were tested. Mineral aggregate mixture was produced by adding 1% filler quantity (measured at sieve 0.09 mm)
D2	for one mixture : 5.9 - 6.2 - 6.5 - 6.8
F1	with four mixtures : - in one case : 5.7 as given by the French Standard. - in three cases : 6.0 with higher filler and lower sand content
H1	for one mixture 6.1 - 6.5 - 6.9
H2	with three mixtures : - with highest sand content : 6.4 - 6.8 - 7.2 - second and third case : 6.0 - 6.4 - 6.8
I1	for one mixture : 4.3 - 4.8 - 5.2 - 5.4
NL2	with six different mixtures : + fractions crushed stone 2/6 mm = 15 % = constant + fractions crushed stone 6/10 mm = 45 % = constant while ratio of sand/filler (S/F) changes gradually : 43.5 / 1.5 6.2 - 6.6 - 7.0 42.0 / 3.0 5.8 - 6.2 - 6.6 - 7.0 41.0 / 4.0 5.4 - 5.8 - 6.2 - 6.6 - 7.0 40.0 / 5.0 5.4 - 5.8 - 6.2 - 6.6 - 7.0 39.0 / 6.0 5.4 - 5.8 - 6.2 - 6.6 38.0 / 7.0 5.4 - 5.8 - 6.2 - 6.6
NL3	with five different mineral aggregate mixtures (F / S / 2/6 / 6/10) 5 / 42 / 25 / 28 : 5.5 - 5.8 - 6.1 0 / 50 / 14 / 36 : 4.4 - 4.7 - 5.0 - 5.3 3 / 57 / 22 / 18 : 5.5 - 5.8 - 6.1 5 / 55 / 22 / 18 : 5.8 - 6.1 - 6.4 7 / 40 / 22 / 28 : 5.8 - 6.1 - 6.4
S1	for one mixture : 5.5 - 6.0 - 6.5 - 7.0 - 7.5
SLO	for one mixture : 5.1 - 5.5 - 5.9 - 6.3 - 6.7
US1	for one mixture : 5.0 - 5.5 - 6.0 - 6.5 - 7.0

4.4.4 MIXING AND COMPACTING

Table 4.5 contains some important data on the fabrication of the asphalt mixes and the Marshall specimens :

- mixing temperatures,
- compaction temperatures,
- number of compaction blows.

In order to calculate void characteristics every participating laboratory measured the density of the asphalt mixtures.

Additional tests to define compactibility of asphalt mixtures were done by :

- CRO, H1 (with changing numbers of Marshall blows),
- CH2, F1 (according to PCG method).

Table 4.5 Mixing temperature and data on Marshall compaction.

Lab-Code	Temperature for mixing the asphalt [°C]	Marshall specimens	
		Compaction temperature [°C]	Number of compaction blows
A1	150	150	2 x 50
B1	150	150	2 x 50
CH1	170	-	2 x 50
CH2 **)	160	-	2 x 50
CRO	150 - 160	145	2 x 50
D2	130 - 170	135	2 x 50
F1 *)	150	150	60 gyrations
H1	165	155	2 x 50
H2	165	155	2 x 50
I1	158	148	2 x 75
NL3	157	147	2 x 50
N1	155	140	2 x 75
S1	aggregates :175 bitumen : 160	165	2 x 50
SLO	155	155	2 x 75
US1	145 (290 °F)	135 (280 °F)	2 x 75

*)   Specimens were made exclusively by PCG method.
**)   Some specimens were made by PCG method.

4.4.5 CHARACTERISTICS AND REQUIREMENTS

1. Requirements by Marshall Method :

    a) Density of asphalt specimens : only A1 required a special optimum value.
    b) Specifications, limits concerning voids are shown in Table 4.6.
    c) Requirements concerning Marshall Stability, Marshall Flow, Stability/ Flow ratio are given in Table 4.7.

2. Requirements by LCPC Method.

    These data are given Table 4.8.

Table 4.6 Voids condition specification at participating laboratories.

Lab-Code	Air voids in Marshall specimens [% by vol]	Voids in mineral aggregate [% by vol]	Voids filled with bitumen [% by vol]	Code, name of specification standard
A1	3 - 5	< 16	70 - 80	RVS 8.627
B1	3 - 5	-	70 - 83	BRRC Recommand.
CH1	3.5 - 5	-	70 - 80	SN 640.431a
CH2	3.5 - 5	-	70 - 80	SN 640.431a
CRO	3.5 - 6.5	-	64 - 80	JUS.U.E.4014/1990
D2	2 - 5	-	-	ZTV StB 84/90
H1	3 - 5	-	75 - 82	MSZ-073210
H2	3 - 5	-	75 - 82	MSZ-073210
I1	4 - 6	-	-	CNR Off. Bull.
N1	*)	**)	**)	
NL3	max. 6	-	max. 82	Standard RAW
S1	2 - 4	-	-	Sw. Road Adm.
SLO	3.5 - 6.5	-	64 - 80	JUS.U.E.4014/1990
US1	3 - 5	< 16		ASTM

*)   Limit values by specification not reported, though test results are given.
**)   Neither limit values nor test results reported.

Table 4.7 Marshall Stability, Marshall Flow, Stability/Flow ratio requirements at participating laboratories.

Lab-Code	Stability at 60 °C [kN]	Flow at 60 °C [mm]	Stab./Flow [kN/mm]	Code, Name of Specification. Standard
A1	*)	*)		RVS 8.627
B1	> 10	2 - 3.5	> 3.5	BRRC Recommand.
CH1	> 10	< 3.5	-	SN 640.431a
CH2	> 10	< 3.5	-	SN 640.431a
CRO	> 7	-	> 2	JUS.U.E.4014/1990
D2	**)	**)		ZTV StB 84/90
H1	8 - 15	-	> 3	MSZ-073210
H2	8 - 15	-	> 3	MSZ-073210
I1	> 11	-	3.0 - 4.5	CNR Off. Bull.
N1	***)	***)		
NL3	> 7.5	2 - 4	> 3	Standard RAW
S1	*)	*)	-	Sw. Road Adm.
SLO	> 7	-	> 2	JUS.U.E.4014/1990
US1	> 8	3 - 5	-	ASTM

*)   Neither specified nor tested.
**)   Not specified, test results not reported.
***)   Neiher name of specification nor its requirements are reported.

Table 4.8 Requirements by LCPC mix design method.

Tested characteristics	Specified limit values	Name, code of specification
Voids Void content of specimens, made with PCG method after 60 girations. Voids filled with bitumen	4-9 % by volume not specified, but test results reported	NF-P-98-252
Duriez test "R" compression strength (at 18 °C) r/R ratio r = compression strength at 18 °C after 7 days of specimens under water	> 6 MPa > 0,75	NF-P-98-251-1
Permanent deformation test Wheel tracking test of 10 cm thick flat specimens (T = 60 °C, P = 0,6 MPa)	after 30.000 cycles < 10 %	NF-P-98-253-1

Other mechanical tests and procedures
Several laboratories also used other test methods and procedures to find the recipe for their optimum mix (A1, B1, CH2, CRO, H1, H2, I1, N1, S1, SLO1). These test and procedures are reported in the next section.

4.4.6   DETERMINATION OF THE OPTIMUM ASPHALT MIXTURE

Some laboratories made several mineral aggregate mixes with one or more bitumen contents. Other laboratories considered it sufficient to design just with one mineral aggregate mix, usually applying 3-5 different bitumen contents. Optimum mixture was then defined by finding the optimum bitumen content.

Most laboratories emphasised producing optimum air voids content in the Marshall specimens. Basically they aimed for the mean air voids content at the extreme limit values given in their national specification.

When determining optimum bitumen content or optimum asphalt mixture, mechanical test results were not directly used. Laboratories rather determined the optimum mixture by traditional tests, and then carried out some additional mechanical tests. Mechanical test results were given without further comments in some cases, while in other cases national specifications were also shown.

Laboratory A1
Three distinct mineral aggregate gradation mixes were tested with different bitumen contents and the following characteristics were defined :

- Bitumen content at optimum (maximum) density of Marshall specimen.

- Bitumen contents at minimum and maximum air voids content limit values (according to national specification).

- Bitumen contents at minimum and maximum voids filled with bitumen values (according to national specification).

In the two latter cases two bitumen content intervals were identified. The centre of gravity of these two intervals was defined as optimum bitumen content. This process was used distinctly with all the three mineral aggregate mixes. These bitumen contents slightly differed from the one received by "maximum, optimum density" method.

Optimum mixture could not be determined as "voids of mineral aggregate" >16 % by volume was required for all three mineral aggregate mixes. (This was caused by the unfavourable quality of the mineral aggregate components).

Mixes also did not prove to be adequate by the Carinthian model mix design method, as "deformation moduli" were smaller than required. According to given traffic and climatic conditions required deformation moduli (T = 50 °C) were calculated. Despite this the mixture having the best Deformation Modulus was sent for testing to EMPA.

- Special test methods and procedures :

Mix was designed by the Carinthian model method (CM method) :

- Filler/bitumen ratio and voids filled with mastix are systematically varied;
- Optimal bitumen content and mastix content is defined;
- Uniaxial static compression test is made (T = +50°C, +30°C, +10°C, -20°C).

Measured characteristics :

a) dynamic deformation modulus [N/mm2]
b) strain [%]

- Required value of dynamic deformation modulus is calculated according to traffic and weather conditions. This value is compared with the test result at 50 °C.
- Critical temperature, connected with winter cracking, is calculated from tested strain values.
- "Adhesion strength" is also defined.

Laboratory B1
The supplied material did not comply with any of the mix compositions of the National Specifications. The BRRC recommendations were applied for two types of compositions :

- a Talbot type of mix,
- a stone mastic type of mix (SMA).

The experimental verification confirms the analytically found binder content for both mixes. However as Marshall stability and flow are inadequate criteria's for SMA, the Talbot mix with B'=6.1 % binder was accepted as optimal.

- Special test methods and procedures (see Appendix 2.2) :

The design method involves two parts :

1. a computer aided analytical study,
2. an experimental verification.

The analytical study (PRADO - Mix design software) is carried out on a volumetric basis and consists of the following steps :

- setting up the aggregate mix according to a fixed grading curve (target),
- calculating the expected voids VQ in the mix skeleton (> 2mm particles),
- evaluating the maximum allowable volume of mastic in VQ,
- designing the mastix composition (filler/bitumen ratio),
- deriving the theoretical binder content for the case of dense bituminous mix.

Experimental verification :

Marshall requirements must be verified for a given aggregate composition, verification is carried out on three binder contents :

B' - 0.3 %, B' and B' + 0.3 %.

Laboratory CH1
There were ten different asphalt mixtures - practically in the full spread of possibilities within national specification. Several variants fulfilled the requirements for all parameters. The optimum asphalt mixture had good air void content, and bitumen filled void content (average of lower and higher limit values), and had the highest Marshall stability value.

Laboratory CH2
There were two mineral aggregates gradations with two and three different bitumen contents (five variants all together). Concerning air void content only one asphalt mixture was adequate to national specification.
  Here are some characteristics of the mixture :

- air voids [% by volume] about average of lower and upper limits in national specification,
- bitumen filled void [%] upper third (33%) in the interval given by limit values of national specification.
- Marshall stability [kN] just fulfilling the requirement, close to the given minimum.
- Flow [mm] close to tolerable maximum.
- Indirect Tensile Test and Wheel Tracking Test results fulfil requirements.
- Special test methods and procedures :

  a)   PCG method (according to LCPC)
  b)   Indirect tensile test (T = 45 °C, 25 °C and -10 °C), limits :
       test temperatures :   + 45 °C   -10 °C;
       - indirect tensile [N/mm2] >0,18   >3,8;
       - deformation 10^{-3} [-]   <34   >4,6.

c)   Wheel tracking test (according to LCPC, T = 60 °C, P = 0,6 MPa)
         limit : rut depth <10% after 30 000 cycles.

Laboratory CRO1
One mineral aggregate gradation with five bitumen contents was tested. Void content
was about average of specified limit values at the mixture declared to be optimum.
Bitumen filled void characteristic is the same. Marshall stability and Marshall
stability/flow ratio are favourably better than the required minimum values.

•   Special test methods and procedures :

a)   Experimental determination of maximum possible concentration
         skeleton with pores in asphalt, sample for stone skeleton mixtures
         according to optimum mix.
b)   Volumetric characteristics of asphalt.
c)   Water absorption in vacuum.

Laboratory D2
Having just one mineral aggregate mix and four distinct bitumen contents, optimum
bitumen content was not determined.

Laboratory F1
There were four different mixtures (one mixture with two bitumen contents and the
other two mixtures with two bitumen contents). Only one of the four mixtures
proved to be adequate by PCG method voids content requirement. This mixture was
also acceptable by the Duriez method, thus this mixture was declared to be optimum
mixture.

Laboratory H1
There was one mineral aggregate mixture with three different bitumen contents. The
second version with medium bitumen content showed a favourable air void content
(about average of lower and higher limit values according to specification). Other
values also showed values fulfilling the requirements. Various additional mechanical
tests produced acceptable parameters as well according to earlier experience of
laboratory staff. Thus optimum asphalt mixture became the version with medium
bitumen content.

•   Special test methods and procedures :

Permanent Deformation Test :

a)   Dynamic creep test (T = + 40 °C, f = 4 Hz, sinusoidal loading,
         compression increased with 0.1 N/mm^2 steps, thus compression stress
         P = 0.6 - 0.7 - 0.8 - 0.9 - 1.0 N/mm^2). Several measured and
         calculated characteristics, parameters are evaluated.

b)   Wheel tracking test (T = + 45 °C and T = + 60 °C, P = 0,4 MPa, number of cycles N = 7560).

Fatigue Tests :

a)   Dynamic uniaxial bending test (T = + 10 °C, f = 4 Hz, sinusoidal loading, bending-tensile stress increased by steps)
b)   Dynamic biaxial bending test (T = 0 °C, f= 1 Hz, f= 10 Hz, sinusoidal loading, bending stress increased by steps).

Test results : Complex modulus = bending stress at N = 10^6 cycles.
Test for behaviour at low temperature :

a)   Indirect tensile test (splitting test) at T = + 5 °C, - 5 °C, - 20 °C
b)   Measurement of linear thermal expansion coefficient (+ 20 °C < T > - 20°C).
c)   Calculation of fictitious cracking temperature of asphalt.

Test results are evaluated in comparison with earlier set of results at Laboratory H 1.

Laboratory H2
Three bitumen contents were used for three distinct mineral aggregate gradation mixes (9 variants all together). Optimum bitumen content was declared when air void content was optimum (about average of national specification limit values) with all three mineral aggregate mixes. For these three mixtures bitumen content values of "bitumen filled void", "Marshall stability", "stability/flow ratio" were also favourable according to national specification. Thus optimum asphalt mixture was finally determined as a result of dynamic creep test. Mixture having the best total deformation value in the dynamic creep test was considered to be optimum.

•   Special test methods and procedures :

Dynamic creep test (T = + 50 °C, 1500N loading, loading cycle 1.7, loading time 0.2s).
Result : total deformation after 10000 loads.

Laboratory I1
There was one mineral aggregate gradation mixture with five distinct bitumen contents. Optimum bitumen content was considered at maximum Marshall stability. However, according to national specification this maximum reached the minimum value only. Beside that, air void content is also in the lower 33% part of the tolerable interval.
  Mechanical tests with the optimum asphalt mixture (indirect tensile test, static creep test) produced adequate results according to national specification requirements .

- Special test methods and procedures :

  a)   Uniaxial static creep test (T = + 10 °C, + 25 °C and + 40 °C)
  b)   Indirect tensile test (T = + 10 °C, + 25 °C and + 40 °C)
       technical specification limits :

test temperature [°C] :	+ 10	+ 25	+ 40
indirect tensile strength [N/mm2]	1.5 - 2.5	0.7 - 1.0	0.3 - 0.6
tensile strain coeff. [N/mm2]	>160	>70	>35

Laboratory N1
A mixture with relatively small air void content was chosen as optimum among the great number of mixtures. During split cylinder test this mixture had the highest splitting strength. Dynamic creep test also produced good result with that mixture (total deformation < 1 mm).

- Special test methods and procedures :

  a)   Dynamic creep test (T = 40 °C, P = 0.6 MPa, 3 s on - 3 s off, test time 24 hours). Result : the total deformation after 24 hours in [mm] Limit in "ZONE A" : <1 mm.
  b)   Splitting test. Test results : split cylinder [kN/m2] and deformation [mm] .

Laboratory NL3
Many mixture variants were made by the basic matrix test method. Several mixtures fulfilled the requirements of national specification. The method is influenced by traffic load. The method regulates mastix quantity and quality and bitumen/filler rate.
  Thus optimum bitumen content (considering the tolerable deviations as well) and filler quantity was determined according to the traffic. As the grading - determined this way - matched the national grading specification, this mixture was declared to be optimum asphalt mixture.

Laboratory S1
There was one mineral aggregate gradation with five distinct bitumen contents. Optimum asphalt mixture was chosen at a bitumen content producing good air void content (average of lower and higher limit values in national specification). Analysing some parameters as function of varying bitumen contents the following statements can be made for the optimum bitumen content and asphalt mixture :

- Indirect tensile strength is a maximum
- Marshall stability and resilient modulus are favourably high.
- Special test methods and procedures :

a)   Incirect tensile test (T = 10 °C). Result : indirect tensile strength [kPa].

b)   Dynamic creep test (T = 10 °C, vertical force 1,8 kN, rise time 0,25). Result : resilient modulus [MPa].

Laboratory SLO1

One mineral aggregate gradation and five bitumen contents were tested. Voids content, bitumen filled voids were about average of specified limit values in the mixture declared to be optimum. Stability and stability/flow ratio are better than the required minimum values.

•   Special test methods and procedures :

a)   Water absorption in vacuum.
b)   Percentage of open voids.
c)   Percentage of closed voids.

Laboratory US1

There was one mineral aggregate gradation with five distinct bitumen contents. Optimum asphalt mixture was chosen at a bitumen content satisfying the following parameters :

•   Maximum Marshall stability.
•   Maximum unit weight.
•   4% air voids.

The characteristic properties of the defined optimum mix with a bitumen content of 6.73 % by mass are not reported.

4.5   Results and discussion of different mix designs

The aim of mix design is to find the best composition, "the optimum mixture", of a bituminous mixture, using given raw materials, with the intention to fulfil given conditions for a certain purpose (traffic, temperature etc.). The results of mix design are expressed by the bitumen content and the percentages of the mineral aggregate fractions. Mixing together these fractions in the intended relation leads to the grading curve of all aggregates.

The discussion of the results of this interlaboratory test program concentrates in a first step on the different fractions of the grading curve and on the characteristics of the grading curves. In the next chapter the optimum mixture - bitumen content, mass- and volumetric relations - will be discussed. Those values which were also tested in the additional investigations by EMPA will be discussed in a special section.

4.5.1 MINERAL COMPONENTS

In Fig. 4.2 the percentage of the four given mineral fractions every laboratory used is given (CH2 and S1 did not give any values). A first glance could lead to the conclusion that the laboratories had found very different optimum mixes.

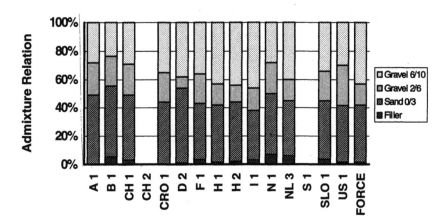

Fig. 4.2 Relation of the mineral components [% by mass].

Fig. 4.3 with the fractions filler (d < 0.09 mm), sand (0.09 < d < 4 mm) and gravel (d < 4 mm), calculated from the effective grading curves of the optimal mixes, illustrates the existing small differences between the gradings of the individual laboratories. The values for the grading used in the FORCE-project and of the Talbot curve are integrated in this figure.

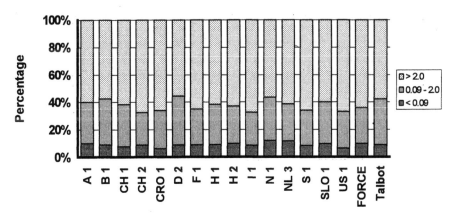

Fig. 4.3 Composition of the optimum mixture [% by mass].

To compare the different gradings there are two possibilities :

Talbot curve
The Talbot curve represents a grain size distribution leading to a composition with minimum of pores in the mineral skeleton. Of course, the composition of real mixes is not identical with the Talbot curve because the volume of the bitumen and the pores in the compacted mixture also needs some space. Nevertheless, it is at least one of the possibilities to carry out an objective analysis.

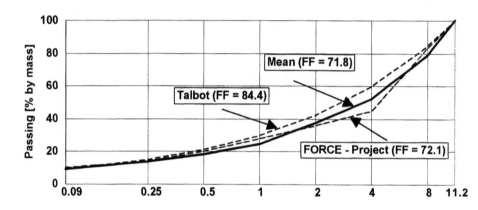

Fig. 4.4 Grading curves and relative fineness factors (FF).

Factor of „fineness"
This factor represents the area between the grading curve and the X-axis, adopting a determined scale. The greater this factor the finer the grading of the aggregate. Fig. 4.4 shows the factors of the grading curves found by the laboratories as well as the factor of the mineral mixture used by FORCE and that of the Talbot curve. The horizontal line in Fig. 4.5 represents the mean value for the results of the participating laboratories (without FORCE and Talbot). It can be seen that only a few gradings (A1, B1, D2, N1 and NL3) are near the factor of fineness of Talbot.   Moreover it can be seen that the factor of the composition used at FORCE is very near the mean value of the laboratories. At the same time it is noticeable that the fineness factor of FORCE is rather far away from the factor for the Talbot curve.

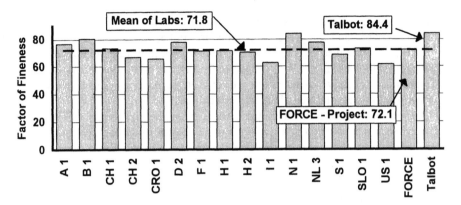

Fig. 4.5 Fineness factors (FF).

4.5.2 BITUMEN CONTENT

A bituminous mixture is characterised especially by the bitumen content in the mixture. Usually one adopts the value based on the mass relation of the bitumen to the mixture. Sometimes alternatively the relation of the bitumen mass to the mass of the mineral aggregate is used. In this analysis the first method is adopted.

Fig. 4.6 shows the individual bitumen contents and the mean (without FORCE). Looking at the distribution of the values it appears that four values are situated outside the range of a one standard deviation (± 0.46) of the mean (6.33 %); only one value is situated outside the double range (± 0.92). Therefore one can conclude very good accordance between the different mixes.

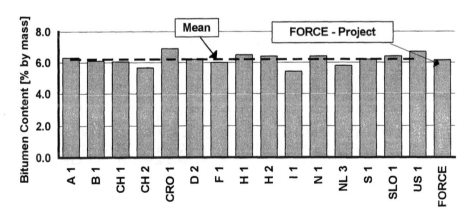

Fig. 4.6 Bitumen content [% by mass].

4.5.3 COMPOSITION OF THE WHOLE MIXTURE

Fig. 4.7 shows the composition on the basis of the mass relation of all components (bitumen, filler, sand and gravel) in the mixture. In Fig. 4.8 the same is shown based on the volumetric distribution. Naturally, the volumetric values were calculated only for the mixture itself, not for compacted specimens. The comparison of the two figures does not show a great difference between the relations. This is caused by the rather good correspondence of the bitumen contents.

Of great importance seems to be the mastic content, i.e. the content of bitumen plus filler in the mixture. Fig. 4.9 shows the mastic content of all optimum mixes as well as that of the mixture used in the FORCE-project. This value varies widely between the laboratories.

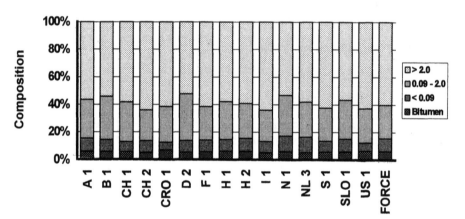

Fig. 4.7 Composition of the optimum mixture (relation by mass).

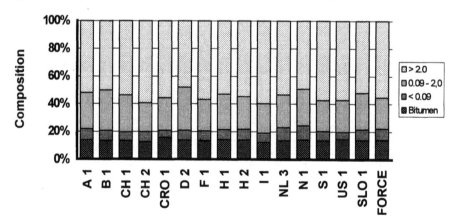

Fig. 4.8 Composition of the optimum mixture (relation by volume).

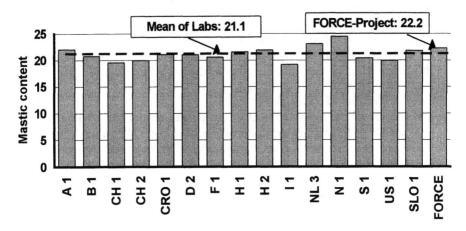

Fig. 4.9 Mastic content of the optimum mixture (% by volume).

4.6   Additional tests by EMPA

4.6.1   PROGRAM

Apart from designing its own optimum mixture EMPA tested a sample of the optimum mixture from each participating laboratory in respect of, EMPA-report, 1994 :

- gradation of the aggregates,
- content of recovered soluble binder,
- density of the mixture,
- behaviour at elevated temperatures (Marshall test),
- behaviour at low temperatures (indirect tensile test),
- properties of the recovered binder,

and compared its results to those measured in the individual laboratories.

4.6.2   TESTS AND PROCEDURES CONDUCTED BY EMPA

General

Testing procedure and calculation of the mixture parameters were based on the Swiss Standards and performed with the precision prescribed in these standards as well as on the internal EMPA standard operating procedures (SOP's) and on the procedures as described below.

Sampling of the mixtures
The mixtures were heated in an oven for 1.5 to 2 h and then, as described in Swiss Standard SN 671 950a, mixed on a metal sheet by shovelling several times and spread out in a circle. Portions for the various test samples were batched with a shovel.

Content of recovered soluble binder
The content of the recovered soluble binder was determined according to Swiss Standard SN 671 955a. The mixture was separated in a hot extractor with toluene. 3 kg of mixture were weighed in. The content of the recovered soluble binder was determined only once.

Density of Mixture
The density was determined according to Swiss Standard SN 671 965a. Toluene was used as solvent. 2 kg of mixture were weighed in. The density was determined only once and expressed in $t \cdot m^{-3}$.

Gradation of the aggregates
The extracted and dried aggregate was sieved dry according to Swiss Standard SN 870 610c. For this purpose, the grid sieves 16, 11.2, 8, 5.5, 4 mm and the mesh sieves 2.8, 2, 1, 0.5, 0.25, 0.125 and 0.09 mm were used.

Marshall test

- *Preparation of test specimens* : For the Marshall test specimens, 1200 g of every single mixture were filled into 4 prewarmed (approx. 160°C) compaction moulds.
  According to Swiss Standard SN 671 969b, Section 4, these moulds consisted of a steel cylinder of internal diameter d = 101.6 mm and a height h = 76 mm. The mixture was compacted by 50 blows at each side. The mass of the falling weight was 4536 g, dropped from a height of 457 mm. The compacted and cooled specimen was pushed out of the mould with a hydraulic ram.
- *Determination of bulk density and void content* : The bulk density in $t \cdot m^{-3}$ of the non-paraffined Marshall test specimen was determined according to Swiss Standard SN 671 967a, Section 7, by immersion weighing. The bulk density is determined as a mean value of four Marshall test specimens.
  The bulk density, together with the density, permitted the calculation of the void content in % by volume.
- *Stability and flow* : Stability (SM) and flow (FM) were determined based on Swiss Standard SN 671 969b as the mean of four Marshall test specimens. The stability is given in kN, the flow in mm. The deformation rate of the Marshall press was 50.8 $mm \cdot min^{-1}$ (0.847 $mm \cdot s^{-1}$).

Indirect tensile test

In the indirect tensile test a cylindrical test specimen is loaded in the radial direction on the opposite sides of the surface until fracture occurs. There is no Swiss Standard for this test.

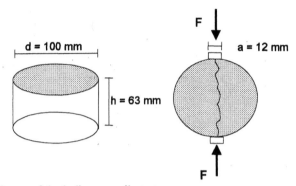

Fig. 4.10 Schema of the indirect tensile test.

The tests were conducted displacement-controlled at a speed of 18 mm/min on a servohydraulic universal testing machine with a force ranging up to 150 kN. The static fracture load (mean) was determined on four Marshall test specimens that had been cooled to -10 °C.

Tests on recovered soluble binder

The binder was recovered with toluene and the following tests were carried out :

- *Penetration at 25°C* according to Swiss Standard SN 671 740a. The mean value of four single measurements is taken. The penetration is given in 10^{-1} mm.
- *Softening point ring and ball* according to Swiss Standard SN 671 743a with an automatic ring and ball machine. The mean value of four single measurements is taken. The softening point is stated in °C.

4.6.3   EMPA TEST RESULTS ON ALL "OPTIMUM" MIXES

All properties of the optimum mixes investigated by EMPA were summarised in a so-called "Identity Polygon". In such a graph all significant mean values of the mixture, the aggregate and the binder results are grouped around a centre point and connected by a polygon. That means that all characteristic properties of the mixture can be seen on one graph (see Fig. 4.11). In Fig. 4.12 the identity polygons of all optimum mixes are assembled.

MG = MIXTURE

MS = AGGREGATES

BM = BINDER

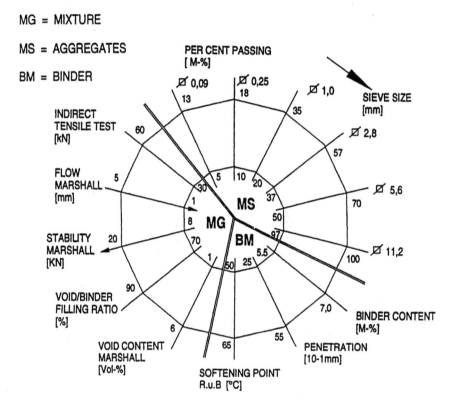

Fig. 4.11 Principle of the "Identity Polygon" (after W. Humm 1993).

From the outfit of these polygons one can read very clearly what characteristics differ and how great the difference or the correspondence between the different laboratories is and in what areas they are. It can be seen e.g. that the gradings of the labs A1, N1 and NL3 are very close together whereas binder content and Marshall values differ widely. The polygons of F1 and H2 are nearly identical in their appearance, but especially the results for the grading differ significantly. It is also very interesting to see that the results of the two Swiss laboratories are clearly different in spite of the fact that they have been based on the same Swiss Standards.

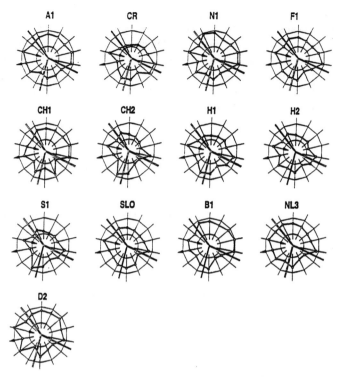

Fig. 4.12 Identity polygons for all "optimum mixtures" determined by EMPA.

4.6.4 RESULTS AND SUMMARY

Gradation of the aggregates
According to Fig. 4.13 the gradation of the aggregates of the various "optimum" mixtures are quite close together.The optimum grading curves produce only 'normal' variations within the framework of standards (the figure shows the limits of the swiss standard).

The filler content varies between 5.9 and 12.5 % by mass. Here three groups can be distinguished :

- Group 1 : B1, N1 and NL3 with a filler content between 12 and 11.2 % by mass.
- Group 2 : A1, CH1, F1 and SLO with a filler content between 10.7 and 9.6 % by mass.
- Group 3 : CH2, H1, H2 and S1 with a filler content between 8.7 and 7.6 % by mass.

The mixture from CR with a filler content of 5.9 % by mass is significantly different.

Binder tests
Fig. 4.14 shows that the *content of recovered soluble binder* is also very close together. The content of recovered soluble binder lies between 5.6 and 6.78 % by

mass. The values for the mixture of CR lies with 0.4 % by mass above all other mixtures. This means a difference of about 1 % by mass, which corresponds to the normal tolerance within the framework of standards. The highest *penetration values* are shown by the mixtures of SLO and A1 at 39.10^{-1} mm. The lowest values are shown by the mixtures of B1 and CH2 at 29.10^{-1} mm. All other mixtures lie between 30 and 34.10^{-1} mm. The extremely high penetration value for the EMPA mixture of 50.10^{-1} mm may be explained by the fact that the "optimum" mixture had been prepared immediately before the test, and hence no hardening of the binder could occur due to rewarming. The *ring and ball values* range from 58 to 61.6 °C, corresponding to the trend of the penetration values. Here the softening point of the EMPA mixture of 54 °C should not be included in a comparison.

Marshall tests
For *mixture density and bulk density*, the values for all "optimum" mixtures are very close.On the other hand, considerable differences are observed for *void content, stability and flow* :

- *Void content* : Very high void contents were found for the mixtures of CH2 with 5.3 % by volume and S1 with 5.1 % by volume. The mixture of D2 with 1.5 % by volume has by far the lowest void content. Roughly comparable are the void contents of A1, N1, NL3 with 2.6 - 3.2 % by volume, of H2 and F1 with 3.2 and 3.4 % by volume, and of B1, CH1, CR, H1, SLO with 3.8 - 4.4 % by volume.
- *Stability* : The highest stabilities occur for the mixture of CH1 (16.7 kN), A1 (16.1 kN) and B1 (15 kN). A further group shows stabilities above 13 kN, namely SLO, NL3 and D2. For the mixtures of CH2, F1, H1, H2 and N1, the stability ranges from 11.1 to 12.2 kN. The lowest stabilities were found with the "optimum" mixtures of CR (10.2 kN) and S1 (9.8 kN).
- *Flow* : The highest flow values were found for the mixtures of D2 with 5.3 mm, N1 with 4.8 mm and A1 with 4.6 mm. The lowest values are shown by CH1 with 3.1.mm and S1 with 3.3mm. The remaining values may be divided into 3 groups :

    - Group 1 : B1, CH2 with 3.5 mm.
    - Group 2 : CR, F1, H2 with 3.8 to 4.0 mm.
    - Group 3 : SLO, NL3, H1 with 4.2 to 4.4 mm..

Indirect tensile test
The highest fracture forces were measured for mixture NL3 at 51.3 kN and A1 at 48.1 kN. Next follow N1 at 46.4 and D2 at 46.1 kN. SLO and B1 show values of 43.3 and 43.1 kN. The mixtures from the remaining laboratories range from 41 to 38.5 kN. The mixture by S1with a fracture force of 34.8kN again shows an extremely low value. Table 4.9 shows the overal results.

Table 4.9 Results of EMPA test on the "optimum mixtures".

Laboratory-code	A1	B1	CH1	CH2	CR	D2	F1	H1	H2	N1	NL3	S1	SLO
Grading [% by mass]													
16 mm	100.0	100.0	100.0	100.0	100.0	100.0	100.0	100.0	100.0	100.0	100.0	100.0	100.0
11.2 mm	99.4	99.9	99.7	99.1	99.7	99.4	99.7	99.5	98.6	99.7	99.3	99.2	99.2
8.0 mm	84.7	88.8	89.6	79.2	80.8	82.6	85.1	78.5	75.9	86.7	83.1	78.2	80.3
5.6 mm	68.3	69.8	68.6	51.2	57.7	63.3	63.7	55.9	51.4	67.2	59.6	54.4	58.7
4.0 mm	59.6	60.5	59.6	44.3	50.8	59.4	54.3	50.3	46.2	58.0	52.6	46.9	49.6
2.8 mm	50.9	53.8	51.5	39.2	43.0	55.0	46.4	44.8	41.8	50.4	46.6	40.7	43.3
2.0 mm	43.3	44.5	43.2	33.2	34.8	46.9	39.1	36.9	35.0	41.5	39.7	32.9	36.5
1.0 mm	31.0	30.2	29.8	22.6	23.1	32.1	27.8	24.1	23.6	28.4	28.2	21.3	25.6
0.5 mm	22.5	22.1	21.0	16.0	15.7	22.7	20.3	16.4	16.7	21.1	20.6	14.7	18.3
0.25 mm	16.8	17.0	15.4	12.2	10.8	16.9	15.3	11.7	12.6	16.8	16.0	10.9	13.8
0.125 mm	12.6	13.5	11.7	9.6	7.4	13.0	11.9	8.8	9.9	13.9	12.9	8.4	10.8
0.09 mm	10.7	12	10.3	8.5	5.9	11.2	10.4	7.6	8.7	12.5	11.5	7.4	9.6
Density [t*m-3]	2.403	2.413	2.423	2.424	2.378	2.393	2.414	2.389	2.388	2.407	2.427	2.402	2.420
Bulk density [t*m-3]	2.339	2.314	2.315	2.297	2.274	2.356	2.331	2.284	2.311	2.341	2.359	2.280	2.327
Void content [% by vol]	2.6	4.1	4.4	5.3	4.4	1.5	3.4	4.4	3.2	2.8	2.8	5.1	3.8
Stability [kN]	16.1	15.0	16.7	11.1	10.2	13.6	12.2	12.0	11.2	11.7	14.1	9.8	13.1
Flow [mm]	4.6	3.5	3.1	3.5	4.0	5.3	3.9	4.4	3.8	4.8	4.4	3.3	4.2
Indirect tensile test [kN]	48.1	43.5	41.3	37.5	38.2	46.1	42.9	40.4	38.3	46.4	51.2	34.8	43.3
Binder content (soluble binder) [% by mass]	6.08	5.83	5.66	5.64	6.78	5.83	6.03	6.39	6.14	6.17	5.75	6.22	5.69
Penetration[10-1mm]	39	29	50	29	33	34	30	31	30	34	31	32	39
Softening point R&B [° C]	58.0	61.6	54.0	59.8	59.2	60.4	59.8	59.3	59.4	58.4	59.3	58.7	59.1

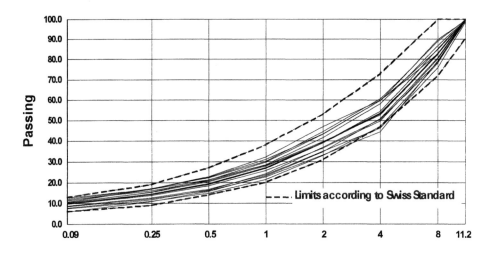

Fig. 4.13 Results of EMPA Test on the Gradation of the Aggregates.

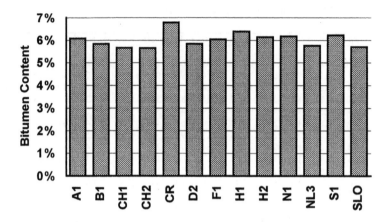

Fig. 4.14 Results of EMPA Test on the Bitumen Content (Content of recovered soluble Binder) [% by mass]".

4.6.5 COMPARISON OF THE VALUES OF THE PARAMETERS FROM THE VARIOUS LABORATORIES WITH THE EMPA RESULTS

In order to determine the correlation between the values of a parameter determined by EMPA and the other laboratories, both values were plotted in the same diagram. As EMPA did not get the test results from all laboratories, these diagrams could only be prepared for the parameters :

Marshall stability, Marshall flow, and Marshall void content. The Fig. 4.15 to 4.17 show the comparisons.

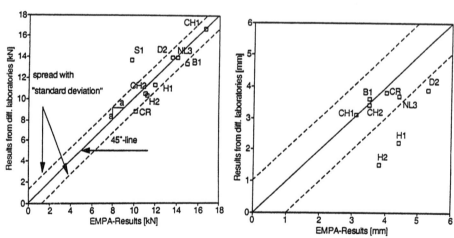

Fig. 4.15 Marshall stability [kN].

Fig. 4.16 Marshall flow [mm].

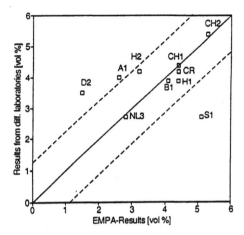

Fig. 4.17 Void content [% by vol].

In the figures, the values of a parameter determined by the other laboratories are plotted on the vertical axis versus the corresponding EMPA values on the horizontal axis. The 45 °-line is the line of equality. The distance of a point from this line is thus a measure of correlation. In addition, lines representing one standard deviation are drawn on both sides of and parallel to the 45 °-line. The values determined by the EMPA for CH1 were situated directly on the 45 °-line (by definition).

It turns out that the deviations of the measured values depend both on the laboratory (e.g. the values for a mixture of B1 and CH2 show relatively small deviations, those of A1, D2 show generally larger deviations), and on the type of parameter (e.g. mixtures of H1, H2 show small Marshall stability and large deviations of Marshall flow). But since the data of only a few parameters determined by the other laboratories were available to EMPA, a statement on the way a laboratory works is not possible in detail.

Overall it is clear that the agreement in the Marshall stability is very good; the values lie, with exception of two of them, either very close to the 45 °-line or even on it. However, for all the other parameters very high deviations and scatter exist.

4.6.6 COMPARISON OF THE OPTIMUM MIXTURE COMPOSITION REPORTED BY THE LABORATORIES AND THE EMPA RESULTS

In this section a comparison of the laboratory results with the results of the additional tests executed by EMPA concerning the mixture composition is carried out.

Bitumen content
For both grading and bitumen content the best way to judge the differences is on the basis of a graph. Fig. 4.18 shows the two results of the bitumen content the optimum mixture, one column for the laboratory value and one for the value by EMPA. One can state a certain difference, but the reproducibility seems rather good.

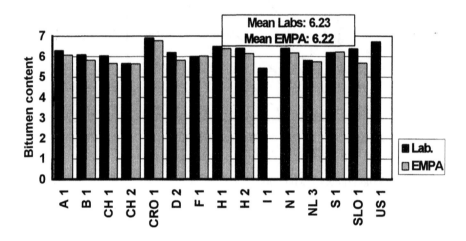

Fig. 4.18 Bitumen content [% by mass].

Grading
To compare the grading the mean values of the gradings and additionally the factors of fineness are used. The Fig. 4.19 illustrates the first comparison. It can be seen that the grading is practically equal. Looking at the fineness factors in the Fig. 4.20 one must state that the mean is more or less the same, but there are some differences between the results reported by the individual laboratory and that tested on the optimum mixture by EMPA.

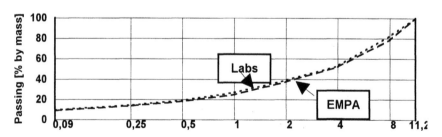

Fig. 4.19 Mean of grading of laboratories and EMPA.

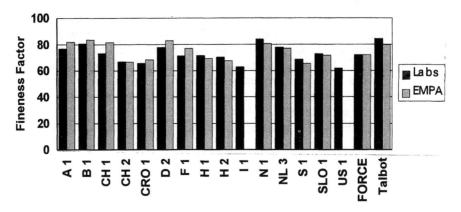

Fig. 4.20 Fineness factors determined by individual laboratories and EMPA.

4.7 Conclusions

Within the framework of the 'Interlaboratory Test Program Mix Design' the participating laboratories had the task of producing and investigating an optimum mixture according to their national standards and regulations using the prescribed materials (aggregate, bitumen). The EMPA received a specimen mixture produced by each participating country in order to carry out comparison tests.

With the exception of France, which used the LCPC method based on the PCG gyratory test, all countries used the Marshall method or some of its elements so that a comparison of basically different compaction methods was not possible. However, it is worth noting that a good comparison between the optimum mixture from France and the others was also possible. The question whether a basic conformity between the different methods exists, however, cannot be answered here.

Also the individual methods used by some laboratories and other additional investigations could not be compared in these investigations due to the wide range of differences.

A glance at the data and results indicates a wide diversity which can be explained by the different working methods employed in the various laboratories.

By first of all observing the main components, the gradation of the aggregate and the bitumen content, good conformity can be established between the individual laboratories, and also to the data produced by the FORCE-project.

Although the mixtures of the 4 prescribed components (filler, sand, gravel 1 and gravel 2) vary by a wide range, the optimum calculated sieve curve produces only 'normal' variations. Also the distribution of the gradings themselves, expressed by the fineness factor are very similar.

The bitumen content of the various 'optimum mixtures' lies, with one exception, between 5.6 and 6.4 % by mass (FORCE-project = 6.05 % by mass) which means a difference of approx. 1 % by mass. This corresponds to the normal tolerance within the framework of standards.

More difficulty is experienced when making comparisons between the test results achieved by the individual laboratories and those from EMPA. For instance laboratory D2 quoted the Marshall flow value at 3.6 while EMPA determined a value of 5.3. Laboratory S1 quoted a Marshall void content of 2.8 % by volume for its optimum mixture, EMPA, in contrast, achieved a value of 5.1 % by volume.

When considering all the comparable results of the investigations carried out by EMPA, both, the Marshall stability determined for all the optimum mixtures ≥ 10 kN, and indirect tensile force at -10°C between 37.5 and 51 kN indicate a high stability under hot conditions and a strong consistency in cold conditions. These results indicate excellent characteristics with respect to the prevention of deformation and cracking when subjected to cold.

Greater differences between the individual mixtures was established for Marshall void content and flow. For the Marshall void content, the laboratories can be divided into three groups : laboratories (D2, A1, N1 and NL3) whose mixes exhibited a very low void content between 1.5 and 2.8 % by volume, laboratories (CH2 and S1) for whose mixtures EMPA determined a void content of over 5 % by volume and laboratories whose mixtures had an average void content of between 3.8 and 4.4 % by volume. The groups are not so clearly discernible with regard to the Marshall flow value. Laboratories whose mixtures had a low void content all had a very high flow value between 4.6 and 5.3 mm, however, the Marshall flow of mixtures from all other laboratories was between 3.1 and 4.4 mm.

A ranking of the various optimum mixes from the individual laboratories cannot be carried out because the results, produced by non-performance assessment, do not provide direct evidence of the mixture's properties in practice (e.g. resistance to rutting or cracking etc.). It is therefore not absolutely certain what is really good, or not so good or bad.

To get more information about the behaviour of the material on the road it would be useful for further interlaboratory test programs to consider also so called "performance" tests as modulus tests, fatigue tests, shear tests and simulation tests as the wheel tracking test.

Overall within the framework of the Interlaboratory Test Program the optimum mixes produced exhibit remarkably similar results so that it can be assumed that a mix design of asphalt concrete surface layers will generally create comparable mixtures.

A comparison of the results of the individual laboratories with those of EMPA indicated the considerable difference between the working methods of the various test and research institutes. In order to improve this situation, it will be necessary to draw up and apply common guidelines and definitions.

4.8   References

OECD full-scale pavement test, *OECD-report,* 1991.

Raab C., RILEM TC 152-PBM, Interlaboratory Test Program : Part 1, Mix.-Design, *EMPA-report FE 146'411,* 1994

Humm W. : "Entwicklung eines Verfahrens das grafische Darstellen der Mischgutquelität", *EMPA-report No. 143'188,* May 1993.

PART THREE

MECHANICAL TESTING OF MIXTURES

5

State of the Art on Stiffness Modulus and Fatigue of Bituminous Mixtures

Hervé di Benedetto and Chantal de La Roche

5.1 Introduction

It is possible to identify three main "typical" behaviours according to the strain amplitude and the number of loadings applied to bituminous mixtures [di Benedetto, RILEM 90] (Fig. 5.1).

- For a small number of loadings and strains of a few percent, the observed behaviour is highly non-linear.
- For loadings including a few hundred cycles and "small" strains ($<10^{-4}$) the behaviour is regarded, to a first approximation, as linear viscoelastic.
- With loadings of several tens of thousands of cycles and small strains, when damage phenomena appear, the material shows "fatigue".

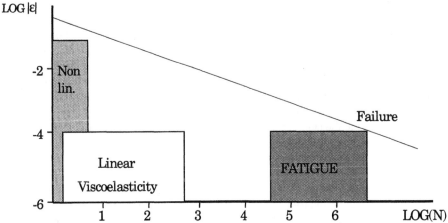

Fig. 5.1 "Typical" behaviours observed on bituminous mixtures (ε) strain - (N) number of loadings.

This chapter concerns only the main developments and tests aimed at characterizing the linear viscoelastic behaviour and the fatigue behaviour of bituminous mixtures.

Bituminous Binders and Mixes, edited by L. Francken. RILEM Report 17. Published in 1998 by E & FN Spon, 11 New Fetter Lane, London EC4P 4EE, UK. ISBN 0 419 22870 5

Using the background of thermomechanics, it is shown that some explanations can be given about the validity of the tests.

5.2   Linear viscoelastic behavior - Complex modulus

5.2.1   ASSUMPTIONS

In the literature, bituminous materials are generally regarded as being continuous, isotropic, viscoelastic, linear media that are thermorheologically simple (i.e. the principle of time-temperature superposition applies). These assumptions apply only under certain conditions, which are briefly discussed. It is clear that the "closer" the behavior of the material to these properties, the more valid the theoretical developments in connection with each of these assumptions.

5.2.1.1 Continuous medium
While their composite granular structure makes the bituminous mixtures currently used on pavements heterogeneous, macroscopic continuity can be assumed on the scale of a pavement layer. For laboratory specimens, it is generally held that a ratio of 10 between the size of the coarsest aggregate and that of the specimen is necessary for the latter to be considered non-heterogeneous [Biarez et al, 1994]. Even though this ratio is not always attained for practical purposes (some specimens tested in fatigue have a ratio of size of grains to size of specimen of the order of 3), this assumption is made. But it cannot be applied to surface dressings, for example.

5.2.1.2 Isotropy
The manner of placement of the bituminous mixture on pavements (spreading of swollen material in layers and compacting by successive passes of compactors on the surface) gives the material some anisotropy (orientation of grains favored by compacting). Similarly, specimens made in the laboratory exhibit properties of this type. Complex modulus measurements in tension-compression on cylindrical samples taken along three core drilling axes by Doubbaneh [1995] in slabs of bituminous mixtures made in the laboratory have shown variations of as much as 20%.
In road engineering calculations, the material is assumed to be isotropic.

5.2.1.3 Viscoelasticity and linearity
In the whole range of solicitations they are submitted to, bituminous mixtures exhibit a complex elastoviscoplastic behavior, but at low strain amplitudes this behavior can be considered as purely viscoelastic and also linear (Fig. 5.1). The modulus of bituminous materials must be measured inside this linear viscoelastic domain. This requires applying "small strains" [Linder, 1977], [Huet, 1963]. A few experimental results presented in this article give an idea of the limits of this domain. It seems to appear for strains of which the amplitude is less than approximately 10^{-4} .
To facilitate understanding the developments that follow, some basic principles of linear viscoelasticity are summarized up in appendix I at the end of this chapter.

5.2.1.4 Influence of temperature

Temperature is a decisive parameter for the behavior of bituminous materials. In effect, like the binder it contains, a bituminous mixture is temperature-sensitive, which means that its response to a given loading is strongly dependent on temperature. The experimental results show that the principle of time-temperature equivalence [Huet, 1963] applies to bituminous mixtures. However, mixtures based on some modified bitumens or very structured bitumens significantly depart from this property [Ramond et al, 1996].

5.2.2 MEASUREMENT OF LINEAR VISCOELASTIC PROPERTIES

To measure the properties of materials having a viscous behavior, in addition to the type of test in question, one must also take into account the evolution of the applied loading versus time.

5.2.2.1 Types of tests

As regards the existing tests for bituminous mixtures, and more generally geomaterials, two main categories can be distinguished: homogeneous tests and non-homogeneous tests.

Homogeneous tests give direct access to the stresses and strains, and therefore to the constitutive law (whether viscoelastic or not). Non-homogeneous tests call for postulating a constitutive law a priori (linear viscoelasticity, for example) and taking account of the geometry of the specimen to obtain, after calculations that are often complex, the parameters of the constitutive law (viscoelastic modulus, for example). Non-homogeneous tests can be used only if the behavior in question is "simple" (isotropic linear elastic or isotropic linear viscoelastic). When the behavior of the material tested departs from the postulated behavior, a large error may be introduced.

Homogeneous tests would therefore seem to be more pertinent for describing linear viscoelastic behavior. In effect, analysis of them makes it easy:

- to define the parameters of the behavior and
- to find the limits of the linear domain.

However, a non-homogeneous test gives good results if the behavior of the mixture tested is linear viscoelastic.

The homogeneous tests used for bituminous mixtures have been reviewed by Di Benedetto [1990].

5.2.2.2 Types of loading

To describe the linear viscoelastic behavior of a material, it is possible to use several loading versus time signals. The three main loadings applied in tests on bituminous mixtures are:

- monotone loadings analyzed by introducing the time (time mode);

- sinusoidal loadings processed using the frequency (Fr) or the pulsation (ω) (frequency mode). The complex modulus (E*) is found directly with this type of loading;
- non-sinusoidal cyclic loadings, such as pulse tests, which require specific analyses.

It is clear that the moduli obtained with the various types of loading signal, which can be very different, are related by more or less complex expressions. These expressions reflect the linear viscoelastic behavior of the material.

The theoretical developments show (cf. appendix I) that the statement of the creep function f(t) or of the relaxation function r(t) or of the complex modulus E*(ω) fully describes the linear viscoelastic behavior.

Remark : It has to be emphasized that the word "dynamic" has to be used only for a test with non negligible inertia effects inside the sample (i.e. when wave propagation is observed). Considering :

- the frequency (or strain rate) of the tests,
- the size of the samples,
- the modulus of bituminous mixtures,

the inner inertia effects can be neglected for classical repeated loading tests conditions on bituminous mixtures. For example, complex modulus or fatigue tests are not dynamic tests but cyclic tests or tests with repeated loading and are therefore interpreted as static tests.

5.2.3 MEASUREMENT IN THE FREQUENCY DOMAIN - COMPLEX MODULUS

5.2.3.1 Definition

The material is subjected to sinusoidal loadings at varied frequencies. Since the measurements are made in the domain of small strains, for which the bituminous mixture behaves mainly like a linear viscoelastic material, the response to a sinusoidal loading is also sinusoidal [Huet, 1963].

The complex modulus E* is the complex number defined as the ratio between the complex amplitude of the sinusoidal stress of pulsation ω applied to the material $\sigma = \sigma_0 \sin(\omega t)$ and the complex amplitude of the sinusoidal strain that results in a steady state. Given the viscoelastic character of the material, the strain lags the stress, which is reflected by a phase angle φ between the two signals: $\varepsilon = \varepsilon_0 \sin(\omega t - \varphi)$. Given this definition, the complex modulus is not a function of time but depends on the pulsation ω (or on the Frequency Fr) for a fixed temperature.

Writing :

$$\sigma(t) = \mathrm{Im}\big[\sigma*(t)\big] \quad with \quad \sigma*(t) = \sigma_0 \, e^{i\omega t}$$

$$\varepsilon(t) = \mathrm{Im}\big[\varepsilon*(t)\big] \quad with \quad \varepsilon*(t) = \varepsilon_0 \, e^{i(\omega t - \varphi)}$$

$$E*(\omega) = \frac{\sigma_0}{\varepsilon_0 \, e^{-i\varphi}} = |E*| \, e^{i\varphi}$$

|E*| is the norm of the complex modulus or stiffness modulus.

φ   is the phase angle of the material (for example used to estimate the energy dissipated in the material).

It is also possible to use the following notations $E* = E_1 + iE_2$

E_1 is the storage modulus.
E_2 is the loss modulus.
$K*(\omega)$ the bulk modulus and $G*(\omega)$, the shear modulus are also defined.

Assuming a linear viscoelastic and isotropic behavior, the relations between these rheological parameters are the following:

$$K^* = \frac{E^*}{3\,(1 - 2\upsilon^*)}$$

$$G^* = \frac{E}{2\,(1 + \upsilon^*)}$$

where υ^* is the Poisson's ratio.

In these relations, υ^* is a priori a complex number. However, direct measurements of υ based on volumic strain measurements in complex modulus tests in tension-compression with [Charif, 1991] or without [Doubbaneh, 1995] a confinement pressure tend to show that its imaginary part is very small. For the bituminous materials tested, it can therefore be treated as real; its value varies between 0.3 and 0.5 depending on temperature and frequency.

5.2.3.2 The various types of test

Generally, for all of the tests, from the values of force F and displacement D applied to the boundaries of the specimen and the phase angle φ between these two signals, the complex modulus of bituminous mixtures can be determined using two factors:

- a shape factor γ that depends on the dimensions of the specimen;
- a mass factor μ that takes into account (if necessary) the effects of inertia related to the mass M of the moving specimen and the mass m of moving parts (attachment helmets, specimen-motor unit link, etc.).

The real and imaginary parts of the complex modulus are then given by [Huet, 1963]:

$$E_1 = \gamma \left(\frac{F}{D} \cos \varphi + \mu \omega^2 \right)$$

$$E_2 = \gamma \left(\frac{F}{D} \sin \varphi \right)$$

where : ω is the pulsation.

A table of the existing modulus types of measurement is given in paragraph 5.2.6.

5.2.3.3 Presentation of the results

The various components of the complex modulus vary with the temperature and the frequency of loading, which are fixed for each elementary test. The experimental results $|E^*|$, φ, E_1, E_2, are usually expressed using the conventional representations shown in Fig. 5.2, 5.3, 5.4, 5.5, and 5.6. The examples presented in these figures are for a standard French semi-granular bituminous mixture (5.4 % binder) in the two-points bending test on trapezoidal specimens.

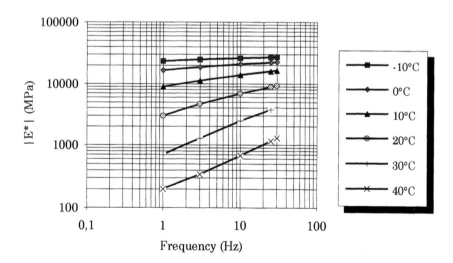

Fig. 5.2 Isothermal curves of the complex modulus.

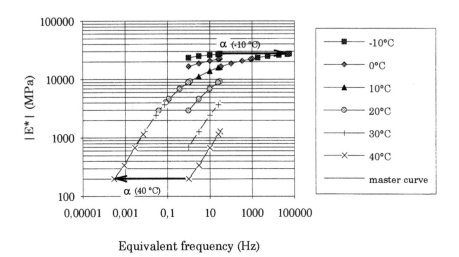

Equivalent frequency (Hz)

Fig. 5.3 Master curve (Reference temperature = 10°C).

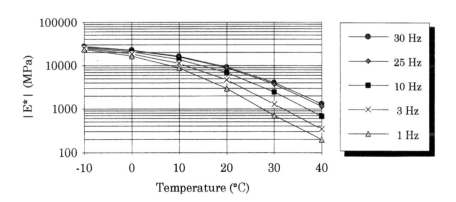

Temperature (°C)

Fig. 5.4 Isochrone curves of the complex modulus.

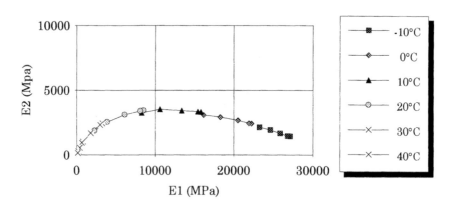

Fig. 5.5 Complex modulus curves in the Cole and Cole plane.

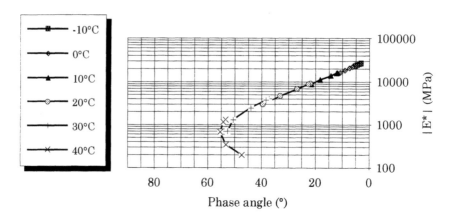

Fig.5.6 Complex modulus curve in the Black space

5.2.3.4 Frequency-temperature equivalence

It can be seen from these curves that the material follows the frequency-temperature equivalence [William et al, 1955] : $E^*(\omega,T)$ can be expressed as a function of a reduced variable $\omega\alpha(T)$, where $\alpha(T)$ is a function of temperature. In particular, the same modulus value of the material can be obtained with different (frequency, temperature) couples. It is possible to use this property of frequency-temperature equivalence to construct a single "master" curve (log $|E^*|$, log Fr) for an arbitrarily chosen reference temperature (T_R).This curve (Fig. 5.3) is obtained by translating each isotherm parallel to the frequency axis, with respect to the isotherm corresponding to the reference temperature, until the points having the same modulus value are superposed.

The coefficient of translation of isotherm T with respect to the chosen reference isotherm T_R is log α_T such that $E^*(\omega,T) = E^*(\omega.\alpha_T, T_R)$ with $\alpha_{T_R} = 1$

Several formula are proposed to represent log α_T. The most commonly used are:

- the "WLF" formula (obtained by William, Landel and Ferry) [1955]:

$$\log \alpha_T = \frac{-C_1(T - T_R)}{(T - T_R) + C_2}$$

where :   C1 and C2   are constants that depend on the material;
          T and T_R are temperatures in °K.

- Arrhénius's equation:

$$\ln \alpha_T = \frac{\Delta H}{R}\left(\frac{1}{T} - \frac{1}{T_R}\right)$$

ΔH represents the apparent energy of activation characterizing the material;

R   is the universal gas constant (8.31 J/mole/°K);

T and T_R are temperatures in °K.

5.2.4   MEASUREMENT OF THE MODULUS IN THE TIME DOMAIN

Another way to determine the modulus of a bituminous mixture is to measure a secant modulus in the time domain during monotone loading. The **secant modulus** S(t) of the material is introduced by:

$$S(t) = \frac{\sigma(t)}{\varepsilon(t)}$$

where : $\sigma(t)$ is the stress at time t
            $\varepsilon(t)$ is the strain at time t

This type of measurement is also used to obtain the modulus of the binder such as in the Bending Beam Rheometer (BBR) test [Anderson et al, 1993].

When the loading rate is zero, one obtains either the relaxation function, r(t) = 1/S(t), if the loading is a strain, or the creep function, f(t) = S(t), if the loading is a stress. The standardized French direct tensile test (standard NFP 98-260-1) developed by Linder [1977] and extended by Moutier [1990] is one example of those tests. It consists, at a given temperature, of subjecting a cylindrical specimen to a tensile loading up to a specified value of axial strain ε_{max} (fixed in the linear domain of the material), following a loading law controlled in strain of the type $\varepsilon = at$. The test is repeated for several loading times t_i (same value ε_{max}).

The range of possible loading rates is limited by the experimental resources. For a fixed temperature, the tests are generally performed for loading times of 1, 3, 10, 30, 100, and 300s. It is repeated at several temperatures: in practice from -10 to +20° C.

The results obtained are used to plot the isotherm of the secant modulus versus the loading time (Fig. 5.7).

The principle of time-temperature equivalence $(S(t,T)) = S(t/\alpha_T))$ can be used to plot the master curve of the material and have access to loading times shorter than those tested. It is the modulus value at 15°C and 0.02 s, obtained from the master curve, that is used for pavement design calculations in France.

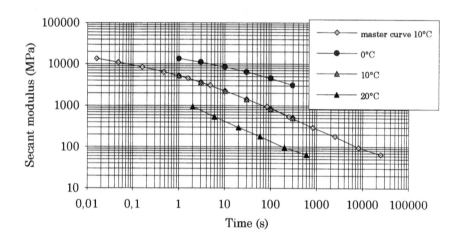

Fig. 5.7 Isotherms of the secant modulus and master curve at 10°C derived from direct tensile tests (semi-granular bituminous mixture, 5.5% binder).

5.2.5 OTHER TYPES OF MEASUREMENT

The notion of resilient modulus has been defined to make better allowance for the conditions of loading on pavements,

The previously described modulus tests often require sophisticated servocontrol systems. For "inspection" measurements, "pulse" tests have been developed. For this type of test, the maximum loading value applied is controlled but there is no true control of a loading setpoint. Non-sinusoidal cyclic loadings are applied.

These tests are used to define a resilient modulus E - or reversible modulus - corresponding to the ratio of the repeated stress to the reversible strain.

In fact, if the applied pulse proves to be a linear loading with respect to time, the resilient modulus is interpreted theoretically as a secant modulus identical to that of the direct tensile test described above. In this case, E is equal to S(t).

If the loading is not linear with time, the modulus can depart substantially from what would be obtained by following a linear path. This explains some errors that may occur in the interpretation of tests of this type.

This type of modulus can be measured for example in a triaxial test (with or without confinement pressure) (Fig. 5.8) or in an indirect tensile test.

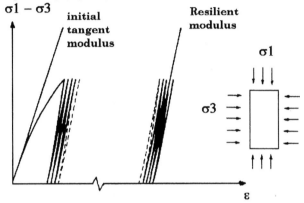

Fig. 5.8 Typical stress-strain curve in a triaxial resilient modulus test (after [Mamlouk et al, 1988]).

The diametrical compression or indirect tensile test used for measurements at small strains on bituminous mixtures consists of applying an impulse loading on two diametrically opposed generating lines of a cylinder of bituminous mixture. The (diametrical) central part of the sample is then subjected to a tensile stress that is in large part constant over the diameter joining the two generating lines loaded (Fig. 5.9).

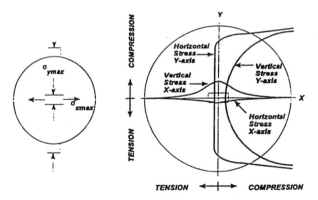

Fig. 5.9 Principle of the diametrical compression test and distribution of stresses in the sample (linear elastic calculation) after [Read et al, 1996].

This test is standardized and is routinely used in the United States to measure the resilient modulus of bituminous mixtures (ASTM standard D 4123, 1982 (revised in 1987)). The loading then consists of repeated force pulses.

It is a non-homogeneous test, interpreted by assuming an isotropic material having a constant Poisson's ratio (ν). The resilient modulus E and the Poisson's ratio are given by:

$$E = P \, (\nu + 0.27) \, / \, t \, \Delta H$$

$$\upsilon = 3.59 \ \Delta H \, / \, \Delta V - 0.27$$

where : P   is the load in N and E in MPa (Fig. 5.9).
       t   is the thickness of the cylindrical specimen (mm).
     ΔH is the change of diameter along x (mm).
     ΔV is the change of diameter along y (mm).

If the vertical displacements are not measured, E is determined using an approximate value of the Poisson's ratio. This is the case of measurements made with the NAT (Nottingham Asphalt Test) [Brown, 1993].

While this type of test can be used as a means of checking the performance of the material, it is also necessary to know the value of the Poisson's ratio υ. Moreover, its very principle may favor creep of the specimen in the course of the test. According to tests performed for the SHRP, Tayebali et al [1994] conclude that this type of test must be restricted to temperatures below 20°C.

5.2.6 RECAPITULATORY OF THE DIFFERENT TESTS FOR MODULUS MEASUREMENT

Table 5.1 groups various existing tests with the diagrams showing their principle, their "shape" factors, and a few associated references to the literature. The formulas given in this table are used for the interpretation of the interlaboratory tests presented in chapter 6 of this state of the art report.

Table 5.1 Recapitulatory of the various existing test methods for modulus measurements (from Francken et al. 1996)

≈ HOMOGENEOUS TESTS

	Principle	Form factor* γ $[L^{-1}]$	Reference (used in chapter 6)	Literature
tension compression (with or without confining pressure)		$\dfrac{h}{\pi D^2}$	CO-CY	[Charif, 1991] [Doubbaneh, 1995]
Shearing test		$\dfrac{1}{h\,e}$		[Assi, 1981] [de La Roche, 1996]
Constant height shearing test		$\dfrac{h}{\pi D^2}$		[Sousa, 1994]
Shearing test Machine		$\dfrac{h}{\pi D^2}$	SH-CY	[Lempe et al, 1992]
		$\dfrac{h}{2Lb}$	SH-PR	
Co-axial Shear test**		$\dfrac{\ln\left(\dfrac{d}{D}\right)}{2\pi h}$	SH-BE	[Gübler, 1990]

* in the case of shearing tests, the shape factor is given for the calculation of G
** the validity of the homogeneity hypothesis depends on the ratio D/*d*

NON HOMOGENEOUS TESTS

Principle		Shape factor* γ	Reference	Literature
2 points bending		$\dfrac{4\,L^3}{b\,h^3}$	2PB-PR	[Francken et al, 1994]
		$\dfrac{12\,L^3}{b\,(h_1 - h_2)^3}\left[\left(2 - \dfrac{h_2}{2h_1}\right)\dfrac{h_2}{h_1} - \dfrac{3}{2} - \ln\dfrac{h_2}{h_1}\right]$	2PB-TR	[Huet, 1963] [Chauvin, 1990]
3 points bending		$\dfrac{24\,L^3}{\pi^4\,b\,h^3}$	3 PB-PR	[Myre, 1992]
Indirect Tensile test		$\dfrac{1}{b}(\nu + 0.27)$	IT-CY	[Brown, 1993] [Kennedy et al, 1994] [Tayebali et al, 1994]
4 points bending		$\dfrac{2\,L^3 - 3L\,l^2 + l^3}{8\,b\,h^3}$	4PB-PR	[Pronk, 1996]

5.2.7 COMPARISON OF RESULTS OBTAINED WITH THE VARIOUS TYPES OF TESTS

5.2.7.1 Complex modulus and Secant modulus

Theory : For a given loading time t from a direct traction (or indirect tension test if the loading is linear) and a given pulsation ω (from a sinusoidal test), the relation between S (t) and |E* (ω)| can be obtained only if the whole creep (or relaxation) function is known. This relation appears also to be dependent on temperature. Considering both a general viscoelastic model (Huet biparabolic (cf. appendix II)) and a time-temperature superposition law, a general calculus giving the ratio R between |E* (ω)| and S (t):

$$R = \frac{|E^*(\omega)|}{S(t)}$$

is proposed by Moriceau [1994] and Di Benedetto [1995].

If t = 0,1 second and ω/2π = 1 Hz, then :

R = 0,87	at 0°C
R = 0,65	at 20°C
R = 0,54	at 40°C

The conclusion is that the correlation between the results of the Indirect Tensile Test and the complex modulus is influenced by the temperature and the whole viscoelastic properties. This correlation will change for different mixes.

The theoretical relation between the complex modulus and the secant modulus is found simply in the context of linear viscoelastic behavior [Linder et al, 1986], [Moriceau, 1994], [Piau, 1996]. By definition (cf. appendix I), the complex modulus is expressed versus the creep function (f(t)) by:

$$E^*(\omega) = \frac{1}{i\omega \int_0^\infty e^{-i\omega t} f(t)\, dt}$$

The secant modulus is in the form:

$$S(t) = \frac{\sigma(t)}{\int_0^t f(t-\tau)\, d(\sigma(\tau))}$$

In what follows, it is assumed that the loading is linear and in the form σ(t) = a t, the secant modulus then becomes:

$$S(t) = \frac{t}{\int_0^t f(t-\tau)\, d\tau}$$

Experimental results : Experimental results have been obtained by Serfass et al [1990], in direct tension (secant modulus) and in two-point bending on trapezoidal specimens (complex modulus), for 18 formulations of bituminous mixtures. These bituminous mixtures differ in type of binder (pure bitumens, bitumens with SBS polymers, hard bitumens, bitumens with additions of polyethylene) and grading formulation. The results obtained, Fig. 5.10, confirm the good agreement between the secant modulus and the complex modulus at the same measurement temperature (10 or 15°C) with pulsation ω and loading time t conforming to the relation ωt = 1.

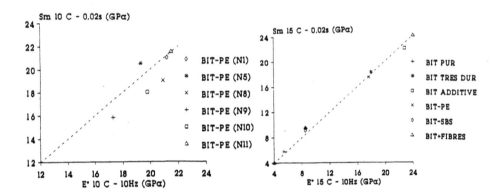

Fig. 5.10 Correspondence between secant modulus Sm obtained in the direct tensile test (10 and 15°C, 0.02 s) and complex modulus |E*| obtained in two-points bending (10 and 15°C, 10 Hz), from [Serfass et al, 1990].

5.2.7.2 Complex modulus found in different tests
Several studies have been performed to compare the results yielded by various types of complex modulus tests. The inter-laboratory tests organized by the RILEM [Francken et al, 1996], which are presented and analyzed in chapter 6, made it possible to compare results on a single 0/14 bituminous concrete in bending (2-, 3-, and 4-points), tension-compression, and shear (3 different types of test). For a given (frequency, temperature) couple, the following trends were found:

- the different **bending tests** give similar values of |E*| and φ [Francken et al, 1996]. Comparative two- and three-point bending tests reported by de La Roche et al [1994] lead to similar conclusions.
- comparison of results obtained in **bending** and in **tension-compression** leads to less unanimous conclusions. For Francken et al [1996], the values of |E*| obtained in tension-compression are twice those obtained in bending, but this claim is based on only one tension-compression result. A more complete study grouping six laboratories made it possible to compare the results obtained in two-point bending tests (three laboratories) and in tension-compression tests (three laboratories) [Aguirre et al, 1981]. For temperatures of 10 and 20°C, the

modulus values obtained in tension-compression and in bending are close, but with a systematic tendency to higher modulus values in tension-compression than in bending, the largest differences between the two groups were found at 0 and 30°C.

In all comparisons of results, it is necessary to make sure that the measurements compared were in fact made under the same frequency and temperature conditions, and also in the linear domain of the bituminous mixture or at the same strain value. When these conditions are satisfied, [de La Roche, 1996] and [Doubbaneh, 1995] have found good agreement between the moduli measured in bending and in tension-compression.

- there are few results in the literature concerning comparisons of **modulus values in bending or tension-compression** and **in shear**, which require prior knowledge of the Poisson's ratio. The SHRP has however compared modulus values obtained in shear (shear test at constant height) and values obtained in bending (four-points) [Tayebali et al, 1994]. Statistical analysis of the results has led to the following empirical relations between the two types of measurement:

$$|E^*| = 8.560 \, (|G^*|)^{0.913}$$
$$E_2 = 81.125 \, (G_2)^{0.725}$$

with $|E^*|$:   magnitude of the modulus in bending (psi),
         E_2:   loss modulus in bending (psi),
         $|G^*|$:   magnitude of the modulus in shear (psi),
         G_2:   loss modulus in shear (psi).

The results obtained by Francken et al [1996], for their part, show modulus values in shear that dependent significantly on the type of test used: for a given frequency-temperature couple, the values of $|G^*|$ vary in a ratio of 1 to 5 according to the type of test. On the other hand, the phase angle values are very consistent across all types of test. By comparison with the bending tests performed on the same material, application of the relation $G^* = E^*/\, 2(1+\nu)$ (a relation usable in the context of isotropic linear viscoelasticity), assuming ν real and equal to 0.35 (a value routinely used for bituminous mixtures) leads to modulus values in bending (E^*), calculated from G^*, varying between 0.3 and 2 times the mean modulus value measured by the bending tests. These results have led to questioning of the quality of the measurements made in shear (the most scattered), since the various results obtained in bending are very consistent.

5.2.8   INFLUENCE OF VARIOUS PARAMETERS ON THE VALUE OF THE COMPLEX MODULUS

The parameters influencing the complex modulus values of bituminous mixtures can be divided into two main categories:

- the loading parameters related to the testing conditions (frequency, temperature, loading level);
- the mix design parameters related to the condition of the material (void content) and to its composition (type and content of the various components: aggregates, filler, binder).

5.2.8.1 Loading parameters

Frequency and temperature : All studies of the question agree that these are the parameters that most influence the modulus value of bituminous mixtures. The influence of frequency on the relative value of $|E^*|$ is greater at high temperature than at low temperature, while the influence of temperature is greater at low than at high frequency, in the relatively narrow window of modulus measurements on bituminous mixtures (temperature between -10 and +50°C, frequency between 1 and 40 Hz). For a 0/14 bituminous concrete, the value of $|E^*|$ varies in a ratio of 1 to 100 between -10 and +40°C, at a frequency of 3 Hz, and in a ratio of 1 to 5 between 1 and 30 Hz at 30 °C [de La Roche et al, 1994]. The phase angle φ also varies with frequency and temperature. It increases with temperature up to a temperature plateau, then decreases beyond it. Given the influence of frequency and temperature on the modulus value of bituminous mixtures, these parameters must be controlled very closely in tests. The temperature, in particular, must be controlled not only in the environment of the specimen, but also in the specimen itself, since any variation may lead to an erroneous interpretation of the observed phenomena.

Loading level : In the standards for tests on bituminous mixtures, it is assumed that the bituminous mixtures exhibit a linear behavior in tension for strains less than or equal to $50 \cdot 10^{-6}$, at the usual temperatures ($-20 \, °C < \theta < 40 \, °C$) (standard NF P 98-260-2). According to Soleimani [1965], the linear range of bituminous mixtures is very limited and decreases when the temperature increases. It is therefore possible to go outside the linear domain in tests at high temperature, even when applying very small strains. According to Charif [1991], the stiffness modulus values measured in the triaxial test with confinement pressures of 0.1 and 0.3 MPa, between 5 and 30°C, at 1 and 25 Hz, on a 0/10 bituminous mixture, remain substantially constant for strain values varying between 10 and $100 \cdot 10^{-6}$ (Fig. 5.11). The domain of linearity therefore seems to be larger than in tests without a confinement pressure.

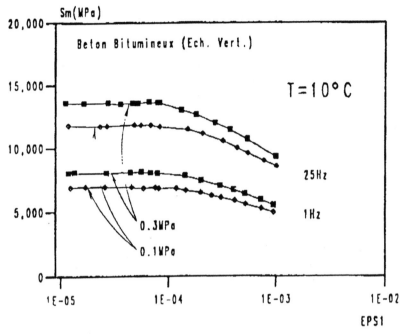

Fig. 5.11 Variation of stiffness modulus Sm versus axial strain ε1 (EPS1) (after [Charif, 1991])

According to Doubbaneh [1995], tension-compression tests performed on a 0/6 bituminous concrete show a reduction of modulus when the amplitude of axial strain increases between 30 and 80 10^{-6}. This reduction becomes larger as the temperature increases and the frequency decreases (3 % at 5°C, 20 Hz and 27 % at 41°C, 1 Hz) (Fig. 5.12).

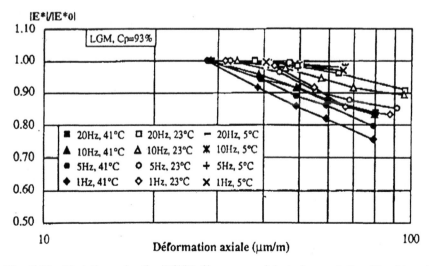

Fig. 5.12 Variation of ratio |E*|/|Eo*| versus axial strain ε_{ol} (after [Doubbaneh, 1995]).

The nonlinearity observed in the behavior of bituminous mixtures can have several causes: Hadrzynski [1995] suggests a physical nonlinearity such as that observed in bitumens in connection with a gradual change of microstructure.

Linder [1977] and Bazin [1967] mention a geometrical nonlinearity related to the presence of faults such as voids, cracks, with the loss of rigidity being explained by the opening of cracks under the effect of a local stress intensity factor.

For comparisons of modulus values, the variation of $|E^*|$ with the loading level makes it necessary to fix the strain level to be applied in the test as low as possible without compromising the quality of the measurements.

5.2.8.2 Mix design parameters

Type of binder [Bazin, 1967],[Francken, 1977], [de La Roche et al, 1994]: Under given temperature and frequency conditions, the complex modulus is largely influenced by the type and therefore the modulus of the bitumen, both in terms of the magnitude of the modulus $|E^*|$, and in terms of thermal and kinetic susceptibility (influence on φ). The harder the bitumen, the higher the magnitude of the complex modulus of the mix. Binder modifiers also influence the stiffness modulus of the mix usually by reducing its thermal susceptibility but no systematic study is available regarding the influence of binder modifiers on the stiffness modulus of bituminous mixtures.

Binder content: This is a mix design parameter that significantly influences the stiffness of bituminous mixtures [Soliman, 1976] [Ugé et al, 1977]. It can not in fact be considered in isolation, since it influences the void content of the final mixture and through it, the modulus. It is however possible to detect the following general tendencies: increasing the bitumen content increases the modulus when starting from very small binder contents, up to an optimal value. Beyond this optimal binder content value, the modulus decreases (fig 5.14).

Filler content : The addition of filler in a bituminous mixture improves the quality of the mastic. For a given bitumen content, the thickness of the film of bitumen on the aggregates depends on the quantity of filler, which has a large specific surface area. The mechanical characteristics of a bituminous mixture are therefore improved by an increase of filler. However, beyond some filler content, these characteristics can decline, especially if the bitumen content of the material is not sufficient (case of bitumen-stabilized continuously-graded aggregates, classical French "grave-bitume") [Soliman, 1976].

Aggregates : The mineralogical type of the aggregates and their shape seem to have little influence on the value of the complex modulus [Soliman et al, 1977], at least when the binder is rigid enough. It must on the other hand be expected that the importance of the aggregate in the bituminous mixture will increase as the behavior of the bitumen becomes predominantly viscous (at low frequencies and high temperatures) [Saunier, 1968], [Ugé et al, 1977]. The influence of the grading curve, for its part, appears in the variations of void content it entails.

Void content : While the void content of bituminous mixtures is not strictly speaking a mix design parameter, it is a characteristic of the mixture that results from both mix design factors and compacting factors, and it influences the modulus of bituminous mixtures: the rigidity of bituminous mixtures increases up to a maximum value when void content decreases [Bazin et al, 1967], [Saunier, 1968], [Soliman, 1977], [Moutier, 1991]. This is a classical result for soils [Tatsuoka et al, 1997]. For a given composition, the decrease of void content obtained by different compacting procedures, also makes bituminous mixtures less sensitive to temperature (Fig. 5.13).

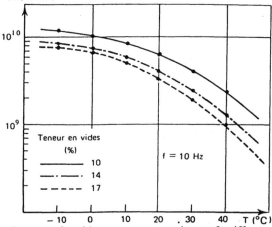

Fig. 5.13 Influence of voids content on values of stiffness modulus measured between -10 and +40°C, after [Soliman, 1977].

5.2.9 EMPIRICAL RELATIONS GIVING THE COMPLEX MODULUS

Since complex modulus tests on bituminous mixtures are long and expensive, many authors have taken an interest in prediction of the complex modulus of a bituminous mixture from its composition, mainly for the purpose of obtaining orders of magnitude usable in pavement design. The relations found are derived from statistical work based on comparisons of results of mechanical tests on many bituminous mixtures having compositions that sweep out the main mix design parameters.

Several types of predictive formula have been developed.

The first approach consists of starting from a given (frequency, temperature) couple (the one used for the pavement design calculation for example), for a given bitumen, and establishing an empirical relation between the composition of the material and the complex modulus value.

The second assumes that the influences of the frequency, temperature, and type of binder parameters are included in the value of the complex modulus of the bitumen and establishes a relation between the modulus of the bituminous mixture and that of the binder it contains, with allowance for the volumic distribution of the various constituents of the mixture.

5.2.9.1 Composition-based relation for a given bitumen

Moutier [1991] has found a relation between complex modulus and composition parameters following a long-term experimental plan of two-point bending tests, processed statistically. The preponderant explanatory parameters are the binder content and a compacity parameter ΔC.

For a conventional semi-granular bituminous mixture, the relation found is the following:

$$|E*|_{15°C,10Hz} = \left(-3,36 + 1,79\ TL - 0,184\ TL^2 + \left(\Delta C \cdot \left(-0,078 + \frac{0,63}{TL}\right)\right)\right) \cdot 10000\ MPa$$

where : TL   is the binder content by weight for aggregates having a density of 2.85. A correction must be applied for different densities.

   ΔC is a parameter reflecting the energy of compaction used on site. $\Delta C = C - Cm$, is the difference between the site compacity and the compacity obtained with "standard" means of compaction in the laboratory (LCPC slab compactor).

This formula has been obtained with binder contents varying from 3.6 to 5.7 % (in weight of aggregates).

The graphic representation of this relation (Fig. 5.14) visualizes the influence of the binder content and energy of compaction parameters already mentioned above.

While this relation yields very precise predictive values for materials identical to those tested in the experimental plan used as basis for the regression, any prediction for other conditions (change of type of bitumen, for example) may give erroneous values.

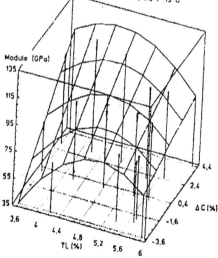

Fig. 5.14 variation of stiffness modulus versus binder content and energy of compaction [Moutier, 1991].

5.2.9.2.Relations incorporating the modulus of the bitumen.

On the assumption that the influences of the frequency, temperature, and type of bitumen parameters are included in the modulus value of the bitumen, several formula have been proposed to calculate the modulus $|E*|$ of a bituminous mixture from the volumic fractions of aggregate (Va), binder (Vb), and voids (Vv) and the modulus of the bitumen (Sb). These various relations have been compiled and analyzed by Francken [1977].

Heukelom and Klomp's relation

$$|E*| = Sb \left[1 + \frac{2,5}{n} \, x \, \frac{Cv}{1-Cv} \right]^n$$

$$with \; n = 0,83 \, lg \, \frac{4 \, x \, 10^4}{Sb} \qquad and \; Cv = \frac{Va}{Va+Vb}$$

where : Sb (MPa)   is the stiffness modulus of the bitumen taken from Van der Poel's nomograph giving the stiffness modulus of the bitumen versus the frequency, the temperature, and the penetration index of the binder [Shell Bitumes, 1991].

This formula can be used to calculate the moduli of a very great variety of compositions, but was established with voids contents that were all close to 3%. Like any statistical correlation, it is primarily valid for the ranges of variation of each parameter included in the regression, and experience has shown that it applies poorly to less compact bituminous mixtures.

This relation has been corrected by Van Draat et al [1965] to apply to less compacted mixes, using C'v instead of Cv defined as

$$C'v = \frac{Cv}{\left(1 + (Vv - 3)\right)}$$

Francken's relation [1996]

With the same notations as above, Francken's model gives the modulus of the bituminous mixture versus that of the bitumen through an expression in reduced form:

$$|E*| = E\infty \, |R*|$$

E∞ depends only on the volumic composition of the mixture.

$$E\infty = 1{,}436 \ 10^4 \left(\frac{Va}{Vb}\right)^{0.55} e^{-5.84 \ 10^{-2} \ Vv}$$

$$\text{and } \log|R*| = \log|B*| \ x \left[1 - \left(F\left(\frac{Va}{Vb}\right) \ x \ H(|B*|)\right)\right]$$

$$\text{with } F\left(\frac{Va}{Vb}\right) = 1.35\left(1 - e^{-0.13 \ x \frac{Va}{Vb}}\right)$$

$$H(|B*|) = 1 + 0.11\log(|B*|)$$

where : $|B*|$ is the reduced modulus of the bitumen ($|B*| \approx S_b/3000$).

This method is broader than Heukelom and Klomp's because it applies in the following domains: $0{,}16 \langle \frac{Va}{Vb} \langle 12$ *and* $1{,}5\% \langle Vv \langle 32\%$, which covers the common compositions used in road techniques.

Ugé's relation [1977]

For a bitumen modulus Sb between 10^7 and 10^9 Pa

$$\log|E*| = \frac{M+N}{2}\left(\log Sb - 8\right) + \frac{M-N}{2}\left|\log Sb - 8\right| + B$$

For a bitumen modulus Sb between 10^9 and $3 \ 10^9$

$$\log|E*| = B + M + (A - B - M) + \frac{(\log Sb - 9)}{\log 3}$$

with $A = 10{,}82 - 1{,}342 \ (100 - Va) / (Va + Vb)$

$B = 8{,}0 + 5{,}68 \ x \ 10^{-3} \ Va + 2{,}135 \ x \ 10^{-4} \ Va^2$

$$M = \frac{1,12\,(A-B)}{\log 30}$$

$$N = 0,6\ x\ \log\left(\frac{1,37\,V_b^2 - 1}{1,33\,V_b - 1}\right)$$

Like Francken's relation, already described, this approach is applicable to all mixtures of bitumens and aggregates, whatever their proportions, provided that the bituminous binder is sufficiently rigid (modulus S_b greater than 10^7). The estimate of the modulus of the bituminous mixture is given to within a factor of 2, which is satisfactory given the great variety of results analyzed.

Other relations

Other authors [Di Benedetto et al, 1996], [Boussad et al, 1996] establish correlation of the type $\log(E^*) = a \log(G^*) + b$ for a given granular mixture.

With : E*   complex modulus of the mix.
     G*   complex shear modulus of the binder.

In all cases, the experimental results depart from the model at low binder modulus values (high temperatures, very low frequencies), which seems logical because in this case the granular skeleton becomes more and more important.

Also interesting is the prediction formula developed by [Fonseca et al, 1996], using the prediction model developed by Witczak. This formula, based on the composition parameters of the mix, also includes a bitumen aging parameter through the value of the bitumen viscosity. It allows to predict the dynamic modulus of in placed aged asphalt mixtures, but is only valid for positive temperature up to 40°C.

5.2.10 CONCLUSION ON MODULUS

Of the various ways of measuring the linear viscoelastic properties of bituminous mixtures, measurement of the complex modulus in the linear domain is quite pertinent. When measuring the modulus of a bituminous mixture, care must be taken to remain in the domain of linearity of the material (small strains) and correctly interpret non-homogeneous tests. These points are developed in another section of this book.

It should also be emphasized that modulus measurement tests must be performed rapidly, since otherwise viscous dissipation might cause heating within the specimen and hence lowers the modulus. This phenomenon seems to be important in fatigue tests and is discussed in the section 5.3 on that subject.

Frequency and temperature are the parameters that most influence the value of the complex modulus of a bituminous mixture. It also depends, of course, on the mix design parameters, which include the types and proportions of the various constituents.

While there are empirical formula that give the complex modulus of a bituminous mixture versus its composition for a given frequency and temperature, some theoretical models, such as homogenization methods, not discussed here, can also be used to determine the modulus of a bituminous mixture from the modulus of the binder and its composition.

It has, for example, been shown in [Boutin et al, 1995], using a homogenization method, that the (time-temperature) translation coefficient α_T of the bitumen would be equal to the coefficient α_T of the bituminous mixture made with this bitumen. This result, obtained by theoretical arguments, seems to be confirmed experimentally, as close values of translation coefficient are obtained for both materials.

5.3   Fatigue of bituminous mixtures

5.3.1   INTRODUCTION

Bituminous materials in roads are subjected to a short-term load each time a vehicle passes. This causes micro-damage that results in a loss of rigidity of the material and can by accumulation, lead, in the long term, to failure. This fatigue is of considerable importance in the field of roads and must be correctly understood to ensure good pavement structural design.

The first fundamental laboratory studies of fatigue were undertaken in 1852 by Wöhler. Today, metals are the materials about which there is the most knowledge of fatigue behavior [Brand et al, 1992], but there have also been many studies on the fatigue of concretes, of road materials, and more generally on a large number of composite materials.

The usual test to characterize fatigue subjects a specimen of material to repeated loadings, generally identical, and records the number of cycles to failure of the specimen (or life duration). Failure is defined by some specific criterion.

The curve representing the life versus the amplitude of the applied loading (stress σ or strain ε) is the material's Wöhler curve (Fig. 5.15). It is usually characterized by a relation of the type:

$$\varepsilon \text{ or } \sigma = A\,N^{-b}$$

The curve is then a straight line in log-log coordinates, but some materials do not follow this law [Peyronne et al, 1984].

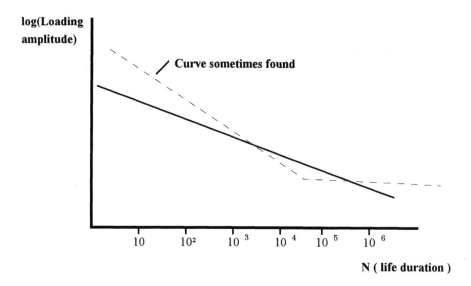

Fig. 5.15 Wöhler curve: Loading amplitude versus number of loading cycles.

The fatigue strength for N loading cycles is the loading value leading to the failure for N cycles.

In reality, conventional fatigue tests give very scattered results. If the same test is repeated several times, the number of cycles to failure can vary by a ratio of 1 to 10 for metals [Brand et al, 1992]. This ratio, which could reach as much as 1 to 30 for bituminous materials in the 80's [Soliman et al, 1977], is now close to 1 to 10 given the improvements made regarding the experimental control and measurements devices. It can reach 1 to 1000 for some materials treated with hydraulic binders.

Cumulative damage laws are used to introduce the varied loading levels applied in reality. The most widely used cumulative damage law is Miner's law [1945], expressed by :

$$\sum_{i=1}^{n} \frac{n_i}{N_i} = 1$$

where n_i is the number of cycles of loading level S_i applied, and N_i the life corresponding to the loading S_i.

This law is difficult to verify for bituminous materials [Doan, 1977, Bonnot 1995]. In spite of this, it is widely used in road techniques because of its simplicity [Peyronne et al, 1984], [Francken, 1979 and 1987].

5.3.2 LABORATORY TESTS TO CHARACTERIZE THE FATIGUE OF BITUMINOUS MIXTURES

5.3.2.1 Introduction

There are full-scale tests - the AASHO test [Highway Research Board, 1966] - and half-scale tests (the LCPC's circular fatigue test track at Nantes, etc.) for investigating fatigue damage of pavements. But in what follows only laboratory tests will be considered.

Three different major types of test can be distinguished:

- bending tests (2-, 3-, and 4-points),
- tension-compression tests (direct and indirect),
- shear tests.

Conventionally, bending tests are supposed to represent the repeated bending forces caused in the pavement by the passage of vehicles, while tensile tests represent the tensile forces induced at the base of the pavement by this bending. Finally, shear tests concern the shear forces induced in the pavement surface layers [Assi, 1981].

However, a classification into 2 main categories, homogeneous and non-homogeneous tests, as proposed by [Di Benedetto, 1990 and 1996a], seems better suited to interpretation in the context of rational mechanics. The tests are described in SHRP (Strategic Highway Research Program) documents [Rao Tangella et al, 1990], [Eurobitume, 1995], [Saïd, 1988], etc., for example.

5.3.2.2 Type of loading

In addition to the test design, the shape of the cyclic loading applied is very important in the analysis of fatigue test results [Francken, 1979]. The shape of the longitudinal and transversal strains as indicated by gauges bonded to the base of a bituminous layer laid on a reconstituted humidified granular material in an experiment on the LCPC's circular fatigue test track [de la Roche et al, 1993] are shown in Fig. 5.16.

This type of loading is difficult to reproduce in the laboratory. Most of the time, the loading cycles applied are all the same. The main forms of loading signal used for fatigue tests are shown in Fig. 5.17.

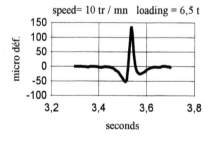
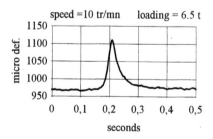

Fig. 5.16 Shape of the longitudinal strain signal.   Shape of the transversal strain signal.

However, out of a concern for simplification and control of the test, the loadings used are generally sinusoidal, and may or may not include a resting time.

Two types of extreme loading are applied:

- sinusoidal force loadings: control of force (of stress if the test is homogeneous),
- or sinusoidal displacement loadings: control of displacement (of strain if the test is homogeneous).

It is generally held that [Doan, 1977]:

- a thin pavement (< 6 cm) is loaded at constant strain (it is the strain of the more flexible lower layer that controls that of the upper layer),
- a thick bituminous pavement (> 15 cm) is loaded at constant stress,
- a pavement of average thickness is loaded in an intermediate mode.

Calculations using Alizé[1] software [Odéon, 1996] and the mode factor MF introduced by Monismith and Deacon [Myre, 1992] confirm this classification.

To better represent the in situ loadings, tests with resting times are sometimes used.

The various studies performed on this subject lead to the following conclusions:

- there is an increase of life between the continuous and discontinuous tests, and the test temperature and the duration of the resting time are factors of influence [Bonnaure et al, 1982 and 1983].
- the higher the temperature [Raithby et al, 1972] [Verstraeten, 1976] or the longer the resting time [Doan, 1977], the larger the gain seems.

In studies done with intermittent loadings, several results show that beyond a certain limit, resting periods yield no additional increase of life. It is often assumed that a coefficient of ten between the duration of the resting periods and the loading periods yields the optimal increase of life [Bonnaure et al, 1983], [Rivière, 1996]. Moreover, the beneficial effect of the resting time is generally greater with softer bitumen and it also increases with the binder content of the mixes [Francken et al, 1987].

---

[1]   *software for calculation of stresses and strains at the base of the layers, based on Burmister's model and used in the French structural design method.*

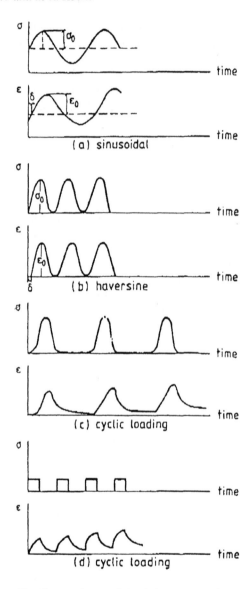

Fig. 5.17 Shape of loading cycles used for fatigue tests, after [Saïd, 1988].

5.3.3   ANALYSIS OF TESTS CHARACTERIZING THE FATIGUE BEHAVIOR OF BITUMINOUS MIXTURES.

5.3.3.1 Classical analysis

Specific failure criterion : An arbitrary specimen fatigue life criterion has been defined; it corresponds to a loss of half the stiffness of the specimen, namely:

Force$_{final}$ = 1/2 Force$_{initial}$ for imposed-displacement tests.
Displacement$_{final}$ = 2 Displacement$_{initial}$ for imposed-force tests.

This criterion makes it possible to plot Wöhler curves (Fig. 5.15) for bituminous materials. It should be noted that the loadings in controlled-force tests and controlled-displacement tests are very different even if the first cycle is identical in force and displacement amplitude (Fo, Do). In fact, the failure cycle for a controlled-force test is characterized by the couple (Fo, 2Do) and that for a controlled-displacement test by the couple (Fo/2, Do). In the former case, the accumulation of damage leads to an increase in the loading applied at each cycle, while in the latter it leads to a reduction of this loading.

For tests at constant force, the lives defined by complete failure of the specimen differ little from those defined by the halving of the stiffness modulus of the specimen [Doan, 1977]; the lower the temperature, the more this is true. On the other hand, in controlled-displacement tests, the criterion is often reached long before the specimen fails.

These findings partly explain the observed differences of behavior between the two loading modes.

It should be emphasized that, so far, no rational analysis has been able to predict the results in one mode from those in the other. This is very surprising, because the material, which does not know in which mode it is being loaded, has an intrinsic fatigue law. In addition, the qualitative ranking of fatigue strength based on the tests are not always borne out on pavements.

These last remarks show the limitations and weaknesses of the usual approach.

Processing of results : The Wöhler curve is generally regarded in the case of bituminous materials as a straight line in a logarithmic coordinate system (it is obtained by linear regression):

$$\log N = b \log \lambda + a$$

where λ is the loading amplitude and N the life (number of cycles) associated with it. λ is either a force (F) (or stress) amplitude or a displacement (D) (or strain) amplitude. In some countries, the allowable loading for 10^6 cycles (written ε_6 or σ_6) is the loading amplitude that must be applied to the material to obtain a fatigue test life of 10^6 cycles. This value is regarded as characteristic of the fatigue behavior of the bituminous mixtures. It is highly dependent on the material tested and on the test conditions used.

Factors influencing fatigue test results :
Test conditions : Many studies intended to characterize the influence of the test conditions on fatigue results have been performed, using the conventional analysis [Pell, 1962 and 1967], [Lucas et al, 1965], [Thibaud, 1976], [Doan, 1977], [Goddard et al, 1977], [Rao Tangella et al, 1990], [Moutier, 1991], [Kim et al, 1992], [Brennan et al, 1992], [de la Roche et al, 1993], [Tayebali et al, 1994 a and b].

Table 5.2 summarizes the main tendencies and the differences observed in "force" and "displacement" tests.

Table 5.2 Comparison of results obtained with imposed force and displacement.

	Controlled-force test	Controlled-displacement test
Evolution during test	Increase of displacement	Reduction of force
Usual failure criterion	Failure of specimen	Loss of half of initial force
Life	Shorter	Longer
Dispersion of results	Lower	Higher
Increase of temperature (> 0 °C)	Reduction of life	Increase in life
Increase of stiffness modulus	Increase of life	Reduction in life
Effect of resting time	Large	Small
Duration of propagation of cracks	Brief	Long
Increase of test frequency	Increase	Reduction
Growth of damage	Fast	Moderate

Among the various tendencies, it should be emphasized that the life at imposed strain can be up to ten times as long as the life in a test at imposed force starting from the same initial strain level.

Several geometries (beam, trapezoidal cantilever beam, cylinder (slender or not), special geometry) relative to different types of loading (bending, tension-compression, shear) are used for fatigue tests.

A few studies have been performed to compare the results obtained with these different types of tests [Aguirre et al, 1981]. It seems that lives are shorter in tension-compression than in bending at the same nominal strain. On the other hand, results obtained in the same modality are mutually consistent (see §6) (Fig. 5.18).

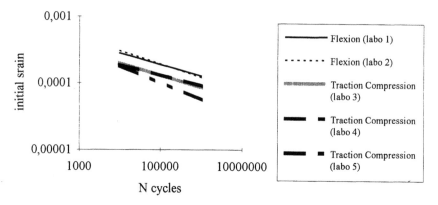

Fig. 5.18 Fatigue behavior at constant stress amplitude (10°C, 10 Hz) [Aguirre et al, 1981].

According to de la Roche et al, [1994], the life is longer in three-point bending than in two-point bending (the ratio ε_6 three-point bending/ε_6 two-point bending ranges from 1.4 to 2).

Mix design parameters : The influence of the mix design parameters on fatigue behavior has also been the subject of many studies [Pell et al, 1969], [Soliman et al, 1977], [Brennan et al, 1990]. The feature common to these studies is an attempt to predict the fatigue behavior of bituminous mixture from their composition. Among the parameters having the most influence, we may mention: the type of binder and binder content, the voids content (or the compactness), the fines content, and the characteristics of the aggregates (mineralogical type, grading curve, etc.). An analysis of the influence of each of these parameters is tricky as they may be interdependent.

The following results can, however, be quoted :

Type of binder *:*The type of binder can have a strong influence. The harder the bitumen, the better the fatigue strength [Bazin et al, 1967]. The class of bitumen affects the slope of the straight fatigue line: in controlled-stress tests, the harder the bitumen, the flatter the slope (except for high strain levels) [Saïd, 1988].

The results obtained in imposed strain in the SHRP are strongly correlated with the type of bitumen, via the modulus of loss of the aged binder [Tayebali et al, 1994]. Similarly, in a given penetration class, differences of type of bitumen can lead to very large differences in life [Moutier, 1991].

Bituminous materials made with modified binders often exhibit better fatigue results than those made with pure binders [Brennan et al, 1992].

Binder content: According to many authors, the binder content is the factor that most influences the fatigue results [Pell et al, 1962], [Saunier, 1968], [Epps et al, 1969], [Saïd, 1988], [Brennan et al, 1990], [Moutier, 1991].

For a given bitumen, there exists an optimal value, below which the life increases with increasing binder content. Above this value, the fatigue resistance decreases as the binder content increases. This optimal content is around 7% binder for Californian bituminous materials and seems to depend on the type of aggregates used [Epps et al, 1969].

Void content *:*The fatigue results are generally better and less scattered when the void content decreases [Bazin et al, 1967]. The influence of the void content varies with the type of material tested and the type of loading imposed [Doan, 1977]:

- at imposed strain, with dense bituminous mixtures, decreasing void content leads to a higher modulus and a shorter life. For bitumen-treated granular materials (low binder content), the life increases when the void content decreases.
- at imposed stress, the life increases when the void content decreases.

Aggregates : The shape of the aggregates does not seem to influence the fatigue strength of the materials [Saunier, 1968], [Brennan et al, 1990], [Pell et al, 1969], [Epps et al, 1967]. Only the solidity of the aggregates and their adherence properties affect the fatigue resistance [Kim et al, 1992].

Filler content : As in the case of the complex modulus, there is an optimal filler content with respect to fatigue strength. It is between 7 and 9% for a bitumen-treated granular material [Soliman et al, 1977]. It seems that it is the filler/bitumens ratio that determines the fatigue resistance, with the filler filling the voids between the aggregates.

Global predictive relations

Several global statistical relations have been formulated to predict the fatigue resistance of bituminous mixtures from their composition. In the context of SHRP, [Tayebali et al, 1994] propose the following relation, obtained from four-point bending tests on rectangular specimens at imposed strain:

$$Nf = 10^5 \; 2{,}738 \; \exp^{0{,}077 \, VFB} \; \varepsilon_0^{-3{,}624} \; E_2^{-2{,}720}$$

with Nf :   fatigue life.
    ε_0 :   initial strain.
    E_2 :   initial loss modulus (psi).
    VFB :   percentage of voids filled by bitumen.

We may also mention:

- the formula of [Myre, 1992] obtained in three-point bending tests on rectangular specimens placed on a rubber support,
- the formula established by [Moutier, 1991] on the basis of results obtained in a long-term plan of two-point bending fatigue tests on trapezoidal specimens at imposed strain,
- and the previous works of [Verstraeten, 1974].

However, like all statistical correlation, these relations are valid for the type of materials on which they have been established, and the predicted values for different materials (hard bitumen or modified binders, etc.) differ substantially from the experimental values measured in the laboratory.

The analysis of these various factors and the proposed statistical correlations must not conceal the fact that the criterion of qualification of fatigue using the conventional analysis is increasingly challenged both by theoretical analyses (heating, etc.) and because of comparisons with actual behaviors observed on pavements [de la Roche et al, 1994].

To remedy these shortcomings, other approaches have been proposed more recently.

5.3.3.2 Analysis in terms of dissipated energy

Several authors have advanced the hypothesis of a unique relation between the fatigue life and the total dissipated energy to failure. This energy approach could lead to more coherent results than the usual fatigue laws [Tayebali et al, 1994] [Baburamani P.S, 1992].

The energy W_i produced by viscous dissipation in the course of cycle i is given by the following expression:

$$W_i = \pi \, \varepsilon_i \, \sigma_i \, \sin(\varphi_i)$$

with ε_i: strain amplitude in cycle i.

       σ_i : stress amplitude in cycle i.

       φ_i : phase difference between stress and strain in cycle i.

This energy varies in the course of the test because the phase angle increases in the course of the test, the stress decreases in controlled displacement (the energy dissipated during one cycle will therefore tend towards 0), and the strain increases in the course of the test in controlled force (the energy will therefore increase).

Van Dijk et al. [1977] propose the following fatigue law:

$$W_N = A \, (N_f)^{\,z}$$

where W_N :   total dissipated energy $= \sum_{i=1}^{N} W_i$

   N_f :   life

   A, z :   coefficients determined experimentally

This relation is regarded by its authors as dependent on the mix design of the mixture but independent of the fatigue test chosen (two- or three-point bending), of the temperature (from 10°C to 40°C), of the loading mode (controlled stress or strain), and of the frequency (from 10 to 50 Hz). This claim seems however to have to be moderated in light of the results obtained in the context of the SHRP (Strategic Highway Research Program) [Tayebali et al, 1994].

Another use of this dissipated energy approach has been developed by Hopman et al, [1989], taken up by Rowe, [1993], and completed by Pronk, [1995 and 1996]. The approach described to define the life of a material is attractive but it is difficult to apply to tests at constant strain.

The foregoing developments assume that the dissipated energy goes to damage the material. But this assumption has been challenged by some authors [Lesueur et al, 1995]. Noting that, as the application of the foregoing theory presumes, the fatigue life is proportional to the loss modulus, some bituminous mixtures with blown bitumen, for example, should give excellent fatigue results, which is not the case. Again, perfectly elastic materials should not fail by fatigue.

In fact, the energy dissipated does not just destroy bonds. It seems to act mainly to heat the specimen [Soltani, 1993], [Moriceau, 1994], [Di Benedetto et al, 1996], [de la Roche et al, 1996].

5.3.3.3 Analysis in terms of damage.

Some authors propose using damage theory [Montheillet et al, 1986] to characterize the fatigue of bituminous mixtures [Alimani, 1987], [Piau, 1989], [Moriceau, 1994], [Di Benedetto et al, 1996], [Ullidtz et al, 1997], [Lee et al, 1997]. This approach, generally applied to the one-dimensional case, seems very promising. In the context of this theory, a damage parameter, d, is introduced. It characterizes a relative loss of modulus between the initial state and the state of the material in cycle N:

$$d = (E_O - |E^*|)/E_O$$

where :   d   is the damage in cycle N
            Eo   the norm of the complex modulus in the initial state
            $|E^*|$ the norm of the complex modulus in cycle N

The damage law states the change of d between two successive cycles (d) :

$$\Delta d = f(d)\ g(\text{loading applied at cycle N}) \qquad 0 \le d \le 1$$

where f is a function of d that may also depend on the temperature and g is a function of the loading applied during cycle N. This function g is, in general, chosen as dependent on the stress and/or strain amplitude in cycle N.

Analysis of the curve $|E^*|$ versus N can yield the damage law. However, according to [Di Benedetto et al, 1996], it is necessary to eliminate the start of this curve (phase I of Fig. 5.19) in the analysis of the damage, because during this phase I, heating of the specimen by internal viscous dissipation, is preponderant. This heating at the start of the test has been clearly demonstrated by [de la Roche, 1996] by measurements with an infrared film camera.

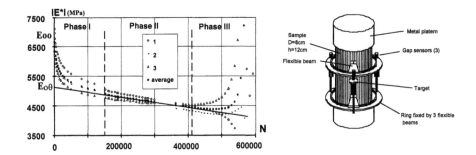

Fig. 5.19 Fatigue test [Di Benedetto et al, 1997] at 10°C and 10 Hz - 3 transducers values and average values. Initial heating in phase I.

a -   principle of measurement with three axial transducers.
b-   results of a fatigue test at imposed strain (141 10⁻⁶m/m).

Recent results obtained in homogeneous tests by [Di Benedetto et al, 1996 and 1997] show that the tests in controlled stress and controlled strain yield results that have little scatter and are mutually compatible when allowance is made for the effects of heating by dissipation.

5.3.4 CONCLUSION

Fatigue is one of the main failure modes of pavement structures. A correct description of this phenomenon in the laboratory is therefore important.

Of the various analysis proposed here, the classical one, widely used in pavement design where it shows its usefulness, also shows its limits in terms of relevance. For the future developments, it seems that the analysis based on damage laws is promising. Whatever the analysis used, the various ancillary phenomena that act in the course of the test and may affect the result must be understood. The most important seems to be internal viscous dissipation, which causes an increase of temperature within the material and so modifies its rigidity.

Analysis of the healing phenomenon is also a means of investigation; it is delicate and there are currently very few experimental data concerning it. This phenomenon is also associated with the reduction of temperature during the rest period, which also induces a recovery of the modulus.

Finally, work must be continued on a better description of fatigue cracking, which appears by coalescence of microdamage; it is not discussed in this state of the art report so is not the fracture mechanism applied to fatigue behavior [Majiddzadeh, 1991]. The relative weights of the initiation and propagation phases are probably highly dependent on the type of test.

These different paths of investigation require determination in the course of the test of the evolution of various rheological parameters of the material (modulus, phase angle, Poisson's ratio, etc.) and of the temperature.

In addition, there must be validations and fittings with observations on real pavements.

5.4 References

Aguirre, Morot, De La Taille, Doan Tu Ho, Bargiacchi, Smadja, Udron, Guay et Roncin : Etude comparée des essais de module complexe et de résistance à la fatigue des enrobés bitumineux. *Bulletin de Liaison des Laboratoires des Ponts et Chaussées*, N° 116, pp.3343, 1981.

Alimami M. : Contribution à l'étude de l'endommagement par fatigue des enrobés bitumineux *Thèse de Docteur Ingénieur* Université Paris VI, 1987.

Anderson D.A. and Kennedy T.W : Development of SHRP binder specification. *Proceedings of the annual meeting of the Association of Asphalt Paving Technologists AAPT*, march 1993.

Assi M : Contribution à l'étude du comportement des enrobés à la fatigue en cission. *Thèse de Docteur Ingénieur*, Ecole Nationale des Ponts et Chaussées, 1981.

Baburamani P.S : The dissipated energy concept in fatigue characterisation of asphalt mixes A summary report. *Research report ARR 235*, Australian Road Research Board , 1992.

Bazin P. and Saunier J.B : Deformability, fatigue and healing properties of asphalt mixes. *Proceeding of the Second International Conference on the Structural Design of Asphalt Pavements, Ann Arbor, Michigan*, 1967.

Biarez J. and Hicher P.Y : Elementary Mechanics of Soil Behaviour Saturated Remoulded Soils. Ed. A.A. Balkema, Rotterdam, Brookfield, 1994.

Boltzmann L : Zur Theorie der Elastischen Nachwirkung Wien Berichte, 1874 ; Pogg. Ann., Bd. 7, 1876.

Bonnaure F., Huibers A. and Bonders A : Etude en laboratoire de l'influence des temps de repos sur les caractéristiques de fatigue des enrobés bitumineux. *Revue Générale des Routes et Aérodromes*, N° 595, mars 1983.

Bonnaure F.P, Huibers A.H.J. and Bonders A : A laboratory investigation of rest period on the fatigue characteristics. *Proceedings AAPT*, Vol. 61, 1982.

Bonnaure F.P., Gest G., Gravois A. and Uge P. A new Method of predicting Stiffness of Asphalt Paving Mixtures. *Journal of the Association of Asphalt Paving technologists AAPT*, Vol. 46, pp.64100, 1977.

Bonnot J. : Rapport du modérateur, thème Interprétation. First Eurobitume Workshop on the Rheology of Bituminous Binders, Brussels, April 1995.

Boussad N. and Dony A : Bitumespolymères : relations entre la rhéologie des liants et la rhéologie des enrobés. *1rst Eurasphalt and Eurobitume Congress*, may 1996.

Boutin C., De la Roche C., Di Benedetto H. and Ramond G. De la rhéologie du liant à celle de l'enrobé bitumineux, théorie de l'homogénéisation et validation expérimentale. *Eurobitume Workshop*, Bruxelles, April 1995.

Brand A., Flavenot J.F. and Gregoire R : Données Technologiques sur le Fatigue. Centre Technique des Industries Mécaniques CETIM, 1992.

Brennan M.J. and Clancy F : A new initiative in measuring the fatigue performance of bituminous materials. *Proceeding of the 7th International Conference on the Structural Design of Asphalt Pavements, Nottingham, UK*, 1992.

Brennan M.J., Lohan G. and Golden J.M : A laboratory study of the effect of bitumen content, bitumen grade, nominal aggregate grading and temperature on the fatigue performance of dense bitumen macadam : *Proceeding of the IVth International Rilem Symposium Budapest*, Ed. Chapman & Hall, 1990, pp.358366.

Brown S.F. and Cooper K.E : Simplified methods for determination of fundamental material properties of Asphalt Mixes. *Proceedings of the International Conference SHRP And Traffic Safety On two Continents The Hague The Netherlands*, 1993.

Charif K : Contribution à l'étude du comportement mécanique du béton bitumineux en petites et grandes déformations. *Thèse de Doctorat*, Ecole Centrale Paris, 1991.

Chauvin J.J : L'essai de module complexe utilisé pour la formulation des enrobés. *Proceedings of the Fourth International RILEM Symposium Budapest*, Ed. Chapman and Hall, pp.367381, 1990.

De la Roche C. and MARSAC P. Caractérisation expérimentale de la dissipation thermique dans un enrobé bitumineux sollicité en fatigue *First International Eurobitume and Eurasphalt Congress* may 1996.

De la Roche C. and ODÉON H : Expérimentation USAP/LCPC/Shell Fatigue des Enrobés Phase 1 Rapport de synthèse. *Document de Recherche LCPC, sujet* n° 2.01.05.2, 1993.

De la Roche C., CORTE J.F., GRAMSAMMER J.C. ODEON H., TIRET L. and CAROFF G. Etude de la fatigue des enrobés bitumineux à l'aide du manège de fatigue du L.C.P.C. *Revue Générale des Routes et Aérodromes*, n° 716, pp. 6274, 1994.

De la Roche C., ODEON H., SIMONCELLI J.P. and SPERNOL A. Study of the Fatigue of Asphalt Mixes Using The Circular Test Track of the Laboratoire Central des Ponts et Chaussées in Nantes, France *Transportation research Record*, n° 1436, pp. 6274, 1994.

De la Roche C : Module de rigidité et Comportement en Fatigue des Enrobés Bitumineux, expérimentations et nouvelles perspectives d'analyse. *Thèse de Doctorat*, Ecole Centrale de Paris, 1996.

Di Benedetto H. Nouvelle approche du comportement des enrobés bitumineux : résultats expérimentaux et formulation rhéologique, *Procceedings of the fourth international RILEM Symposium, Budapest*, Ed Chapman and Hall, p 385401, 1990.

Di Benedetto H. Reflection about the possible explanation for the test differences and recommendations, work report for TG3, *RILEM Technical Committee 152: Mechanical tests for bituminous mixes*, Zurich, 1995.

Di Benedetto H. and DES CROIX P : Bindermix rheology: limits of linear domain, non linear behaviour. *Irst European Eurasphalt and Eurobitume Congress*, may 1996 c.

Di Benedetto H., Ashayer Soltani A. and Chaverot P : Fatigue damage for bituminous mixtures : a pertinent approach. *Journal of the Association of Asphalt Paving technologists AAPT*, 1996 b.

Di Benedetto H., SOLTANI A., CHAVEROT P. : "A rational approach for fatigue damage", *1rst European Eurobitume and Eurasphalt Congress*, Strasbourg, 1996a.

Di Benedetto H., Soltani A., Chaverot P. : "Fatigue damage for bituminous mixtures", *5th Int. Conf. Rilem : Mechanical Tests for Bituminous Materials*, Lyon, may 1997.

Doan T.H : Les études de fatigue des enrobés bitumineux au LCPC. *Bulletin de Liaison des Laboratoires des Ponts et Chaussées*, 1977, N° Spécial V, pp.215228.

DoubBaneh E : Comportement mécanique des enrobés bitumineux des petites aux grandes déformations. *Thèse de Doctorat*, Institut National des Sciences Appliquées de Lyon, ENTPE, 1995.

Epps J.A. and Monismith C.L : Influence of mixture variables on the flexural fatigue properties of asphalt concrete. *Journal of the Association of Asphalt Paving technologists AAPT*, Vol. 38, pp.423458, 1969.

EUROBITUME "The rheology of bituminous binders", *Eurobitume European workshop*, Brussels, April 1995.

EUROBITUME Rheology of bituminous binders. Glossary of rheological terms. *Edited by Eurobitume*, 1996.

Fonseca O.A. and Witczak M.W : A prediction methodology for the dynamic modulus of in place aged asphalt mixtures. *Journal of the Association of Asphalt Paving technologists AAPT*, Vol. 65, pp.532565, 1996.

Francken L. and CLAUWAERT C. Caracterisation and structural assessment of bound materials for flexible road structures. *Proceedings of the VIth International Conference on the structural design of Asphalt Pavements, Ann Arbor, Michigan*, 1987.

Francken L., Partl M. and RILEM TC 152 PBM. Complex Modulus Testing of Asphalt Concrete, Account of a Rilem Interlaboratory Test Program. *Paper presented to the Transportation Research Board Meeting, Washington*, 1996a.

Francken L : Fatigue d'un enrobé bitumineux soumis à des conditions de sollicitations réalistes. *La Technique routière Bruxelles*, Vol 24, n° 4, 1979.

Francken L : Fatigue performance of a bituminous road mix under realistic test conditions. *Transportation Research Record, Washington*, n° 719, p 3037, 1979.

Francken L : Module complexe des mélanges bitumineux. *Bulletin de Liaison des Laboratoires des Ponts et Chaussées*, N° Spécial V, pp.181198, déc.1977.

Goddard R. and PowelL W.D : Résistance à la fatigue des enrobés denses. Influence des facteurs de formulation et température. *Bulletin de Liaison des Laboratoires des Ponts et Chaussées*, 1977, N° Spécial V, pp.247254.

Gubler R : Méthode d'essai par oscillation axiale pour la détermination des caractéristiques mécaniques et du comportement à la fatigue des asphaltes. *Proceedings of the Fourth International RILEM Symposium Budapest*, Ed. Chapman and Hall, pp.432444 , 1990.

Hadrzynski F : Etude de la rhéologie d'un bitume modifié par des polymères et prévision du comportement mécanique des mélanges avec fines, sables et graviers. *Thèse de Doctorat*, Université de ParisNord, 1995.

HIGHWAY RESEARCH BOARD. Essai AASHO. *Bulletin de Liaison des Laboratoires des Ponts et Chaussées* N° Special E, Paris, 1966.

Hopman P.C., Kunst P.A.J. and PronK A.C : A renewed interpretation method for fatigue measurements : verification of Miner's rule. *Proceedings of the 4th Eurobitume Symposium Madrid*, 1989, Vol. 1, pp.557561.

Huet C : Comportement viscoélastique en régime dynamique : modules et complaisances complexes. *Conférence au Groupe Français de Rhéologie*, 1966.

Huet C : Etude par une méthode d'impédance du comportement viscoélastique des matériaux hydrocarbonés. *Thèse de Docteur Ingénieur*, Faculté des Sciences de l'Université de Paris, 1963.

Huhtala M : The rheology of bituminous mixtures. Proceedings of the First European Workshop on the rheology of bituminous binders Brussels, April 1995.

Kennedy T.W. and Huber G.A : The SuperpaveTM System Superior Performing Asphalt Pavements. *Strasse und Verkher / Route et Trafic*, n° 6. Ed. VSS Union des Professionnels Suisses de la Route, pp.347354, juin 1994.

Kim Y.R., Kim N. and Khosla N.P : Effects of aggregate type and gradation on fatigue and permanent deformation of asphalt concrete. *ASTM STP 1147 Effects of aggregates and mineral fillers on asphalt mixture performance*, Ed. Richard C. Meininger, American Society for Testing and Materials, Philadelphia, 1992.

Lee H.J. and Kim Y.R : Prediction of fatigue damage in asphalt concrete using a viscoelastic continuum damage model. *Proceedings of the 5th Int. Conf. Rilem : Mechanical Tests for Bituminous Materials*, Lyon, may 1997.

Lempe U., Leykauf G. and Neumann U : Participation to the RILEM Interlaboratory Test Program Part 2 "Dynamic tests". *Research Report*, Technische Universität München, 1992.

Lesueur D. and Dekker D : Fatigue resistance : What's wrong with the dissipated energy ?. *Eurobitume Workshop The rheology of bituminous binders Brussels*, April 1995.

Linder R., MOUTIER F., Penet M. et Peyret F. La machine asservie d'essais rhéologiques MAERLPC et son utilisation pour l'essai de traction LPC sur enrobés. *Bulletin de Liaison des Laboratoires des Ponts et Chaussées*, N° 142, pp.132138, 1986.

Linder R : Application de l'essai de traction directe aux enrobés bitumineux. *Bulletin de Liaison des Laboratoires des Ponts et Chaussées*, N° Spécial V, pp.255274, 1977.

Lucas M., Bazin P. and Saunier J : Essais de fatigue sur enrobés bitumineux. *Revue Générale des Routes et Aérodromes*, N° 404, 1965.

Madjiddzadeh K, Ramsamooj D.V. and Kaufmann E.M. Application of Fracture Mechanics in the Analysis of Pavement Fatigue, *Journal of the Association of Asphalt Paving technologists AAPT*, Vol. 50, 1971.

Mamlouk M.S. and Sarofim R.T : Modulus of Asphalt Mixes An Unresolved Dilemma. *Transportation Research Record n° 1171*, Transportation Research Board, 1988.

Mandel J : Cours de mécanique des milieux continus, Tomes 1 et 2 Ed. Gauthier et Villars, Paris,1966.

Miner M.A. Cumulative damage in fatigue *Journal of applied mechanics*, vol. 12 n° 3, 1945.

Montheillet F. and Moussy F : Physique et mécanique de l'endommagement. , Ed. Les Editions de Physique, 1986.

Moriceau L : Etude du comportement à la fatigue des bétons bitumineux. *Rapport de Travail de Fin d'Etudes*, Ecole Nationale des Travaux Publics de l'Etat, 1994.

Moutier F. et Delorme J..L : Le contrôle de la qualité des enrobés à l'aide de la Machine Asservie d'Essais Rhéologiques. *Proceedings of the 4th International RILEM Symposium*, pp.234266, Ed. Chapman & Hall, 1990.

Moutier F : Etude statistique de l'effet de la composition des enrobés bitumineux sur leur comportement en fatigue et leur module complexe. *Bulletin de Liaison des Laboratoires des Ponts et Chaussées*, n° 172, pp.3341, mars 1991.

Myre J : Fatigue of asphalt materials for Norwegian conditions. *Proceeding of the 7th International Conference on Asphalt Pavements, Nottingham*, 1992, Vol. 3, pp.238251.

NF P 982601. *Norme Française*. Essais relatifs aux chaussées. Mesure des caractéristiques rhéologiques des mélanges hydrocarbonés. Partie 1 : Détermination du module et de la perte de linéarité en traction directe, AFNOR, 1992.

NF P 982602. *Norme Française*. Essais relatifs aux chaussées. Mesure des caractéristiques rhéologiques des mélanges hydrocarbonés. Partie 2 : Détermination du module complexe par flexion sinusoïdale, AFNOR, 1992.

Odeon H : Mode factor. Fautil effectuer des essais de fatigue à déformation ou à contrainte imposée? *Communication interne*. Laboratoire Central des Ponts et Chaussées, janvier 1996.

Pell P.S. and TAYLOR I.F : Asphaltic road materials in fatigue. *Journal of the Association of Asphalt Paving technologists AAPT*, Vol. 38, pp.371422, 1969.

Pell P.S : Fatigue characteristics of bitumen and bituminous Mixes. *Proceeding of the First International Conference on the Structural Design of Asphalt Pavements, Ann Arbor, Michigan*, 1962.

Pell P.S : Fatigue of Asphalt Mixes. *Proceeding of the Second International Conference on the Structural Design of Asphalt Pavements, Ann Arbor, Michigan*, p 577593, 1967.

Peyronne C. and Caroff G : Dimensionnement des Chaussées, *Cours de Routes*, Ed. Presses de l'Ecole Nationale des Ponts et Chaussées Paris, 1984, 216 p.

Piau J.M. :"Modélisation thermomécanique du comportement des enrobés bitumineux",*Bulletin de Liaison des LPC*, n° 163, pp 4154, septembreoctobre 1989.

Piau J.M. and Heck J.V : Comportement des enrobés bitumineux, passage du domaine fréquentiel au domaine temporel. *Raport Interne*, Laboratoire Central des Ponts et Chaussées Nantes, 1996.

Pronk A.C : Evaluation of the dissipated energy concept for the interpretation of fatigue measurements in the crack initiation phase. *Research Report n° P. DWW95.001*, Ministerie Van Verkeer en Waterstaat Directoraat Generaal Rijkswaterstaat PaysBas, 1995, .

Pronk A.C : Theory of the Four Point Dynamic Bending Test. *Research Report*, Ministerie van Verkeer en Waterstaat Dienst Weg en Waterbouwkunde The Netherlands, 1996.

Raithby K.D. and Sterling A.B : Some effects of loading history on the fatigue performances of rolled asphalt. *TRRL Report L.R. 496 U.K.*, 1972.

Ramond G., Pastor M., Durrieu F. and Giavarini C. Méthodologie d'étude de la modification des bitumes par ajout de polymères. *1rst European Eurasphalt and Eurobitume Congress*, may 1996.

Rao Tangella S.C.S., Craus J., DeacoN J.A. and Monismith C. L : Summary report on fatigue response of asphalt aggregate mixtures. *Strategic Highway Research Program Project A003A.*, Institute of Transportation Studies. University of California., 1990.

Read J.M. and Brown S.F : Practical Evaluation of Fatigue Strength for Bituminous paving Mixtures. *Proceedings of the 1rst Eurasphalt and Eurobitume Congress, Strasbourg*, 1996.

Rivière N : Comportement en Fatigue des Enrobés Bitumineux. *Thèse de Doctorat* Université de Bordeaux I, 1996.

Rowe G.M : Performance of Asphalt Mixtures in the Trapezoidal Fatigue Test. *Journal of the Association of Asphalt Paving technologists AAPT*, 1993, Vol. 62, pp.344-380.

Saïd S.F : Fatigue Characteristics of Asphalt Concrete Mixtures. *Research Report N° 413 005038*, Vägoch Trafikinstitutet, Sweden, 1988, .

Salençon J. Cours de calcul des structures anélastiques Viscoélasticité. *Presses de l'Ecole Nationale des Ponts et Chaussées* Paris, 1983.

Saunier J : Contribution à l'étude des propriétés rhéologiques des enrobés bitumineux. *Thèse de Doctorat ès Sciences appliquées*, Faculté des Sciences de l'Université de Paris, 1968.

Sayegh G : Contribution à l'étude des propriétés viscoélastiques des bitumes purs et des bétons bitumineux. *Thèse de Docteur Ingénieur*, Faculté des Sciences de Paris, 1965.

Serfass J.P. and Van Belleghem S : Utilisation de la traction directe pour l'étude de la valeur structurelle des enrobés bitumineux. *Proceedings of the Fourth International RILEM Symposium Budapest*, Ed. Chapman and Hall, pp.290305, 1990.

SHELL BITUMES. Bitumes, Techniques et Utilisations. , Ed. Société des Pétroles Shell, 1991.

Soleimani P : Etude sur le comportement viscoélastique des matériaux bitumineux par la méthode de fluage. *Thèse de Docteur Ingénieur*, Faculté des Sciences de Paris, 1965.

Soliman S. and DOAN T.H : Influence des paramètres de formulation sur le module et la résistance à la fatigue des gravesbitumes. *Bulletin de Liaison des Laboratoires des Ponts et Chaussées*, 1977, N° Spécial V, pp.229246.

Soliman S : Influence des paramètres de formulation sur le comportement à la fatigue d'un enrobé bitumineux. *Rapport de Recherche des Laboratoires des Ponts et Chaussées*, n° 58, 1976.

Soltani M.A.A : Comportement à la Fatigue des Enrobés Bitumineux. *Rapport de Travail de Fin d'Etudes*. Ecole Nationale des Travaux Publics de l'Etat, 1993.

Sousa J.B : Asphalt Aggregate Mix Design using the Simple Shear Test Constant Height. *Journal of the Association of Asphalt Paving technologists AAPT*, 1994, Vol. 63.

Stefani C : Etude thermique des phénomènes de fatigue dans les matériaux composites bitumineux. *Journées de Physique LCPC*, Les Arcs, déc. 1981.

Tatsuoka A., Jardine R., Lo Presti D, Di Benedetto H, Kodaka J : "Soil testing and characterisation of prefailure deformation properties of geomaterials. XIVth Int. Conf. Soil Mech. and Found. Engin., theme lecture, Hamburg, 1997.

Tayebali A.A., Deacon J.A., Coplantz J.S., Finn F.N. and MONISMITH C.L : Fatigue Response of Asphalt Aggregate Mixtures. Part II Extended Test Program. *Strategic Highway Research Program*, Project A 404. Asphalt Research Program. Institute of Transportation Studies, University of California Berkeley. 1994 b.

Tayebali A.A., Deacon J.A., Coplantz J.S., Harvey J.T. and Monismith C.L : Fatigue response of asphalt aggregate mixtures. Part I Test Method Selection. *Strategic Highway Research Program*, Project A 404. Asphalt Research Program. Institute of Transportation Studies, University of California Berkeley. 1994 a.

Tayebali A.A., Tsai B. and Monismith C.L : Stiffness of Asphalt Aggregate Mixes. Report n SHRP A388, *Strategic Highway Research Program*, National Research Council, Washington DC, 1994c.

Thibaud O : Influence de la température sur la fatigue d'une GraveBitume. *Rapport de Travail de Fin d'Etudes*, Ecole Nationale des Ponts et Chaussées, 1976.

Ugé P., Gest G. and Gravois A : Nouvelle méthode de calcul du module complexe des mélanges bitumineux. *Bulletin de Liaison des Laboratoires des Ponts et Chaussées*, N° Spécial V, pp.199213, déc.1977.

Ullidtz P., Kieler T.L. and Kargo A : Finite element simulation of asphalt fatigue testing. *Proceedings of the 5th Int. Conf. Rilem : Mechanical Tests for Bituminous Materials*, Lyon, may 1997.

Van Dijk W. and Wisser W : The energy approach to fatigue pavement design. *Journal of the Association of Asphalt Paving technologists AAPT*, Vol. 46, pp.140, 1977.

Van Draat W.S. and Sommer P : Ein Gerät zur Bestimmung der Dynamischen Elastizitäts Module von Asphalt. *Strasse und Autobahn*, Heft 6, p 201206, 1965.

Verstraeten J : Aspects divers de la fatigue des enrobés bitumineux. *Rapport de Recherche N° 170/JV/1976*, Centre de Recherches Routières Bruxelles, 1976.

Verstraeten J : Loi de fatigue en flexion répétée des mélanges bitumineux. *Bulletin de Liaison des Laboratoires des Ponts et Chaussées*, 1974, mars, N° 70, pp.141156.

William M.L., Landel R.F. and Ferry J.D : The temperature dependence of relaxation mechanisms in amorphous polymers and other glassforming liquids. *J. of American Chemistry Society, July, N°20*, 1955.

6

Interlaboratory test program on complex modulus and fatigue

by L. Francken and J. Verstraeten

6.1 General principles, materials and participating laboratories

6.1.1 INTRODUCTION

The historical background of the RILEM international testing program on significant testing methods for asphalt pavements has already been given in chapter 4 when dealing with the mix design interlaboratory test. We must recall here that the test circuit of the "OECD-FORCE" project constructed at Nantes in France was a large scale experiment focused on the analysis and evaluation of flexible road structure designed according to the fatigue criteria. At the end of these experiments in 1988 it had been subjected to more than 4 million axle loads (*OCDE 1991*). The EMPA - Swiss Federal laboratories were then assigned the task of extracting paving pieces from the circuit for further modulus measurements and fatigue tests to be carried out in the frame of the second part of the RILEM interlaboratory test.

6.1.2 WORKING PROGRAM

The aim of this program was to compare different test methods to study or test the dynamic behaviour of a compacted bituminous mix with the final purpose to release recommendations for test methods having a high degree of significance and reliability in the determination of fundamental characteristics (*Eustacchio et al. 1994*).

Two properties were considered in this programme : complex modulus and fatigue law. The tests were realised by 15 participating laboratories among those cited in Table 4.1 of chapter 4. The main results of this experiment are presented in this chapter. The conclusions and future prospects given in 6.4 were derived after further discussions of the summary report with the partners of the testing programme.

This intensive effort has allowed several participants to make substantial improvements in their procedures and equipments. Recommendations derived from this experience are expected to induce further progress in the harmonisation of different test methods.

Bituminous Binders and Mixes, edited by L. Francken. RILEM Report 17. Published in 1998 by E & FN Spon, 11 New Fetter Lane, London EC4P 4EE, UK. ISBN 0 419 22870 5

6.1.3 PRINCIPLES OF THE TESTING PROGRAM

1. Allowance was given to the participating laboratories to choose their usual testing procedures and physical conditions of temperatures and loading frequencies,
2. four conditions resulting from a combination of two temperatures (0 and +20°C) and two loading frequencies (1 and 10 Hz) were recommended as reference conditions,
3. it was asked to the laboratories to provide, in addition to their usual testing conditions, measurements for at least one of these four reference conditions,
4. each laboratory was requested to give the technical features of the testing facilities together with a description of the operating procedures followed.

6.1.4 MATERIALS USED IN THE TESTS

The material to study was a French bituminous concrete (BB 0/10) containing a 50/70 pen grade bitumen with a mean content of 6.5% per 100% aggregates. The base components were identical with those delivered for the "mix design" (Test described in Chapter 4). The gravel and sand materials were all crushed and distributed in 3 fractions resulting in the final mixture grading presented in Fig. 6.1. Filler (<.08 mm aggregates) was also provided. The binder was a bitumen with following conventional characteristics :

- Penetration : 53.6 10^{-1}mm.
- Ring and Ball softening point : 53.1°C.

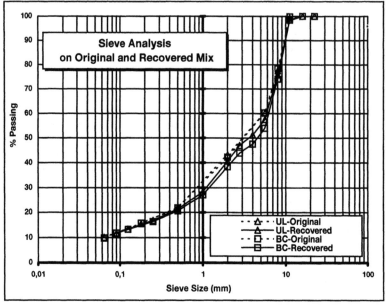

Fig. 6.1 Gradation of the mix.

The material was sampled in 400 x 600 x 120mm blocks taken from a test section of the FORCE project (LCPC's Fatigue test track at Nantes). Two blocks were delivered to each participator who was free to adopt his own working procedure for manufacturing the testing samples out of these blocks. This pavement consisted in two layers (of about 60 mm thickness each) which were, in principle, made of the same composition. It was expected in this way to allow a comparison in which specimen composition, preparation and compaction did not come in as variables. Controls and analysis made on the material lead to conclude that the composition of the upper and lower layers (UL and BC) can be considered as identical for what concerns the gradation (Fig. 6.1) and the binder content.

Table 6.1 Density distribution of the investigated mix.

Lab.	Block mber	UL = Upper layer			BC = Base Course		
		Nb	Bulk density (t/m^3)	Std Dev	Nb.	Bulk density (t/m^3)	Std.Dev
B1	1	10	2.265	.07	10	2.340	.01
	2	10	2.344	.01	10	2.369	.01
CH1	1	2	2.266	--	2	2.310	--
	2	2	2.335	--	2	2.359	--
CH2	1	5	2.278	.024	--	--	--
	2	12	2.281	.025	--	--	--
D2	1	4	2.275	.005	6	2.318	.007
	2	4	2.252	.013	7	2.308	.008
H1	1	13	2.337	.01	13	2.355	.004
I2	1	14	2.320	.006	15	2.353	.006
NL4	1	5	2.262	.013	5	2.325	.004
	2	5	2.284	.015	5	2.337	.008
S2	1	8	2.302	.017	7	2.351	.007

However, the analysis of the sample density distribution, carried out during the final interpretation revealed that this condition was not fulfilled: the bulk densities were distributed over two distinct populations in spite of the fact that the grading curves were fairly constant.

Table 6.1 gives for eight laboratories and per layer (UL = Upper layer; BC = Base Course) the mean value and standard deviation of the bulk densities of the specimens. The distribution of the densities over the 201 samples tested in eight laboratories is presented on Fig. 6.2. It must be noticed that :

- the standard deviation of the whole set of specimens (0.038) is much higher than that of the individual groups taken per block and per layer as in Table 6.2,
- the upper layer presents the highest standard deviations,
- the distribution is skew with a tail on the low density side.

This shows that the scatter obtained in the field can be wide even in a closely controlled situation such as the FORCE project.

A closer analysis of the data reveals that this skewness is due to the presence of three different specimen populations.

Different laboratories have indeed noticed a marked difference between the upper layer (UL) and the bottom layer (BC). As can be verified in Table 6.2, about one half of the blocks (25% of the specimens) had an upper layer with a lower density.

The overall distribution is in fact hiding three distinct populations visible on Fig. 6.2:

1.   The base course material (BC).
2.   High density upper layer material (HUL >2.3t/m^3).
3.   Low density upper layer material (LUL <2.3t/m^3).

The identification of these three materials can be readily determined by examining the average densities presented in Table 6.1. This leads to a redistribution of the specimens into three populations described in Table 6.2.

Table 6.2 Separation of the three specimen populations.

	BC	HUL	LUL	TOTAL	BC + HUL
Average (t/m^3)	2.339	2.328	2.277	2.318	2.334
Std.Dev (t/m^3)	0.025	0.015	0.034	0.038	0.023
Number	91	49	61	201	140
% of Total	45.3	24.4	30.3	100	70

A further analysis of these distributions shows that the difference between BC and HUL is not significant and that hence the specimens of the two groups may be taken together. On the other hand the LUL group is significantly different from the two others. We have already seen that this difference cannot be explained by differences in the grading nor in the binder content. It seems that the effect must be ascribed to variations in the compaction of the mix during laying on site.

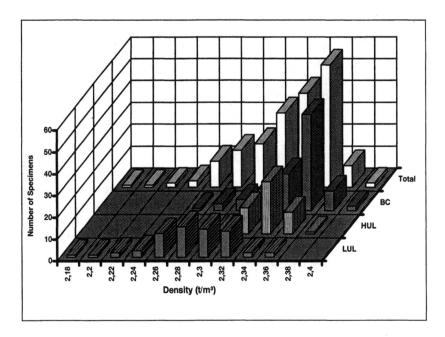

Fig. 6.2 Density distribution per layer (BC = base course ; HUL = high density upper layer; LUL = low density upper layer).

6.1.5 COMPLIANCE WITH THE ISO 5725 STANDARD

We will try as far as possible to rely on the ISO Standard method in order to derive from the mass of the collected results :

- some main points allowing to set up recommendations,
- quantitative evaluations of the repeatability and reproducibility for complex modulus measurements and fatigue testing.

This ISO standard gives guidelines to carry out this type of investigation. Some of the basic conditions required by this standard cannot be met in this case :

1) The participating laboratory have worked on a voluntary basis and hence they do not represent a realistic cross section of laboratories.
2) The test method to be investigated must be one that has been standardised.

A second problem which was raised here before is the uniformity of the test samples. Whatever the reason, this heterogeneity is a factor which has influenced the scatter of some of the results commented hereafter. This problem was unexpected when starting this interlaboratory test.

Consequently the whole set of data was analysed by redistributing the results of moduli and fatigue tests in accordance with two sample populations A and B, with average bulk densities respectively higher or lower than 2.3 t/m³.

1)   Material A : High density material extracted from the base course or taken from the high density upper layer.The criterium to define this material is a minimum of 2.3 t/m³ for the average bulk density in the upper layer.
2)   Material B : Low density material from the upper layer. The criterium used to define this material is a maximum average density of 2.3 t/m³.

6.1.6 PARTICIPATING LABORATORIES

Fifteen laboratories mentioned in chapter 4 (Table 4.2.1) have responded positively to the proposal. Depending on their facilities and availability the results supplied were relative to modulus measurements (15 laboratories), fatigue results (10 laboratories) or both. Table 6.3 gives an overall view of the different types of tests which were carried out.

Table 6.3 Participation in the testing program.

Lab code	Moduli		Fatigue mode**		Method **		Loading Conditions **		
Lab.-Code	E^*	φ	D	F	Test Geom.	Spec. Type	Wave Form	Offset	His.
B1	X	X	X	X	2PB	TR	SI	0	C
CH1	X	X			SH	CY	SI	0	C
CH2	X		X		2PB	TR	SI	0	C
DK2	X	X			3PB	PR	SI	0	C
D1	X	X			SH	PR	SI	0	C
D2	X	X			2PB	PR	SI	0	C
F1	X	X	X		2PB	TR	SI	0	C
H2	X			X	3PB	PR	HS	V	C
I2	X	X	X	X	3PB	PR	SI	0	C
NL2	X	X	X		4PB	PR	SI	0	C
NL4	X	X	X		4PB	PR	SI	0	C
NL5	X		X		4PB	PR	SI	0	C
S2	X				CO	CY	SI	0	P
US1	X	X			IT	CY	SQ	1	P
US2	X	X		X	SH	CY	SI	0	C

*   This laboratory was involved in the measurements carried out in the frame of the OECD FORCE project. These results have been joined to those collected in this interlaboratory test.
**   The meaning of the symbols used is given in 6.1.7 ; see also chapter 5 Table 5.1 :

D =   Fatigue at controlled displacement.
F =   Fatigue under controlled force.

Table 6.4 Shape and dimensions of the specimens.

Lab. Code	Test Method *		Dimensions in mm			
	Testing Geometry	Shape of Specimen	Thickness b or D	Length L	Depth h or h1	Depth h2
B1	2PB	TR	25	250	56	25
CH1	SH	CY	250		25	
CH2	2PB	TR	25	250	56	25
DK2	3PB	PR	40	80	80	
D1	SH	PR	40	160	40	40
D2	2PB	PR	50	300	50	50
F1	2PB	TR	25	250	56	25
H2	3PB	PR	80	150	50	50
I2	3PB	PR	30	340	30	30
NL2	4PB	PR	50	450	50	50
NL4	4PB	PR	50	450	50	50
NL5	4PB	PR	50	450	50	50
S2	CO	CY	100	100		
US1	IT	CY	63	100		
US2	SH	CY	73	50		

*the meaning of the used symbols is given in 6.1.7.

6.1.7 TEST METHODS

The type of test used by each of the fifteen laboratories is determined by the shape of the specimens used, the testing geometry and the way in which the forces are transmitted to the specimen. The information supplied in Tables 6.3 and 6.4 in this respect are in accordance with the following conventions :

6.1.7.1 Shape of specimens

- **PR** : Prismatic (b = thickness, L = length, h = depth).
- **TR** : Trapezoidal (b = thickness, L = length, h1 = depth 1, h2 = depth 2).
- **CY** : Cylindric (b = thickness, D = Diameter).

According to the laboratory, the size of the specimens was sometimes different for a given shape .

6.1.7.2 Testing geometry

- **2PB** : 2 point bending; PR or TR cantilever beam loaded at its extremity.
- **3PB** : 3 point bending; PR specimen resting on two fixed supports at a fixed distance (span) centrally loaded.

- **4PB** : 4 point bending; PR specimen resting on 2 supports ; load is transferred in two points to the specimen.
- **SH** : shear testing procedure ; two parallel PR or CY test pieces are submitted to shear stresses through a plate enclosed in between in a sandwich configuration.
- **CO** : compression ; axially loaded cylindrical specimen.
- **IT** : indirect tensile test ; laterally loaded cylindrical specimen.

6.1.7.3 Loading conditions

The applied load is characterised by its wave form, offset and history. The codes used in Table 6.3 to define the loading conditions of the different laboratories are :

- Wave forms : SI =   Sinusoidal.

  HS =   Haversine.

  SQ =   Square.
- Offset :   0 =   No offset .

  1 =   Offset of 1 amplitude (Haversine).

  V =   Variable offset (varies freely during the test).
- History :   C =   Continuous.

  P =   Pulsed.

6.2   Complex modulus

6.2.1   DEFINITIONS AND INTERPRETATION OF THE MEASUREMENTS

Measurements of moduli were made at different temperatures T and frequencies Fr (or angular frequencies $\omega = 2\pi.Fr$) by using the various facilities and sample geometries presented in Table 5.1. Depending on these facilities, the results supplied for the complex modulus were relative to the norm of the modulus $|E^*|$, the phase angle φ or both (see Table 6.5).

The measurements readily obtained during any of these tests are force F, displacement D and their phase angle Φ. The mechanical material characteristics must be derived from these measurements through formulas given in chapter 5.2.3 assuming the material to be linear elastic and isotropic (cf. 5.3.2.1).

6.2.2   RESULTS AND DISCUSSION

The evaluation will be based on the measured values of the two components of the complex modulus : $|E^*|$ and φ. When the data given by some laboratories did not correspond to the reference set of temperature-frequency conditions (this happened in different cases) interpolated values were derived by using the frequency-temperature equivalency principle. When these conditions were out of the range of experimental values or when the number of measurements was too small to allow the interpolation no values were given.   The values obtained in this way at the different reference conditions of temperatures and frequencies are given in Table 6.5.

6.2.2.1 Evaluation based on the stiffness modulus |E|*
Table 6.5 shows that there are large differences in the values obtained for the stiffness modulus. The ranking of the different laboratories remains independent of the conditions of temperature and frequency. The comparison with the average values leads to distinguish 3 categories of results :

1)   Higher than average : CH1, the Swiss bending-shearing and S2 the Swedish test (compression on cylindrical cores) are yielding modulus values twice greater.
2)   Close to the average : B1, CH2, D2, DK2, F1, H1 , I2, NL2, NL4 and NL5 presented values which are falling more closer together for the two imposed temperatures of 0 an 20°C. All these results were obtained by bending tests.
3)   Lower than average : D1 the lateral shearing of prisms used in Germany leads to values lower than average by a factor of 2 (having already taken the correction of $2.(1 + \nu)$ to convert $|G*|$ modulus to $|E*|$).

It appears clearly that all the results obtained by bending tests fall much closer together than the other test methods. The absolute values of the modulus as well as the phase angles are in good agreement irrespective of the type of test or specimen.

The resilient moduli obtained by S2 are also in good agreement with the former ones if one assumes that their pulsed loading frequency corresponds to 1Hz sinusoidal loading.

The H1 average results obtained during controlled stress fatigue tests are also in good agreement, but owing to their very large scatter they could not be taken into account in our further analysis.

There is clearly a difference in the magnitude of the results for what concerns the following tests : CH1, D1, S2.

The explanation of these large differences remains open to discussion. Following points are among the questions to investigate :

•   distribution of stresses in the sample,
•   validity of the shape factor,
•   starting procedure of the test,
•   preconditioning of the samples
•   anisotropy.

6.2.2.2 Evaluation based on the phase angle φ
Table 6.5 displays the results obtained by 10 laboratories. In spite of the fact that this characteristic is generally difficult to determine experimentally with high accuracy, it can be stated that the agreement is much better in this respect than in the case of the stiffness modulus.

In this case, the CH1 results are in accordance with the bending test results where D1 and I2 are on the low side and NL2 on the high side.

Unlike what was stated for the stiffness modulus one may conclude that the type of test does not influence the results in a significant way.

Table 6.5 Complex Moduli measured by 15 laboratories for a set of 4 reference conditions (T = 0, 20°C combined with Fr = 1, 10 Hz).

Lab. Code	Type	Spec.	Stiffness modulus |E*| (MPa)				Phase angle φ			
			0/1	0/10	20/1	20/10	0/1	0/10	20/1	20/10
B1	2PB	TR	12957	17920	2865	6056	10.9	8.0	31.2	24.6
CH1	SH*	CY	23580	29730	6750	13350	8.9	7.3	30.4	22.2
CH2	2PB	TR	12260	16500	2995	5626				
D1	SH*	PR	5100	6000	1650	2850	8.6	7.5	25.5	19.3
D2	2PB	PR	12137	16385	3150	5746	11.0	7.5	30.1	21.9
DK2	3PB	PR	12282	14946	2983	5713	8.4	5.7	28.2	22.7
F1	2PB	TR	14702	18251	3511	6563	9.7	7.6	27.7	21.2
H1	3PB	PR	11146	16636						
I2	3PB	PR		19073		8933		8.8		18.7
NL2	4PB	PR	12219	16507	2130	4794	14.9	11.1	36.0	29.3
NL4	4PB	PR	13303	16952	2814	5683	10.4	8.3	31.1	24.7
NL5	4PB	PR	13500	17269	2874	5282				
S2	CO	CY	24707	31147	7603	12040				
S2	IT	CY	9507		3363					
US1	IT*	CY			2483					
US2	SH*	CY			2903			31.4		
ALL TOGETHER										
Average			13723	18255	3434	6886	10.4	8.0	30.2	22.7
Std. Dev.			5001	6043	1603	2919	2.0	1.4	2.8	3.0
% Variation			36	33	47	42	19	17	9	13
BENDING TESTS										
Average			12751	17019	2921	6139	10.9	8.1	30.7	23.3
Std.Dev.			950	1095	363	1119	2.0	1.5	2.7	3.1
% Variation			7	6	12	18	18	18	9	13

*   \|E*\| modulus derived from shear modulus by assuming \|E*\| = 3.\|G*\|.

6.2.3   GENERAL STATEMENTS ON ALL THE RESULTS

A first screening of the results presented by 15 laboratories for the reference conditions (Table 6.5) led to the conclusion that :

- the bending tests results performed by nine laboratories are in closer agreement than all other tests. This results in a much lower variation as shown on Table 6.5,
- the resilient moduli obtained in the IT test (S2) are in good agreement with the moduli obtained in bending at a frequency of 1 Hz,
- results of SH and CO test methods are lying out by a factor of 2 irrespective of the temperature and frequency range,
- reproductibility of the measured phase angle is much better than that of the stiffness modulus \|E*\|. This holds for all the testing procedures and testing conditions of temperatures and frequencies.

6.2.4 ANALYSIS OF THE DETAILED RESULTS OBTAINED FROM BENDING

It is clear from the first analysis based on the reference conditions that the bending tests present a consistent set of results. The amount of data available in addition to the reference conditions was large enough in these cases to allow a deeper analysis of the complex modulus.

6.2.4.1 Temperature - Frequency Equivalency Principle
The values of the complex modulus are generally given for several combinations of temperatures and frequencies. These results can be presented graphically under the form of isotherms of the modulus, but most of the laboratories have presented their results under the form of a master curve of the modulus for an arbitrary reference temperature Ts (10°C for instance).

It has been shown indeed by these laboratories that such a master curve can be built up from the modulus values obtained at different temperatures and frequencies by shifting the results along the frequency scale by a factor log (α_T). One single independent variable :

$$X = \log(\alpha_T . Fr)$$

can then be used instead of temperature and frequency (see also the Introduction).

The advantage of this procedure is that once the master curve is established, it is possible to derive an interpolated value of the stiffness modulus for any combination of temperature T or frequency Fr inside the range covered by the measurements. This gives in addition the possibility to compare the results obtained by two laboratories at different sets of conditions, which avoids the recourse to standardise temperature and frequency conditions.

The two different formulas (WLF and Arrhenius equations) generally proposed to compute the shifting factors were given in chapter 5 (§5.2.3.4).

The master curve is then a function of a reduced parameter :

$$X = \log(\alpha_T . Fr)$$

Fig. 6.3 presents the evolution of the WLF and Arrhenius shifting factor obtained by different laboratories as a function of the testing temperature, for a reference temperature Ts of 40°C.

Only one participating laboratory used the WLF formula. In this case the parameters C_1 and C_2 were determined by a fitting procedure allowing other parameters defining the shape of the master curve to be calculated for a reference temperature of 40°C.

Widely different values were obtained for the same specimen when submitted to similar testing conditions.

This indicates that the WLF approach does not allow the representation of different test results on a single scale of reduced variables. It thus appears that the C_1 and C_2 parameters of WLF equation are not material constants in this case.

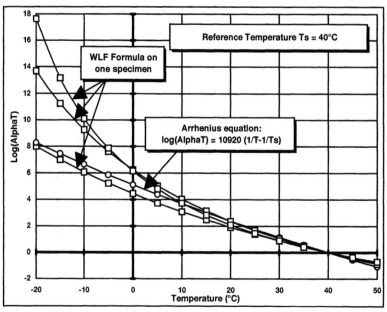

Fig. 6.3 Shifting factors versus temperature. Comparison of Arrhenius equation used by 4 labs and WLF relation used by one lab on 1 specimen tested 3 times.

On the other hand, four laboratories used an Arrhenius equation to adjust their values on a master curve for Ts = 10°C and they came out with almost the same value of 210 to 220 kJ/mol for the apparent activation energy δH. There seems thus to be a clear evidence of the validity of Arrhenius equation. This approach has two advantages over the WLF formula :

1)   There is only one parameter (δH) to be adjusted in the fitting procedure.
2)   This factor is a material constant which is independent of the testing procedure.

Fig. 6.3 also shows that in the range of temperatures higher that 20°C there is a close agreement between both equations.

By using the reduced frequency concept we can now display in Fig. 6.6 the different X (T, Fr) ranges covered with the different types of bending tests considered.

Finally the validity of the Arrhenius approach is clearly demonstrated on Fig. 6.4 where the set of results obtained by six laboratories for two temperatures and different frequencies are presented together in function of the reduced frequency parameter X for a reference temperature of 15°C. Fig. 6.5 represents the Black diagram corresponding to the results of the 6 laboratories having measured the two parameters of the complex modulus over a wide range of temperatures and frequencies.

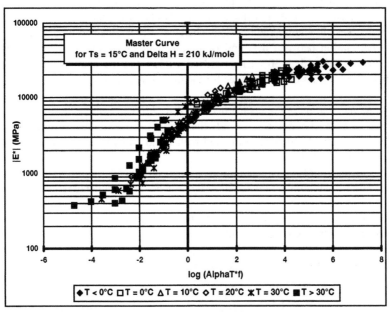

Fig. 6.4 Master curve obtained by application of Arrhenius equation to the bending test results presented by six laboratories.

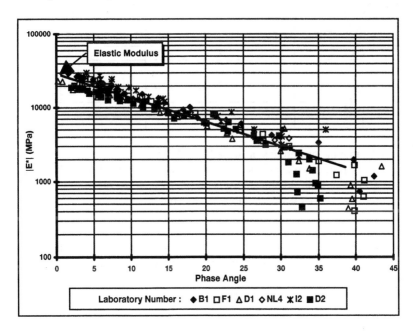

Fig. 6.5 Black diagram for six laboratories.

In spite of the fact that there is more scatter in this type of presentation, it is possible to derive an estimation of the value of the purely elastic modulus E_∞ displayed by the material at very low temperatures. This characteristic is obtained by determining the linear correlation between $|E*|$ and φ for the couples of values corresponding to temperatures lower than 30°C. It is indeed visible that the relationship is no more linear for high values of φ and hence for the high temperatures.

6.2.4.2 Redistribution of the results between A and B materials

H2, NL2 and S2 laboratories gave only the results for the minimal recommendations of the experience (0 and 20°C combined with 1 and 10 Hz). They will not be further taken into consideration.

The eleven remaining laboratories have produced additional results for temperatures covering the ranges $-20 < T < +50°C$ and frequencies $.002 < Fr < 160$ Hz

These results are distributed in an uncontrolled way over the two materials A and B defined in the introduction (see Tables 6.1 and 6.2).

Table 6.6 Tests performed on materials A and B.

Laboratory	Material A Density > 2.3 t/m^3	Material B Density < 2.3 t/m^3
B1	X	X
CH1	X	X
CH2		X
D1	X	
D2		X
DK2	X	
F1	X	
I2	X	X
NL4	X	X
NL5	X	
US2	X	

This results into the two groups presented in Table 6.6.

Nine sets of results are available for material A, but three of them (CH2, D1 and US2) concern the shear modulus G* and are lying out of the other results. Six sets of results of stiffness moduli are available for material B.

6.2.4.3 Comparison procedure

The comparison of the results presented by the different laboratories must be made for the same combinations of temperatures and frequencies (reference T-Fr couples). This is impossible to do in a direct way because the combinations of frequencies and temperatures are different in each case.

Table 6.7 gives an overview of the temperature and frequency ranges covered by the different laboratories who have carried out measurements out of 4 reference conditions of the interlaboratory test. These conditions correspond to a range of X values also mentioned in this table.

They were obtained from equation 6.3 in which an apparent activation energy δH of 209,3 kJ/mole was assumed (δH/R = 10920K).

Table 6.7 Range of conditions used in the modulus measurements.

Lab	Temperature (°C)		Frequency		X = log (α$_T$.Fr)		Strain (μStr)
	From	To	From	To	From	To	
B1	-20	30	1	100	-1.877	7.245	25< <200
CH1	0	20	0.3	10	-1.170	3.083	---
CH2	-10	45	8	50	-2.674	5.303	40
D1	-10	50	1	50	-4.109	5.303	---
D2	-20	40	2	120	-2.727	7.325	---
DK2	-10	50	0.01	256	-5.808	6.012	---
F1	-10	40	1	30	-3.028	5.081	<50
I2	-10	40	10	100	-2.028	5.604	7.5< <24
NL4	0	20	0.5	56	-0.948	3.832	50
NL5	-10	30	1	30	-1.877	5.081	---
US2	20	40	0.01	5	-5.028	0.052	---

6.2.4.4 Defining reference master curves for materials A and B

The properties of the modulus master curve have been exploited to derive two representative master curves for each laboratory (one for material A and another one for material B).

Overall reference curves were also derived for the two materials. These curves were obtained by fitting a power function to the aggregated results of all the laboratories. These curves were obtained for laboratories who determined the stiffness modulus on the basis of bending tests. The G* values given by CH1 and D1 were kept aside to do this.

The result of these operations is presented in Fig. 6.7 and 6.8 for materials A and B respectively.

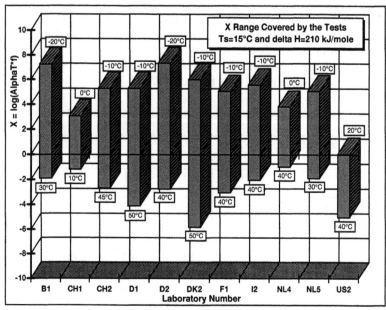

Fig. 6.6 Range of X values covered in the tests (X = log (αT.Fr)).

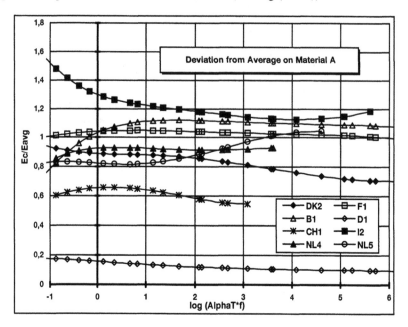

Fig. 6.7 Comparison of the individual laboratory result and overall average master curve obtained for material A (Ec = fitted values for one laboratory Eavg = average value).

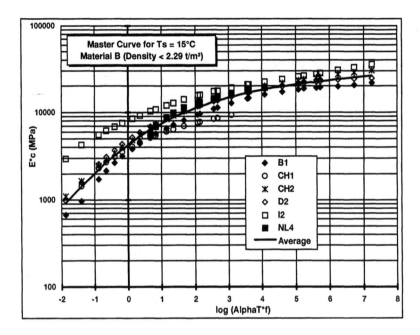

Fig. 6.8 Comparison of the individual laboratory result and overall average master curve obtained for material B.

6.2.5 VARIABILITY IN TEST RESULTS - CAUSES AND WAYS TO IMPROVE

6.2.5.1 Influence of the mix composition
As already mentioned above, the analysis had to take into account the presence of two material types : high density material A and low density material B. Individual master curves were derived from the results given by the different laboratories for these two materials. Also average master curves were derived from all the available results for the materials A and B.

The deviation observed on the master curves between high density (A) and low density (B) materials is illustrated on Fig. 6.9. It can be stated that the ratio of the moduli of material A to material B is almost constant (at a value of about 1.20) over a large range of $X = \log(\alpha_T.Fr)$.

It can be proved that this 20% difference in modulus in favour of material A is to be attributed to the difference in density.

The stiffness modulus $|E^*|$ can indeed be expressed under the form of a product (*Francken and Clauwaert, 1987*):

$$|E^*|(X) = R^*(X) . E_\infty$$

in which : R^* is a reduced modulus or normalised dimensionless, varying from 0 to 1 which determines the shape of the master curve.

E_∞ is the elastic part of the modulus which determines the magnitude of the modulus.

R* is mainly dependent on the bitumen characteristics and E_∞ on the mix composition.

The difference in the |E*| modulus is essentially given by a difference in the purely elastic modulus E_∞ which is related to the mix composition by the relation :

$$E_\infty = 14360 . (\frac{V_a}{V_b})^{0.55} . \exp(-5.84 . \frac{Vv}{100})$$

in which :
E$_\infty$ is given in MPa.
V_a is the volumic concentration in aggregates.
V_b is the volumic concentration in binder.
v is the void content (in %).

Hence any variation in void content, and consequently in density, induces a constant relative variation in the modulus over the full range of X. Using equation 6.5 and 6.6 one can indeed expect between materials A and B a relative variation of :

$$\frac{E_A}{E_B} = \exp (5{,}84 . \frac{\rho_A - \rho_B}{\rho_{max}})$$

Using the density values of A and B (Density A : ρA = 2.334 t/m^3, Density B : ρB = 2.273t/m^3, Max Density : ρ max = 2.5 t/m^3) yields a ratio of 1.205 which confirms the experimental observation.

This result provides the evidence that the stiffness modulus is influenced by the composition in such a way that the modulus can be used as a good indicator of constance and quality.

6.2.5.2 Intra-laboratory variations (repeatability)
It is clear from what was observed that the variability in the mix has largely influenced the scatter of the results, but it is difficult to disclose this effect from experimental errors and inaccuracies.

One laboratory (F1) has shown that a standard deviation of 5% can be obtained for the intra-laboratory scatter provided that representative samples are selected out of the population on the basis of the bulk density and the form factor .

6.2.5.3 Inter-laboratory variations (reproducibility)
A standard deviation of 10 to 20% of the average values was obtained for the interlaboratory scatter on stiffness modulus |E*| measured in bending.

Fig. 6.7 obtained for eight laboratories on material A shows that the overall scatter (ratio of individual values to average) is less than 20% over a large range of X (T, Fr).

The limited number of results available for the other testing methods did not allow this evaluation.

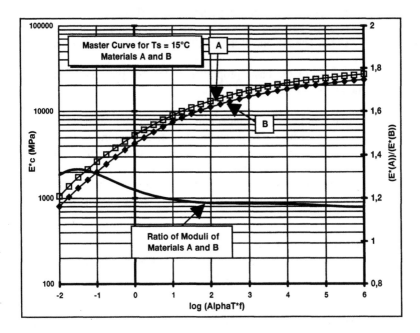

Fig. 6.9 Ratio of measured moduli to average values.

6.2.6 CONCLUSIONS

1. The range of material variation experienced in this inter-laboratory test leads to distinguish two materials A and B with bulk densities lower or higher than 2.3 t/m^3. The heterogeneity of the mix makes the accurate determination of moduli a difficult or even impossible task. A 5% intralaboratory deviation seems to be a limit. One should, therefore, not seek to achieve higher accuracy than required by the application (in view for example determining layer thicknesses for pavement design).

2   A first screening of the results presented by 15 laboratories led to the conclusion that :

    -   the bending tests results presented by ten laboratories are generally in good agreement, especially in the medium range of temperatures and frequencies,
    -   the resilient moduli measured by one participant under pulsed loading are in good agreement with the moduli obtained at 0 an 20°C at a frequency of 1 Hz,
    -   the results of test methods producing shear modulus values G* are beyond or below the average results by a factor of 2 irrespective of

the temperature and frequency range. There seems to be no simple relation between values derived in bending and those measured in shear. The reasons of these differences have to be clarified further.

3. A verification of the basic properties of the complex modulus allowed to ascertain two important facts :

   1) the validity of the temperature-frequency equivalency principle based on the Arrhenius equation to calculate the shifting factors $\log(\alpha_T)$.

   2) the relationship between the stiffness modulus $|E^*|$ and the phase angle φ (Black diagram) determines a typical curve which is independent of temperature and frequency. This approach can be used for the determination of the purely elastic component E_∞ of the complex modulus.

4. The measurements of moduli made in bending by the different laboratories are in agreement within a range of 10 to 20% around the average master curve. Efforts remain to be done in order to track the reasons of some systematic inter-laboratory deviations.

   The closeness of the result must be improved by setting up guidelines for the test facilities, specimen preparation, testing conditions and calibration procedures.

   Experimental reasons for systematic deviations are :

   - material non-linearity; linearity tests must be recommended in order to define the maximum allowable stress and strain levels for complex modulus measurements,
   - lack of accuracy on force measurements in the high temperature range and lack of accuracy of the displacements in the low temperature range. The accuracy of the equipment measuring devices must be in accordance with the range of values to be measured,
   - wrong specimen fitting and too low stiffness of the apparatus when high specimen stiffness are to be measured.

The results obtained in this test indicate that there are many points of agreement between the participators but we have to conclude that a 20% interlaboratory variation does not yet allow a reliable exchange of information on a given material.

6.3 Fatigue tests

6.3.1 DEFINITION AND DERIVATION OF THE PARAMETERS

6.3.1.1 Mode of loading

Material properties such as stiffness are changing in the course of a fatigue test, while applied conditions of loading must be kept constant. However as applied forces and displacements are not independent, a choice must be made between two control modes (see 5.3.3.1):

- D mode = controlled displacement.
- F mode = controlled force.

It is common practice to speak of stress and strain although these parameters are not well defined for non homogeneous test geometries. Unless the test is homogene, one should normally only speak of force and displacement (see 5.3.2.1).

D mode : Displacement controlled mode of loading
In the displacement controlled test the maximal displacement recorded in the test is maintained constant during the test. The strain is the independent variable of the D mode fatigue test and the number of loading cycles corresponding to fatigue failure is based on the evolution of the stress.

F mode : Force controlled mode of loading
In the force controlled test it is the amplitude of the force which is kept constant and the failure criterium is then based on the magnitude of the displacement which increases mostly at the end of the test. The stress is the independent variable of F mode tests and the number of loading cycles corresponding to fatigue failure is based on the evolution of the strain.

6.3.1.2 Statistical analysis
The interpretation is based on the assumption of a linear relationship between the logarithm of the independent variable (initial value of the strain ε_i) and the logarithm of the number of cycles N at failure :

$$Y \equiv \log(N) = A \, . \, \log(\varepsilon_i) + B$$

in which the parameters A and B are determined by a linear regression analysis over the experimental couples of points (log ε, log N).
   This is identical with the well known fatigue law :

$$\frac{N}{10^6} = (\frac{K}{\varepsilon_i})^{\frac{1}{a}} \Leftrightarrow \varepsilon = K \, . \, (\frac{N}{10^6})^{-a}$$

K, sometimes noted by ε_6 (*Moutier, 1990*), corresponds to the critical stress or strain for a fatigue life N of one million load cycles. a is the slope of the linear relationship in a log-log scale. These two parameters are determined from the A and B regression parameters of equation (6.8) through the relations :

$$K = 10^{\frac{(6-B)}{A}}$$

$$a = - \frac{1}{A}$$

6.3.1.3 Meaning and practical importance of the parameters

The strain K determines the quantitative influence of the applied loads on the fatigue phenomenon.

The parameter a is the power factor which governs the form of the load equivalency law used for the definition of the number of standard axle loads for pavement design purposes.

The standard deviation S(N) of the independent variable is a third statistical parameter which deserves some attention in this study.

It allows to make an assessment of the accuracy of the fatigue law for the prediction of the number of cycles N.

It is well known that the distribution of N is log normal. Hence the evaluation of S(N) is a statistical indicator integrating the combined effects of the heterogeneity of the material and the accuracy of the testing method.

If the material is considered as a reference, then S(N) is an indicator of the reliability of the different tests considered in this study (*Moutier, 1990*).

6.3.2   WORKING PROGRAM

The fatigue test programme was carried out at the same time and on the same testing material as the complex modulus measurements described in 6.2.

As this was the first interlaboratory test on fatigue involving such a wide variety of testing facilities it was decided that each participant should use his usual procedures. Fixed conditions of temperatures (0 and 10°C) and frequencies (1 and 10 Hz) were recommended for fatigue but not considered as mandatory in this case. Ten of the laboratories participated in the fatigue programme.

Table 6.8 Overview of the testing conditions.

LAB	B1	CH2	F1	H2	I2	NL2	NL4	NL5	US1	S2
Test Type	2PB	2PB	2PB	3PB	3PB	4PB	4PB	4PB	IT	IT
Wave Form	SI	SI	SI	HS	SI	SI	SI	SI	SQ	HS
Specimen	TR	TR	TR	PR	PR	PR	PR	PR	CY	CY
Mode (D or F)	D F	D	D	F	D F	D	D	D	F	F
T (°C)	10	10	10	10	0 10	10 0	0 20	0 10 20	0 21	10
Fr (Hz)	25	25	25	25	1 10	30 30	10	30 30 30 30 30	10 10	**
Material	A B	A B	B	*	A A	A A	A	B B	B B B	? ?	A A
Nb.tests	15 5	15 5	17	36	12 12	11 19	4	8 6 5 6 5	3 3	7 8

*   Specimens of A and B materials were mixed.
**   Pulsed loading (0.1s loading time, 0.4s rest time).
?   Unknown.

6.3.3   TEST CONDITIONS AND EXPERIMENTAL PROCEDURES (see Table 6.9).

The experimental facilities used by the participants in fatigue testing were generally the same as those used in the measurement of the complex modulus. The symbols used to define conditions such as test type, wave form specimen geometry and physical conditions are given in the introductory part of this chapter (see Tables 6.3 and 6.4).

The lack of a standard procedure for fatigue testing resulted in the 20 sets of conditions presented in Table 6.8. They are differing from one another by the following points :

- type of test (2PB, 3PB, 4PB, IT),
- number of specimens (from 3 to 36),
- specimen size and geometry (TR, CY, PR),
- wave form (S = sinusoidal, HS = Haversine, SQ = square),
- mode of loading (D mode = controlled displacement, F mode = controlled force) (see 6.3.1.1),
- temperature (0, 10 or 20°C),
- frequency (1, 10 ,25 or 30Hz),
- failure criteria (% decrease of stiffness, rupture of the specimen),
- material density (materials A or B).

Some of these points are further commented in the following.

6.3.3.1 Heterogeneity of the investigated material
As already mentioned in 6.1.4, it was noticed by different laboratories that the bulk density of the wearing course material was sometimes lower than that of the base course. In accordance with these observations, it was decided to consider that two types of materials in the interpretation of the fatigue test results :

- Material A : with density higher than $2.3t/m^3$ taken either from the base course or the high density upper layer.
- Material B : with density lower than $2.3t/m^3$ taken from the low density upper layer.

This information was not available for some of the reports.

6.3.3.2 Initial strain value
In many cases, the results are interpreted by taking the initial strain as the independent variable of test whatever the mode of loading.

This initial strain can be derived from the initial value of the stiffness modulus and the stress :

$$\varepsilon_{in.} = \sigma_0 \, |E^*|$$

The initial stiffness value should however be taken after the number of transitory loading cycles which are necessary to bring the specimen at equilibrium temperature. This number is depending on many factors (experimental set up, temperature control system, magnitude of applied forces, specimen size, etc...). This factor which is sometimes variable from one laboratory to another has a strong influence on the final results.

6.3.3.3 Failure criteria
The fatigue law is derived from the relationship which can be established between the independent variable of the test, (i.e. strain or stress for controlled displacement or controlled force respectively) and the number of loading cycles N at failure.

For the laboratories using D mode, failure is defined to occur at the moment where the modulus of the specimen decreases to 50 % of its initial value.

For the laboratories having applied F mode fatigue the failure criterium is defined in different ways :

- B1 considers failure to occur when the displacement increases by a factor of 4.
- I2 considers the moment when the stiffness decreases by 50%.
- H2 takes the inflection point of the permanent strain curve as indicator of the fatigue fracture.

In some cases 3 or 4 point bending (3PB or 4PB) tests can lead to a permanent deformation in addition to the elastic or reversible deformation. This is the case of H2 where the permanent strain is given as an additional information.

6.3.3.4 Distribution of the test specimen over stress/ strain levels
Not only are the number of tested specimens variable, but the way the specimens are distributed over two (minimum) or more strain levels is very different :

- For F1, I2 , NL4 and NL5 the specimen are distributed over two or three levels. This method gives a better estimation of the statistical distribution of the results at the tested strain or stress levels (see Fig. 6.10).
- For B1 and CH2 the specimen are continuously distributed over different levels. This method allows to verify the form of the fatigue law (see Fig. 6.11).

6.3.4   PRESENTATION OF THE RESULTS

As mentioned in Table 6.8, seven laboratories (B1, CH2, F1, I2, NL2, NL4 and NL5) have carried the test at constant displacement and five have made it at constant force (B1, H2, I2, US1, S2).

The overall test results are given in Tables 6.9 to 6.11 are illustrated by Fig. 6.12, 6.13 and 6.14.

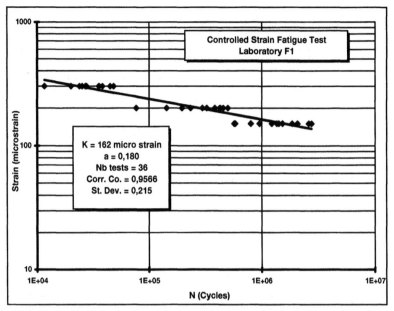

Fig. 6.10 Controlled displacement test results on material A. Example of 3 strain level distribution.

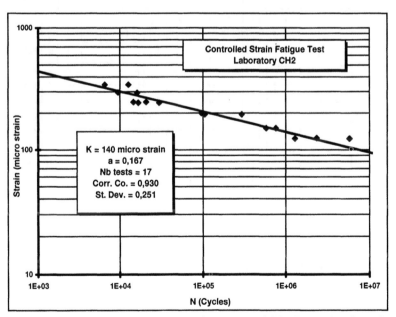

Fig. 6.11 Controlled displacement test results on material B. Example of continuous strain distribution.

Table 6.9 Controlled displacement fatigue results obtained on material A.

Lab	Conditions		Fatigue results					Modulus	
	T (°C)	Fr (Hz)	Nb.	K	a	C.Corr	S(N)	|E*| (MPa)	Std.Dev (MPa)
B1	10	25	15	187	.231	.866	.199	13180	1636
F1	10	25	36	158	.180	.915	.215	11337	375
I2	10	30	19	125	.179	.948	.163	12692	1475
NL2	0	10	4	141	.120	.915	.183	16760	609
*	10	25/30	70	146	.233	.722	.376	11755	1365

*   Average values for B1, F1, I2 grouped.

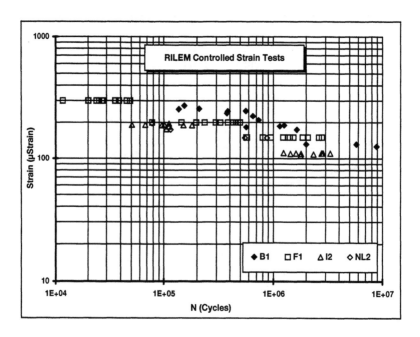

Fig 6.12 Controlled displacement fatigue results obtained on material A (4 laboratories) at two temperatures.

Table 6.10 Controlled displacement fatigue results obtained on material B.

Lab	Conditions		Fatigue results					Modulus	
	T(°C)	Fr(Hz)	Nb.	K	a	C.Corr	S(N)	|E*| (MPa)	Std. Dev. (MPa)
B1	10	25	5	162	.127	.982	.136	10949	1541
CH2	10	25	17	140	.168	.931	.251	11850	703
NL4	0	30	8	151	.179	.976	.116	17663	794
NL4	20	30	6	203	.188	.912	.220	6894	394
NL5	0	30	5	131	.161	.954	.181	15120	1023
NL5	10	30	6	125	.199	.976	.082	10336	478
NL5	20	30	5	187	.154	.931	.143	5068	176
*	0	30	13	139	.185	.859	.267	16683	1365
**	10	25/30	28	140	.166	.919	.236	11666	1119
***	20	30	11	195	.169	.862	.242	6064	962

*   :   NL4, NL5 (grouped values).
** :   B1, CH2, NL5 (grouped values).
*** :   NL4, NL5 (Aggregated data).

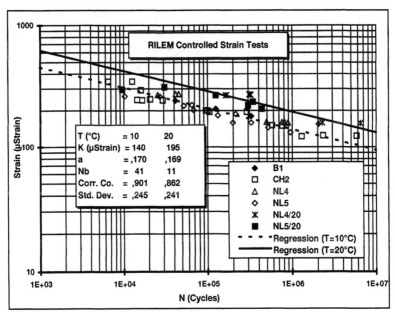

Fig. 6.13 Controlled displacement test results on material B (4 laboratories) at three temperatures.

Table 6.11 Controlled force fatigue results on material A.

Lab	Conditions			Fatigue results				Modulus	
	T(°C)	Fr(Hz)	Nb.	σ (N=10⁶)	a	C.Corr	S(N)	|E*| (MPa)	Std Dev (MPa)
B1	10	25	15	2.128	.176	.823	.267	14137	932
H2	0	1	12	2.490	.143	.821	.332	11146	3613
H2	0	10	12	3.87	.098	.736	.345	16636	2505
I2	10	30	19	1.782	.150	.922	.187	13646	643
S2	10	*	15	1.668	.184	.931	.190	7952	735

*   Pulsed loading (0.1s loading time, 0.4s rest time)

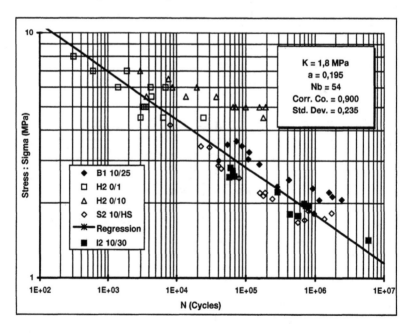

Fig. 6.14 Controlled force fatigue results on material A (5 laboratories). (H2 0/10 not considered in the linear regression).

6.3.5   RESULTS OF THE COMPARISON

Owing to the great variety of conditions and materials, comparisons can only be made on a limited number of cases.

The analysis of these data was made on the basis of the five basic parameters defined above. The interlaboratory variation can be estimated   by comparing the standard deviation S(N) of log(N) obtained by each laboratory with the S(N) corresponding to the aggregated results of all the laboratory results corresponding to comparable situations and conditions. In this way three groups of data were separated by taking into account the material type (A or B) and the mode of loading (D or F).

6.3.5.1 Controlled displacement tests on material A
Four series of test results are available for material B. There are in this case large differences in the results.

The analysis of the aggregated results obtained by the four laboratories is characterised by a very large standard deviation on log(N) which corresponds to a factor 2.4 on the fatigue life.

This is indicative of clear disagreements between the results. In the case of B1 we see on Fig. 6.12 that these results correspond to fatigue lives which are much higher than all the others. The reason of this difference could be explained later on the basis of some defects in the experimental set up.

During the 2PB bending tests used by B1 at high strain levels the displacement measured at the top of the specimens included a part of the deformation of the base plate. Hence the actual strain was lower than the measured one, which led to overestimated fatigue lives. The results given by F1 displayed in Fig. 6.10 give an example of a wide dispersion of the mix properties. As this set of specimens was taken during the initial phase of the OECD-FORCE project it is not sure whether we have to do with a homogeneous material. It could be that in this case we have to do with a mixture of the two materials A and B. Whatever the reason we find in this case a scatter corresponding to a standard deviation on log(N) of .215 which corresponds to a factor 1.6 on the fatigue life.

The fatigue law derived from the I2 results gives a lower strain value than F1 at 1 million cycles, but in this case the scatter is much lower. The data reported by I2 allowed to clearly identify the homogeneity of the specimen population, and this is probably the reason of a lower S(N).

The NL2 results are given for information in this case. They cannot be considered in the comparison because they correspond to a lower testing temperature and hence to a higher stiffness.

These results remain close to those obtained by F1 although the number of four tested specimens seems too low to give statistically significant information on the scatter and the slope of the fatigue law. However the correlation coefficient (0.915) as well as the standard deviation of log(N) are good.

6.3.5.2 Controlled displacement tests on material B

In the case of material B (density lower than 2.3t/m3) we have identified seven sets of results.

But the comparison still remains very difficult in that we have to do here with different sets of testing conditions of temperatures and frequencies.

The examination of the complete set of results presented in Fig. 6.13 allows to distinguish two groups: the results obtained at 20°C and those obtained at 0 and 10°C.

The aggregation of these data per temperature groups indicates indeed in Table 6.10 that 0 and 10°C correspond to similar fatigue laws. The slopes a are rather close to one another for the three cases.

At 10°C we can see that unlike what was the case for material A the three sets of results are in closer agreement with a lower standard deviation of the aggregated results.

This last statement leads to conclude to a better homogeneity. In controlled strain tests there is evidence from the NL4 and NL5 results that the fatigue law is temperature dependent in the range 0-20°C. This cannot be ascribed to a difference in procedure or apparatus. For a same testing procedure, device and mix composition lower temperatures will result in lower K values.

6.3.5.3. Controlled force tests on material A

The parameters given in this case on Table 6.11 correspond to a representation of the fatigue results in function of the applied stress levels.

The results of four laboratories are available. The comparisons of B1 and I2 is directly possible since the test temperature is the same (10°C) and the frequencies (25 and 30 Hz) are not very different (the difference has hardly affected the results, because of the limited influence of this parameter).

As already mentioned in 6.3.5.1, the B1 results correspond to longer fatigue lives with a higher slope. This confirms a systematic experimental deviation due to an overestimation of the strains.

H2 results were obtained at 0°C and for a range of stresses which lies on the high side. In this case the accuracy of the modulus measurements is too low to allow an interpretation of the results in a strain-cycle diagram.

In a stress-cycle diagram only the results obtained by H2 at the lowest frequency of 1 Hz are in agreement with the B1 and I2 results (see Fig. 6.14).

The data given by S2 in the indirect tensile mode under pulsed loading are also presented on this graph. Although we do not know exactly in this case to which testing frequency this corresponds, we see that the results are in fair agreement

6.3.5.4. General observations relevant to both modes of loading

The high scatter sometimes observed is to be ascribed to the fact that the samples consisted in a blend of two different compositions. After elimination of the low density specimens S(N) generally dropped below 0.2. This indicates that :

1)   the fatigue test is very sensitive to mix composition,

2)   the scatter of the results is mostly an indication of the heterogeneity of the tested material.

If one takes care to discard the specimens having an abnormally low density, the range of variation of this parameter goes from 0.115 (NL4) to 0.251 in the worst case (CH2).

  This means that a factor of variation around the mean value of the fatigue life N lies between 1.3 and 1.8.

  These values must be compared with those usually obtained in each laboratory for a comparable mix composition.

6.3.6 PHYSICAL PHENOMENA OBSERVED DURING FATIGUE TESTS

6.3.6.1 Influence of the mode of loading
The influence of the mode of loading (controlled strain D or controlled stress F) can only be evaluated on the basis of the B1 and I2 results by taking two by two the fatigue laws and calculating the ratio of the fatigue lives obtained in the two modes for a given initial strain (200 µstrain for example).

  This leads to the following ratios :

- B1 ratio = 2.58
- I2 ratio = 0.97

One might conclude from this that the mode of loading has no effect at all in the case of I2. On the other hand the difference derived from the B1 results is closer to the factor of 3 which generally is claimed for the temperature range of 10°C (*Van Dijk, 1975*).

6.3.6.2 Energy dissipation
The energy dissipation during a fatigue test is obtained by integrating the energy dissipation per loading cycle over the total number of cycles :

$$Wc = \pi\sigma\varepsilon \sin \varphi$$

If the test is carried out in the controlled stress mode (mode F with a constant stress σ_0) then the cumulated dissipated energy will be :

$$W_f = \pi \sum_{N=1}^{N_f} \frac{\sigma_0^2 \sin(\varphi(N))}{|E*|(N)}$$

In the D mode with constant strain this becomes :

$$W_f = \pi \sum_{N=1}^{N_f} |E^*|(N)\, \varepsilon_0^2\, \sin(\varphi(N))$$

In both cases the final result will mainly depend on the way the stiffness modulus and the phase angle are varying in function of the number of cycles N during the test.

NL4 has found a correlation between the energy dissipated (W_f) during controlled strain fatigue tests and the number of cycles to failure (N_f), at the temperatures of 0 and 20°C and the frequency of 30 Hz.

- It can further be noted that NL4 results show that :

  at 0°C : $W_f = W_c \times N_f$
  at 20°C : $W_f = 1.24 \times W_c \times N_f$

  where W_c is the energy dissipated during the first loading cycle,
  and N_f is the fatigue life.

This is an indication that the variation of $|E^*|$ and φ must be small during a large part of the test. In many cases it has been observed indeed that the energy dissipated per cycle remains almost constant during 80% of the total fatigue life.

Fig. 6.15 aggregates the results of the six sets of controlled strain tests (88 tests) for various couples of temperature and frequency (respectively 10°C and 25 Hz, 10°C and 30 Hz, 0° and 20°C and 30 Hz, 0° and 10 Hz), while assuming that $W_f = W_c \times N$ also holds true at 10°C (which according to *Van Dijk, 1975* seems plausible since the modulus is about 18,000 MPa at 0°C and about 14,000 MPa at 10°C).

It should be observed in this figure that the five B1 points correspond with a lower dissipation of energy; these lowest points relate to 5 specimens with bulk densities ranking among the lowest for the 72 specimens tested in controlled strain by all the laboratories. Some of the results reported by CH2 are in the same case.

Fig. 6.15 does not necessarily prove that the results of the tests performed by the six laboratories are equivalent, but nevertheless indicates some consistency to be ascribed to the fact that the integrated energy is almost linearly depending on the fatigue life N.

6.3.6.3 Thixotropy and healing

Fig. 6.16 represents the Cole-Cole diagram (E_2 versus E_1) obtained from the test results obtained by B1 at 15°C and NL4 (at 0°C and 20°C). The figure represents the evolution of the two components of the complex modulus of the material during a D mode fatigue test.

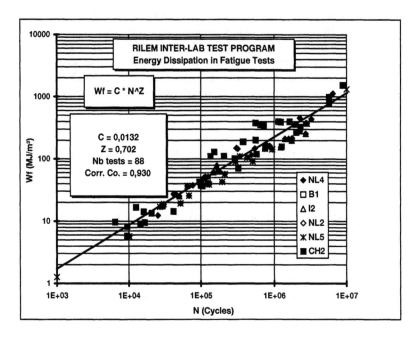

Fig. 6.15 Total energy dissipation versus fatigue life observed by different laboratories under different testing conditions and modes of loading.

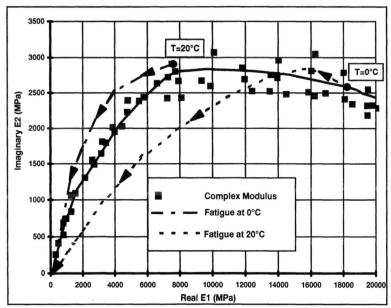

Fig. 6.16 E1-E2 relationship of the material and its evolution during the fatigue process.

It can be seen that at 20°C couple (E$_2$, E$_1$) follows the trend of the Cole-Cole diagram derived from the complex modulus measurements, whereas at 0°C it deviates constantly from it. As suggested by *Verstraeten (1991) and observed by de la Roche (1996)*, this seems to confirm the existence of two distinct fatigue mechanisms : thixotropic effect and possible recovery at high temperature; development of structural faults and partial or virtually no recovery at low temperature.

6.3.7 CONCLUSIONS DERIVED FROM THE DIFFERENT FATIGUE TEST RESULTS

Most of the fatigue tests were performed in 2PB or 4PB test arrangements. If one excludes a few outlyers, the fatigue laws derived by the different laboratories using these methods are rather similar for comparable test conditions and the same material.

- However the variations observed on the **K value** entail large differences in terms of fatigue lives N. At 10°C (see Table 6.10) it may vary between 168 and 199 μstrain which means a factor 2.33 on N if a slope of 0.2 is assumed.
- The results of **slope a** were generally in good agreement between all the laboratories irrespective of temperature and/or mode of loading provided that a minimum number of samples was tested over a wide range of strain or stress levels.
- The standard deviation on log(N) was generally of the order of .2 for the individual sets of fatigue results, this corresponds to a variation of a factor 1.6 on N (the fatigue life).
- The **aggregated results** given at the bottom of Table 6.10 show that the standard deviation is larger and the correlation coefficient lower than the individual results. The number of cycles needed for crack initiation depends on the specimen shape and surface texture; these factors can be responsible for a large part in the interlaboratory variations.
- **Influence of the heterogeneity of the material**. The fatigue of bituminous mixes is a statistical phenomenon. Hence, even if the experimental errors are minimised the remaining standard deviation S(N) can be considered as an indicator of the material heterogeneity. A sufficient number of specimens is, therefore, required in any experiment to correctly determine a fatigue law.
- **Temperature dependence of controlled displacement tests.** The influence of temperature on the fatigue law is visible in the tests carried out at 0, 10 and 20°C (see Table 6.10 and Fig. 6.13). While the fatigue laws derived from the aggregated results of different laboratories are almost identical at 0 and 10°C the results obtained at 20°C lead to much longer fatigue lives. Hence this effect only becomes noticeable for temperatures higher than 10°C.

- **Thermal effects and energy dissipation.** The viscoelastic character of the materials induces energy dissipation with local temperature increases modifying the material characteristics in an unrealistic way during the fatigue tests. Infra Red thermography has given evidence for this art effect which can be considered as responsible for the decrease of the specimen stiffness in the first loading cycles (*de la Roche and Marsac, 1996*). This phenomenon influences directly the failure criterium and the related number of cycles.

- The values of the |**E***| **modulus** obtained in the fatigue tests at 0 and 20°C as well as their standard deviations are in agreement with the results presented on Table 6.5 for the bending test results. This is an indication of the reliability of these methods for the measurement of the complex modulus.

6.4 References

de la Roche C. : Module de rigidité et comportement en fatigue des enrobés bitumineux. Expérimentation et nouvelles perspectives d'analyse. Thèse de Doctorat. Ecole Centrale Paris.

de La Roche C. and Marsac P. : Caractérisation expérimentale de la dissipation thermique dans un enrobé bitumineux sollicité en fatigue. First E & E Congres, Strasbourg, May 1996.

Eustacchio E., Francken L. and Fritz H. : Mix design and repeated loading tests Proceedings of the conference "Road safety in Europe and Strategic Highway Research program" Lille, France, September, 1994.

Francken L. and Clauwaert C. : Characterisation and structural assessment of bound materials for flexible road structures. Proceedings of the VIth Int. Conf on the Structural Design of Asphalt Pavements. Ann Arbor Michigan, 1987.

Fritz H. and Eustacchio E. : Mechanical tests for bituminous Mixes. Proceedings of the 4th RILEM International symposium, Budapest, 1991. Vol 2.

International Standard, ISO 5725, 1986.

Moutier F. : L'essai de fatigue LCPC : un essai vulgarisable ? Budapest 1990 RILEM Proceedings 8.

O.C.D.E. FORCE Project Report of La Baule Workshop, 1991.

Van Dijk W.: Practical fatigue characterisation of bituminous mixes. Association of Asphalt Paving Technologists, Vol. 44, 1975.

Verstraeten J. : Fatigue of bituminous mixes and bitumen thixotropy. Oral presentation at the XIXth PIARC World Road Congress, Marrakech, 1991 (Session TC 8 : Flexible roads).

Younger, K.D., Partl, M.N., Fritz, H.W. and Gubler, R. (1997) Asphalt Concrete Shear Testing With the Co-Axial Shear Tester at EMPA, *RILEM- MTBM97*, Lyon.

Final Conclusions and Future Prospects

by M. N. Partl and L. Francken

1   General

Interlaboratory tests and activities described in this report show that the research initiatives of RILEM initiated by TC-101BAT (Fritz *et al.*, 1991, Eustacchio *et al.*, 1994) and continued by TC-152PBM have attracted much attention. Many laboratories accepted to participate on a voluntary basis and to contribute to this important work without return of investment. This is certainly partly due to the fact that RILEM is a well established platform for prenormative research and technical exchange on a unbiased technical level, thus contributing to an international synthesis of experience in a very early stage of development, validation and verification. In this respect activities of RILEM have, proven very valuable for the laboratories to detect a potential of synergies and to optimize their testing equipment as part of their quality assurance. In fact, some of the test results presented herein did already affect certain test procedures and were already used by the individual laboratories to implement improvements of their test methods (Younger *et al.*, 1997).

Twenty laboratories from over fourteen countries took the opportunity to participate in the interlaboratory binder testing program, fifteen laboratories from thirteen countries were involved in the interlaboratory mix design tests and another fifteen laboratories from ten countries contributed to the interlaboratory mechanical testing program. The financial amount spent on a voluntary basis is in the region of several hundred thousand dollars. RILEM, of course, is not the only organization active on bituminous road materials. In addition, other national and international organizations and research programs are or were recently dealing with asphalt road related activities. These various initiatives show that subjects on the field of bituminous materials and asphalt pavements are still far from being solved.

However, it cannot be ignored that much considerable progress has been made in the last decades. Thanks to those efforts of technical understanding and material optimization roads are now available, carrying amounts of loads in an extend which would have never been considered to be possible some years ago. Nevertheless, traffic still increases dramatically, and so do technical requirements. But the success which has been reached so far gives also strong confidence that actual and future road research will also pay-off in economics if it is done properly and founded with sufficient resources. The most recent example of such a research effort is the five year Strategic Highway Research Program SHRP in the USA which was accomplished 1994 and led to performance based mix design and test methods for asphalt concrete. As SHRP was conducted parallel to this RILEM program, the US laboratories participating either in the RILEM mix design or repeated loading interlaboratory test program were not able to apply the SHRP procedures yet. This,

Bituminous Binders and Mixes, edited by L. Francken. RILEM Report 17. Published in 1998 by E & FN Spon, 11 New Fetter Lane, London EC4P 4EE, UK. ISBN 0 419 22870 5

of course, is regrettable, but waiting for SHRP, on the other hand, would have delayed the RILEM activities in an intolerable way. Obviously, SHRP procedures will have to be considered in more detail by future RILEM work. Not only in this context, it is strongly appreciated that more laboratories and countries from overseas will participate in future RILEM research.

Results and conclusions in this report deal with binders and mixtures as well. *Binders* are dealt with from two aspects. First, the state of the art report on polymer-modified binder test methods is given, secondly the results from the RILEM interlaboratory binder test are described. The conclusions on each of the topics are given in the individual chapters and are not repeated here in detail. Some of the general conclusions and recommendations were already presented by Francken *et al.* (1995) at the first European workshop on rheology of bituminous binders in Brussels 1995.

Mix design and *repeated loading* tests were planned in close coordination, using the same material from the OECD-FORCE project. Again, the conclusions on each of the topics are given in the individual chapters and are not repeated here in detail. Some of the results were also presented earlier by Francken *et al.* (1996).

Although of the same pavement, the samples taken from the OECD-FORCE test track in Nantes for the repeated loading interlaboratory tests were not homogenous enough to be treated as one single population. This demonstrates that it is very difficult to get homogeneous material for comparison tests and that it is difficult to find synergies between two different projects (like OECD-FORCE and RILEM) when the goals are different and when it is not planned from the very beginning which data and measurements are requested by each individual project. Hence not much profit resulted from the fact that the material investigated in the interlaboratory test was tested in the OECD-FORCE project and the synergy between both projects is very limited.

2 Binders

2.1 POLYMER-MODIFIED BINDERS (PMB)

Traditional specifications of plain bitumen are generally not sufficient for polymer modified bitumen and should be supplemented by other test parameters such as tensile strength, elastic recovery, compatibility and storage stability tests. Unfortunately, these traditional empirical test methods, have only poor relation to performance and provide only little information on the rutting, low temperature and mechanical cracking (e.g. fatigue) behavior of PMB's in the critical working ranges of temperatures.

A major step towards better understanding of the function of bitumen and PMB has been taken by SHRP where performance based test methods were proposed including rheological parameters, failure properties as well as short and long term aging properties (Harrigan *et al.*, 1994). Hopefully, the new SHRP specifications and test methods, which are intended for both plain bitumen and PMB, will be more suitable in predicting binder performance on roads than earlier tests.

To characterize aging properties and resistance to climatic damage of PMB chemophysical test methods, such as chromatography, as well as adhesion/cohesion tests should be used and further developed.

The development of PMB specifications is a complicated task and it is very difficult to make a single specification valid for every product, due to the complexity of performance of the various modified bitumen as a function of bitumen nature and the type as well as the polymer content. Hence, many years of experience are needed before a more definite opinion on this matter can be stated.

2.2   BINDER INTERLABORATORY TEST

Nineteen laboratories using eight different types of rheometers participated in the interlaboratory binder tests to determine complex moduli and phase angle of four different binders (one straight run reference bitumen and three PMB's) in their original condition at three frequencies and five temperatures. The results showed that when using different selected equipments (outlyers being removed) the variation on the E* modulus is of the order of 20% at low and medium temperatures and rises to 40% at high temperature. Different laboratories mentioned a large scatter on the results for tire rubber modified bitumen and one of them was unable to make valuable tests for this reason.

The stiffness modulus results for the reference bitumen can be well compared with the values calculated on the basis of Van der Poel's nomogram. Owing to the enormous range covered by the modulus (almost 5 decades) there seems to be a generally good agreement. This is also true for the measured phase angles.

The number of parameters involved is so large and their effect may be such that even if we are aware of all the possible sources of errors and try to limit them, their cumulated effects generally result in a high scatter of the final output. Errors which may be responsible for inaccuracies and lack of reproducibility of rheological binder test results can be classified into different categories, i.e.

- errors due to handling of the binders and sample preparation,
- errors inherent to the type of binder and its evolution,
- errors due to the measuring equipment and transducers,
- errors in the interpretation and data reduction.

The first two categories are often related to storage and reheating. It is well known that some binders can present a long term evolution due to segregation or phase separation even if they are kept in stock at rather low temperatures. Due to this, the binder properties become dependent of the depth of sampling in the binder container. As far as reheating is concerned, the number of melting operations should be kept to a minimum and the thermal history should be controlled as well as possible. Slow heating and carefully homogenization before sampling is recommended.

Sample preparation has been found to play an important role in getting repeatable results. Not only should the procedure and the amount of binder be constant from one experiment to another but also the sample geometry and the centering of the sample in the equipment should be as perfect as possible. Pre-casting the samples in flexible molds, as recommended by SHRP, has proven to be the most efficient way to achieve this goal. This method is also very convenient for molding cylindrical specimen for tension-compression tests.

In order to obtain realistic information on binders in a bituminous mixtures it is now generally considered mandatory to proceed to a preliminary aging procedure. Rolling thin film oven tests RTFOT or thin film oven tests TFOT tests simulate binder hardening during the mixing process while other treatments such as pressurized aging vessel PAV are intended to simulate the long term aging of the binder in service conditions. According to the SHRP this second category of aging procedures is intended for use with the residue of RTFOT or TFOT tests.

Common sources of errors are measuring equipment and transducers. If we want to characterize the rheological properties of a bituminous binder over the full range of service temperatures and loading frequencies we have to deal with complex modulus values covering at least 6 decades (from 1kPa up to over 1000MPa). At low temperature the modulus of bituminous binders is so high that it is possible for the effective stiffness of the specimen to be equal to or greater than the effective stiffness of the drive shaft of the test equipment. This can result in wrong values which become readily apparent if the measured limiting glassy modulus is not approximately 1000MPa. The sample geometry must in any case be adjusted so that its effective stiffness is lower than that of the drive shaft. This requirement holds for any type of rheometer. Furthermore, it is imperative :

- to keep the temperature constant with high accuracy ($\pm 0.1°C$),
- to achieve a homogeneous temperature distribution, avoiding gradients in the sample,
- to calibrate the absolute value of the temperature over the full range covered.

In some vibrational rheometers, the stress is measured in terms of the total force applied to the drive shaft. In such case the inertia of the drive must be included in the equation of motion of the apparatus. Very large errors can also result in the real part of the complex modulus when the frequency comes close to the device resonance frequency. For example, resonance may occur at high frequencies, when the wavelength of the propagated shear wave in the sample approaches the sample dimensions.

Non linear behavior of the material can be the source of unreliable results, especially in the high temperature range. One should therefore perform a linearity test on the binder prior to any extensive measurement program. Such a test might be the measurement of the complex shear modulus G^* at 50°C for increasing stress or strain values. A limitation of the magnitude of stress and strains to be used in the measurements could then be defined in order to achieve a better precision. One should also precisely mention the stress and strains at which each measurement was made.

In order to spot some measuring anomalies data should be evaluated using a Black diagram representation of the complex modulus. It is known that the elimination, of the temperature and frequency in the stiffness modulus - phase angle relationship results in one single curve for conventional binders. Any change in this property indicates either an abnormal measure or a variation of the binder composition and structure.

3   Mix Design

Mix design procedures are very dependent on the type of mixture considered and on the way and conditions how compaction is performed in the field. Asphalt concrete needs a different mix design than other types of mixtures such as Gussasphalt, Stone Mastix Asphalt SMA (or Stone Matrix Asphalt - as it is called in USA), open graded asphalt or overlays etc. Thus, conclusions of this mix design interlaboratory test are restricted to asphalt concrete for surface courses. The optimum of the mix design and the key values are also dependent of traffic, temperature regime, design life and structure of the pavement.

Today, the mix design philosophy in the different countries is still mostly inspired by the AASHTO road test. Hence mix design performed by most laboratories was based on Marshall mix design. Only one laboratory did not use Marshall compaction but gyratory compaction. As SHRP came up with the gyratory compactor for the volumetric mix design, gyratory compaction will probably become more common in future. However, it should be verified by in situ experience which compaction method is the best for which mixture. Mixtures which are not compacted by pneumatic tire roller compactors in the field should probably not be compacted with gyratory. Similar considerations hold for overlays, Gussasphalt and SMA. Question remains of ratio between maximal aggregate size and mold dimensions.

Different ways to deal with limiting values were observed. There are countries, such as Germany which give absolute boundaries for the gradation which are not only true for the target but also for the real gradation. Other countries such as Switzerland give two types of boundaries. The first boundary defines the range where the target values may be fixed, the second boundary indicates the tolerance which is allowed to meet the values. Hence, the first boundary gives the design limits, the second the quality limits.

The concept to perform mix design on the same material, packed and delivered by one laboratory proved to be a good way and is recommended also for similar future work. However, in spite of the fact that the same material was taken as for the OECD-FORCE project, the material characterization of both, the binder and the minerals, was not precise and well enough defined to guarantee that exactly the same material was used. Obviously the OECD-FORCE project did not focus on the characteristics of the individual materials and hence not provide a thorough material description. As this was even done in a research project where a more complete data set was expected than for a "routine" plant this example shows, that in field quality assurance based on a complete documentation including reference material is desperately needed on a routine base to get a full picture in case of problems.

The interlaboratory tests on mix design show that the state of the art of mix design of the different laboratories and the standards their mix design was based on is quite similar. Hence, overall, no big differences between the optimum mixtures of the different participants were observed. In addition, the results generally were very close to the mixture used for the OECD-FORCE project.

For comparison, most of the optimum mixtures were re-evaluated by one laboratory. This showed in some cases considerable differences between the data by the individual laboratory and the re-evaluated data. This demonstrates that a recommendation should be made to provide a better description and definition of test facilities, specimen preparations, testing procedures and calibration procedures to improve reproducibility and to allow a reliable general exchange of information on a given material.

From all the individual mix design optima of the different laboratories it was not possible to evaluate one single absolute "optimal mixture", i.e. to assess which laboratory produced the best mixture over all. To do this, in situ performance of each mixture should have been determined, which, of course, was not possible due to time and financial restrictions and was not the aim of the mix design interlaboratory test.

4 Repeated Loading Testing of Mixtures

4.1 MODULUS AND PHASE ANGLE

Most of the fifteen participating laboratories used bending tests to determine modulus properties. Three laboratories performed shear, two indirect tension and one compression tests.

Complex moduli from bending tests were found to be in good agreement, especially in the medium range of temperatures and frequencies, but the variation may reach 20% at temperatures higher than 20°C. Complex moduli from indirect tensile tests and bending were in good agreement at 0°C and 20°C and at a frequency of 1Hz, whereas complex moduli from shear or compression may differ from bending test results by a factor of two irrespective of the temperature and frequency ranges. There was no evidence for a simple fundamental relation between the different testing modes and loading types, hence theoretical and experimental reasons have to be clarified further.

Phase angle data were in better agreement than complex moduli for all testing procedures and conditions. This holds in particular for all testing modes where phase angles were reported by the different laboratories.

Guidelines for test facilities, specimen preparations, testing procedures and calibration procedures are needed to improve reproducibility and to allow a reliable general exchange of information on a given material. These guidelines should address:

- Material non-linearity: Linearity tests must be recommended to define the maximum acceptable stress and strain levels for complex modulus measurement. To avoid errors due to non-linearity and non-homogenous testing strains should be around $\varepsilon \leq 10^{-5}$ (Doubaneh, 1995).
- Accuracy of force measurements at high temperatures and of displacement measurements at low temperatures: The accuracy of the measuring devices must be in accordance with the range of values to be measured; specimen installation and stiffness of the apparatus must have negligible effect even on high specimen stiffness measurements.
- Specimen dimensions and mechanical modeling.
- Heterogeneity of bituminous mixtures: Determining moduli of a bituminous mixture with great accuracy is difficult or even impossible and no higher accuracy than required by the application should be requested.
- Heating due to repeated loading: amplitudes and number of load cycles should be small to avoid thermodynamic effects on modulus measurement.

The temperature-frequency equivalency principle based on the Arrhenius equation is suitable to calculate the master curve of the complex modulus.

The Black diagram, showing the relation of complex modulus and phase angle, follows a curve which is independent of temperature and frequency and suitable to determine the purely elastic component of the complex modulus at very low temperatures.

4.2 FATIGUE

Most of the fatigue tests were performed in bending test arrangements either on trapezoidal or prismatic specimens. Overall, the fatigue laws derived by the different laboratories using these methods are rather similar for comparable test conditions and the same material.

Expressing the fatigue law by a power function of the number of cycles N with the two parameters K (initial strain or stress at first cycle) and a (slope of the fatigue line in logarithmic scale) showed that variations on the K value entailed large differences in terms of fatigue lives N, whereas the results of slope a were generally in good agreement irrespective of temperature and/or mode of loading provided that a minimum number of samples was tested over a wide range of strain or stress levels.

The number of cycles needed for crack initiation depends on the specimen shape and surface texture; these factors can be responsible for a large part in the interlaboratory variations.

The fatigue of bituminous mixtures is a statistical phenomenon which is influenced by the heterogeneity of the material. Hence, even if the experimental errors are minimized the remaining standard deviation can be considered as an indicator of the material heterogeneity. A sufficient number of specimens is, therefore, required in any experiment to correctly determine a fatigue law.

The influence of temperature on the fatigue law is visible in the tests carried out at 0°C, 10°C and 20°C. While the fatigue laws derived from the aggregated results of different laboratories are almost identical at 0°C and 10°C the results at 20°C lead to much longer fatigue lives. Hence this effect only becomes noticeable for temperatures higher than 10°C.

The viscoelastic character of the materials induces energy dissipation with local temperature increases modifying the material characteristics in an unrealistic way during the fatigue tests. IR-thermography has given evidence for this effect (de La Roche *et al.*, 1996) which can be considered as responsible for the decrease of the specimen stiffness in the first loading cycles. This phenomenon influences directly the failure criterion and the related number of cycles (Di Benedetto *et al.*, 1996).

The values of the |E*| modulus obtained in the fatigue tests at 0°C and 20°C as well as their standard deviations are in agreement with the results for the bending test results of the complex modulus interlaboratory tests. This is an indication of the reliability of these methods for the measurement of the complex modulus.

5   Future Prospects and Activities

The RILEM Technical Committee 152 PBM "Performance of Bituminous materials" is now in charge of the releasing recommendations in the field of binder testing, mix design and relevant dynamic testing procedures. Its main objectives are to use the information provided by the interlaboratory testing programs in order to harmonize the existing test methods in such a way that more accuracy and reproducibility can be achieved in the future. The closeness of the result must be improved by setting up guidelines for the test facilities, specimen preparation, testing conditions and calibration procedures.

5.1   BINDER

From the interlaboratory test on the rheology of binders follows, that an agreement should be made between the builders of rheological binder test devices and the users so that for any combination of rheometer and sample geometry, clear recommendations could be issued on the limits of the different experimental parameters within which reliable and accurate measurements can be made.

In this context it is suggested to deal with the search for a standard reference material with known and traceable properties which compare to bituminous binders. this is necessary to improve the repeatability reproducibility data on asphalt binders and may be used to verify the calibration of rheometers.

Future work by RILEM on binders has to focus on performance based test methods and on comparison of the fundamental tests proposed by SHRP (Harrigan *et al.*, 1994) with other tests, keeping in mind that certain SHRP tests, such as direct tension test, are still under discussion and considering that some characteristics proposed by SHRP, such as $|G^*|\sin(\delta)$ or $|G^*|/\sin(\delta)$, are based on simple assumptions and hence are no universal material properties.

With respect to polymer-modified binders one of the open questions is the range of polymer content where the new methods may be applied. This is especially important for special binders, such as joint sealing materials with almost double polymer content as polymer-modified binders for pavements (Partl *et al.*, 1996). Next question is how aging and durability tests can be further optimized. Another binder related topic is the interaction between binder and aggregate and a better understanding of the filler/binder matrix. Recommendations are needed on testing and characterization methods, some of which are still under development.

So far, activities by RILEM and SHRP concentrated on original binders, i.e. non recovered from a real mix. However, it is well known, that recovered binders from a mix may show properties which are significantly different to the original state. As long term pavement performance LTPP observation studies have also to rely on recovered binder data, the reason for these differences should be further investigated. In particular it should be clarified, how these properties compare to RTFOT and PAV aged binders. The question should be answered by those investigations, whether in addition to RTFOT and PAV a standard mix and binder recovery procedure for the assessment of binder properties should be recommended.

5.2   MIX DESIGN

Future work should incorporate mix design developed by SHRP (Harrigan *et al.*, 1994) consisting of the volumetric mix design gyratory and the so-called level II and level III mix design with all the mechanical tests necessary to do this. It should also include other mixtures than asphalt concrete , e.g. SMA or overlays.

SHRP was specifically designed for the US situation, and due to the short time schedule the test methods and mix design methodology developed in this program still leave many questions open. Nevertheless, it has to be acknowledged that SHRP was certainly a major step in developing and implementing fundamental test methods for bituminous road materials based on performance related mechanical and physical properties. This is certainly the right way to go and also in coincidence with the philosophy of RILEM 152 PBM. However, further work should concentrate on determining these fundamental mechanical and physical properties also with different test methods and not necessarily with the indirect tensile tester or the shear tester proposed by SHRP. This will give the laboratories worldwide the option to keep on working with their own equipment. Besides, this has the great advantage to maintain an innovative variety of testing equipment and the possibility for interlaboratory crosschecks with reduced risk of undetected equipment dependent systematic errors. This is a field for future work of RILEM and has also been addressed in the interlaboratory test on repeated loading.

5.3   MECHANICAL TESTING OF MIXTURES

The RILEM interlaboratory tests on repeated loading reported here are one contribution within the long catalogue of topics in the field of mechanical testing of mixtures still to be covered by worldwide research efforts. Fig. 1 gives a lay-out of a research space which has been designed by RILEM 152 PBM as a basis for future actions in that field. It shows that RILEM will also have to deal with low temperature cracking and permanent deformation and that long term pavement performance (LTPP) studies will have to be involved. To improve international exchange on future research on mechanical testing of mixtures it is recommended that researchers identify the position of their project systematically within this research space, indicating if the research is performed under the aspect of evaluation and comparison of theoretical models (MDL) and methods (TST) for testing, assessment and interpretation of test results (INT) or models and testing with performance in the field (PRF).

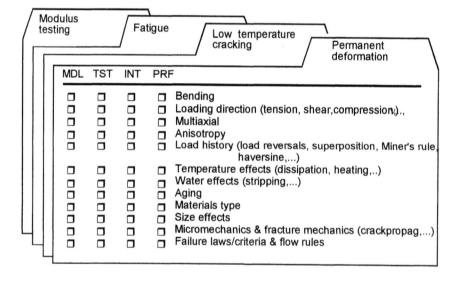

Fig.1.   List of issues for research on mechanical testing. *MDL:* Evaluation and comparison of theoretical models for testing; *TST:* Evaluation and comparison of test methods; *INT:* Assessment and interpretation of test results; *PRF:* Performance, i.e. link of models and testing with in field situation.

One basic reason why only limited correlation between different test methods and test procedures is found today follows from the fact that asphalt concrete is normally assumed to be stressed and heated up homogeneously under constant test conditions ignoring scaling and size effects (Fig. 2). This is not only true for fatigue but also for other mechanical testing. In the present situation mechanical testing is not a priori conducted on the right scale to allow material homogenization and should be done introducing probabilistic and micromechanical models. In addition, as mechanical testing normally is done on a similar scale to the layer thickness, there is a risk that for some types of asphalt layers a homogenization may not be an appropriate design assumption and therefore will only lead to a rough empirical estimate of the behavior.

To overcome this problem and to learn to cover the whole chain of knowledge (Fig. 2), we should compare different test methods and conditions either on a larger scale of test specimens or on a basis of a model material with smaller maximum particle size. In the meantime it is imperative to work with standardized test methods in order to get comparable empirical results. And, of course, there is the need to find the link between a more fundamental testing and the situation on the road. Today, most tests on asphalt pavement materials are conducted on a 1:1 field scale or on a medium sized laboratory scale whereas the intermediate parts of the chain of knowledge are hardly considered.

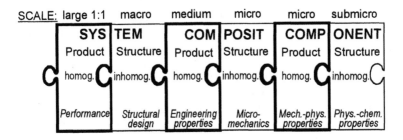

Fig. 2. Chain of knowledge (thick framed elements are comparatively well known).

Another topic to be addressed in future is the influence of aging, water and other substances on the mechanical properties of bituminous mixtures. These durability phenomena are still open to research and discussion and were not part of the RILEM interlaboratory test program. These phenomena are very much influenced by diffusion effects (diffusion velocity etc.), the load history and the contribution of other effects (e.g. loading/crack propagation and oxidation, moisture flow, stripping, densification) and have to be studied on an in field specimen scale. However, working with situations close to reality does only allow qualitative comparisons and - what is even worse- it is slow, because possibilities to accelerate these effects are very limited. Simply using higher temperatures or applying air or water pressure to accelerate the effect may lead to wrong conclusions, because this might activate other effects which are not active in the field. One way to study durability effects would be to perform testing on the smallest possible micromechanical scale but under real temperature and pressure conditions. To understand the behavior in situ, the link between this more fundamental testing and the situation on the road has to be established very carefully.

5.4   COMPLEX MODULUS AND PHASE ANGLE

The results obtained from complex modulus testing indicate that this property may be considered as a physical characteristic which can be determined by reliable procedures. The fundamental knowledge is sufficient to avoid standardization of temperature and frequency conditions. Some recommendations can now be issued to improve repeatability and reproducibility of the tests to such a degree that they could be recommended for material assessment and control. Bending tests appear suitable for this purpose.

However, other testing modes such as shear, tension and compression have to be considered and compared systematically following the work already done by RILEM and also by Tayebali *et al.* (1994) within the framework of SHRP. So far, there is no evidence for a simple relation between the different testing modes. Hence theoretical and experimental relationships and parameters should be further investigated.

The interlaboratory tests demonstrate that the modulus depends on the loading direction and multiaxial stress-strain state (e.g. of the lateral confinement during testing). The test indicate that linear elastic modelization of asphalt with simple models may be not accurate enough and has to be further improved by developing models, which consider inhomogeneities, micromechanics etc.. These models should be general enough to be also valid for different types of mixtures and not only for dense graded asphalt concrete. However, as long as these models are not formulated and validated yet, it is well possible to consider a performance based approach by accepting different problem-oriented loading systems, e.g. using flexural modulus for multilayer analysis stiffness with respect to fatigue, tension modulus to modelize low temperature behavior or shear modulus with dilatancy confinement for high temperature situation.

5.5   FATIGUE

Fatigue is a performance phenomenon which must be considered in combination with the complex modulus to make damage predictions or structural design calculations.

The RILEM interlaboratory test program has revealed that improvements are still needed to allow a better reproducibility between the different test methods and laboratories. Unlike what was concluded for the complex modulus, it is impossible to avoid at this moment the standardization of test procedures and conditions in order to get comparable results and a comparable assessment.

But even under the best possible experimental conditions, the scatter of the results must be considered as unavoidable to some extent. This scatter must rather be taken as the major indicator of the composite character of the material, allowing the application of probabilistic models to predict the development of fatigue cracks.

Although the present state of the art is performance based, it has to be admitted that it is empirical and hence of restricted generality. Even if tests are standardized, a correlation to real behavior can only be expected to give a prediction within the right order of magnitude.

The way to express fatigue as a simple function of load cycles may be dangerous and misleading. It assumes that time has no effect on fatigue behavior and that short and long loading periods have the same effect. There is no theoretical evidence that this can be done. The essential remark to be made is that fatigue is not understood in its physical principles.

In addition, future work on fatigue should consider among other topics:

- *The evolutive nature of fatigue:* During a fatigue experiment at least two processes take place in the asphalt specimens. First during the largest part of the fatigue life the material weakens, but cracks do not exist yet. In a second phase cracks are growing until they join together to finally cause the collapse of the specimen after N_f cycles.

  By loading a structure like asphalt concrete, an inhomogeneous stress/strain distribution occurs with locally very strong/stiff and weak/viscous points and locally different temperatures which are not constant during load application due to stress/strain and temperature redistribution and heating effects. During loading it is inevitable that sooner or later some parts within the material are "overloaded" and microcracks occur which in turn lead to a new stress/strain and temperature distribution. This may result in new local damage but also to some sort of healing and densification, e.g. by closing previous cracks and renewing adhesion and cohesion, which, however, during further loading has only slight effect on the progressing fatigue damage of the whole material.

  By considering only the final number N_f as indicator of fatigue the present traditional approach ignores the details of what really happens in the specimen. The local characteristics of the material and the temperature evolution should therefore be monitored continuously during tests, in order to better understand the fatigue mechanisms and hence to better control them in practical applications.

- *Micromechanics (scaling and size effects, probabilistic modeling):* Tests are usually performed assuming homogeneous stress/strain distribution in the specimens. Hence, more fatigue testing and interpretation should be done using probabilistic and micromechanical models taking into account the composite nature of the material to understand mechanical behavior on a fundamental basis explaining how and why failure occurs and how it is affected by the different material components. This, of course, is different from a view based on global engineering stresses and strains.

- *Influence of aging and durability on fatigue:* Fatigue assumes that no changes occur in the material beyond the damage induced by the applied loads. Model predictions should take into account the simultaneous effects of the different processes in the real world, combining fatigue with other effects such as oxidation, moisture sensitivity, etc..

- *Influence of the loading history (rest periods, healing and heating effects, multiaxial stresses, densification):* Research needs to be done on loading history and on the question how testing can be optimized to separate fatigue phenomena best from other phenomena and to reduce internal heating effects (Di Benedetto *et al.*, 1996; de La Roche *et al.*, 1996).

Fatigue involves either stress or strain induced stress reversals. Simple stress controlled loading and unloading tests, such as indirect tensile tests, lead to a creep and fatigue induced accumulation of permanent deformation and are therefore not suitable to determine pure fatigue properties easily. The phenomenological difference between creep and damage related permanent deformation can be demonstrated by simple mechanical models (Partl, 1991).

Resting periods between loading impulses as e.g. produced by haversine/ recovery cycles in the laboratory or by road traffic are considered favorable to healing and heating effects, as it gives time :

- to close and bond microcracks and to mobilize viscoelastic recovery effects,
- to internal redistribution of stresses/strains, especially by reducing stress/strain concentrations,
- to reduce local heating and hence to create a more equal temperature distribution by temperature flow.

A good characterization of fatigue can be obtained from the homogeneous tests after stabilization of the temperature. However, a fundamental explanation of the rest period effect is necessary, and there is still discussion on how long rest periods in fatigue tests should be to produce a similar effect on the behavior as in the field.

Under traffic, material properties in the pavement are also influenced by multiaxial stresses (e.g. dilatancy effects) and material densification (which is not simulated by simple bending). Further research and improvements of test procedures should be done to simulate and modelize these influences more accurately.

- *Fatigue laws based on dissipated energy approaches* or tensile strain considerations are based on physically plausible considerations. They are, however, not general enough in their present form to deal with the different test methods and to be applied in a general and accurate form for structural pavement design.
- *Interlayer fatigue:* Fatigue behavior is not only to be considered as a material property but also as a system property. So far, research and modeling of interlayer shear and fatigue is only at the very beginning. As these properties are important input parameters for structural design and hence necessary to understand pavement performance, research on this topic should be promoted and intensified.

6   References

de La Roche Ch. and Marsac, P. (1996) : Caractérisation expérimentale de la dissipation thermique dans un enrobé bitumineux sollicité en fatigue, Paper No E&E.4.053, *Eurasphalt & Eurobitume Congress*, Strasbourg, France.

Di Benedetto H., Soltani A. and Chaverot, P. (1996) : Fatigue Damage for Bituminous Mixtures: A Pertinent Approach", *AAPT Journal* No. 65.

Doubaneh E. (1995) : Comportement mécanique des enrobés bitumineux en "petites" et "moyennes" déformations, *Thèse de Doctorat ENTPE-INSA*.

Eustacchio E., Francken L. and Fritz H.W. (1994) : Mix design and repeated loading tests, *Proceedings of the conference "Road safety in Europe and Strategic Highway Research program"* Lille, France.

Fritz H.W. and Eustacchio E. (1991) : Mechanical tests for bituminous Mixes, *Proceedings of the 4th RILEM International Symposium* Budapest, Hungary, Vol 2.

Francken L. and TC152PBM (1995) : RILEM interlaboratory Test Programme on Bituminous Binders. Progress Report and First Conclusions, *First European Workshop on the Rheology of Bituminous Binders*, Brussels.

Francken L., Hopman P., Partl M. N., de La Roche, C (1996): RILEM Interlaboratory Tests on Bituminous Mixes in Repeated Loading. Paper No E&E.4.099, *Eurasphalt & Eurobitume Congress*, Strasbourg, France.

Harrigan E.T., Leahy R.B. and Youtcheff J.S. (editors) (1994) :The Superpave Mix Design System Manual of Specifications, Test Methods, and Practices, *Strategic Highway Research Program, NRC,* SHRP-A-379.

OECD-FORCE full-scale pavement test (1991) *OECD-report*.

Partl M.N. (1991) : Caractérisation et modélisation du fluage des matériaux bitumineux soumis à des charges de traction/compression. *Comptes Rendus Journée Technique* LAVOC-EPFL, Lausanne, Switzerland.

Partl M.N. and Hean S. (1996) **:** Evaluation of Mechanical Tests for Polymer-Bitumen. Paper No E&E.5.147, *Eurasphalt & Eurobitume Congress*, Strasbourg, France.

Tayebali A.A., Tsai B. and Monismith C. L. (1994) : Stiffness of Asphalt -Aggregate Mixes, *Strategic Highway Research Program, NRC*, SHRP-A-388.

Younger K.D., Partl M.N., Fritz H.W. and Gubler R. (1997) : Asphalt Concrete Shear Testing With the Co-Axial Shear Tester at EMPA, *RILEM- MTBM97*, Lyon.

APPENDICES

Appendix 1

Description of non-standardised test procedures used for characterisation of polymer modified bitumens

U. Isacsson and X. Lu

1.1 Compatibility/storage stability test

1.1.1 TUBE TESTS

A. An aluminium foil tube is filled with polymer modified bitumen and placed in an upright position in a 180 °C oven for 14 days, after which it is allowed to cool at ambient temperature. The foil is removed from the specimen, which is laid on its side on a glass plate and put into an oven at 50 °C. The sample is then visually inspected for any non-uniform slump or flow from end to end at intervals over a 24 h period. If the polymer and bitumen are incompatible, the polymer will migrate to the top of the tube while it is in the 180 °C oven, which will cause viscosity differences over the length of the tube and result in visually detectable differences in flow of the material. If the polymer and bitumen are completely compatible, the flow will be uniform (Muncy et al., 1987).

B. An aluminium foil tube (3 cm diameter and 16 cm height) is filled with approximately 75 g polymer modified bitumen. After closing the tube without air enclosures at a temperature between 50 °C and 80 °C, the tube is stored vertically at 180 °C for 3 days, and then cooled to at least below 30 °C. The aluminium foil is then peeled off and the PMB specimen cut horizontally into three equal sections. The top and bottom sections are used to evaluate the compatibility/storage stability of PMB by testing their softening point (Breuer, 1988c) and/or penetration.

C. Several variants of these two test procedures are used, two of which are described in references (Berenguer et al., 1989; Fetz et al., 1985).

1.1.2 UV FLUORESCENT MICROSCOPY (Dong, 1989; Grimm, 1989; Lomi et al., 1989)

This method is used to observe the compatibility and appearance of the dispersion in polymer modified bitumens. At microscopic level, corresponding to enlargements ranging from 500 to 1000, most of the polymer-bitumen binders observed exhibit a

Bituminous Binders and Mixes, edited by L. Francken. RILEM Report 17. Published in 1998 by E & FN Spon, 11 New Fetter Lane, London EC4P 4EE, UK. ISBN 0 419 22870 5

pronounced heterogeneity. Among the polymer-bitumen mixtures, a distinction is made between (1) binders whose continuous phase is a polymer matrix in which the bitumen globules are dispersed, and (2) binders whose continuous phase is a bitumen matrix in which the polymer particles are dispersed. The characteristics of PMB, such as the particle size distribution and rate of swelling of the polymer, can be evaluated by means of image analysis including mathematical treatment. This method is also used to observe the micromorphological changes that occur due to thermic effects.

1.1.3   CRUSHING TEST (Thyrion et al., 1991)

A disc of a bitumen-polymer blend, 20.5 mm in diameter and 2 mm thick, is placed between sheets of absorbent filter paper and placed under a weight of 100 g in an oven at 135 °C for 15 hours. The deformation of the pellet is recorded and the stability of the mix evaluated by observing the oil migration.

1.2   Rheological tests

1.2.1   FLOW BEHAVIOUR TESTS

A. Flow test acc. to Thompson and Hagman (1958). Specimens, 6 cm x 4 cm x 0.32 cm, are casted in a stainless steel or brass form on a rigid stainless steel panel using a release agent on the form. The assembly is conditioned for 1 hour at room temperature, the form removed and the sample on the panel conditioned further in a water bath for 1 hour at 25 °C . The specimen on the panel is then exposed for 1 hour at 38 °C at an angle of 75° relative the horizontal. The change in length of the specimen in centimetres is reported as flow.

B. Apparent viscosity using sliding plate rheometer (Denning and Carswell, 1981). Apparent viscosity is measured using the sliding-plate rheometer in which a sample of binder 30 x 20 mm and 3-10 mm thick is sheared between two aluminium plates. Apparent viscosity is the ratio of shear stress to shear rate.

C. Absolute dynamic viscosity using coaxial cylinders (Sybilski, 1993). In this test, viscosity is measured at increasing shear rates (shear stresses) using a viscometer with coaxial cylinders. Because of the non-Newtonian, pseudoplastic behaviour of polymer modified bitumens, the decrease in viscosity with increasing shear rate is observed. Absolute (zero-shear rate) viscosity is calculated according to Cross equation:

$$\frac{\eta_0 - \eta}{\eta - \eta_\infty} = \left(K\dot{\gamma}\right)^m$$

or its simplified form:
$$\frac{\eta_0 - \eta}{\eta} = (K\dot{\gamma})^m$$

where η - dynamic viscosity, γ - shear rate, η_0 and η_∞ - dynamic viscosity in the first and second Newtonian range, respectively at very low and high shear rate, K and m - constant parameters.

D. Double ball softening point test (Texas Transportation Institute, 1983). The test is a modified version of the ASTM Ring and Ball Softening Point test. The apparatus consists of two 9.5 mm diameter stainless ball bearings cemented together with the test material. One of the ball bearings is fixed to the ring holder of a standard ring and ball assembly, while the other ball is suspended from the first by the test material. Heat is applied to the immersed assembly in the manner described by ASTM D 36. As temperature rises in the apparatus, the weight of the lower ball begins to stretch the specimen. Double ball softening point is recorded as the temperature in the bath between the specimens as each suspended ball reaches the bottom plate of the assembly.

E. Dropping ball test (Harders, 1988). 8 g of the binder is poured into a container with a diameter of 54 mm. Using a distance plate, a steel ball of 55 g weight is partly immersed into the binder. After adjustment to a temperature of 15 °C below softening point for 1 hour, the device is turned upside-down and the ball allowed to fall. The time taken by the ball to fall 8 mm is recorded as t_1, and the time to fall the total distance of 30 cm is recorded as t_2. The ratio of t_2/t_1 is calculated. Normal road grade bitumens have values slightly over 1 and polymer modified binders mostly give values over 2.

1.2.2   ELASTIC RECOVERY TESTS

A. Elastic recovery using ductilometer. The brass plate, mould and briquet specimen are prepared according to standardised test procedures such as ASTM D 113 or DIN 52013. Several variants of the procedure for estimating elastic recovery properties have been described below.

German procedure (Breuer, 1988b). The specimen is elongated to 20 cm at 25 °C, after which the thread is cut in half using a pair of scissors. The elastic recovery is defined as the distance between the two "half-threads" after 30 min in % of initial elongation (20 cm).

Procedure acc. to Muncy et al. (1987). The ductilometer and the specimen are conditioned at 10 °C for 85-95 min. The specimen is elongated to 20 cm and held for 5 min, after which it is cut in half and left undisturbed for 60 min. The percentage elongation recovery is calculated as described above (German procedure).

Procedure acc. to Öster et al. (1989). The bituminous sample is cast in a ductility mould and left at ambient temperature for 30 min, after which it is transferred to a water bath where it is conditioned for 30 min at 13 °C. Excessive sample is cut off and the sample is conditioned for another 30 min. The sample is inserted in the ductility apparatus and elongated to 50 cm. After cutting in half, the sample is left undisturbed for 30 min and the distance between the threads is measured. The elongation recovery is calculated as a percentage.

B. Elastic recovery using sliding plate rheometer (Tosh et al., 1992). In this test, the Sliding Plate Rheometer developed by Shell is used (compare method 2.1 B above, "Apparent viscosity"). A sandwich of the binder is prepared between aluminium plates (20 mm x 30 mm x 3 mm). One of the two plates is fixed in the apparatus and the other is subjected to shear stress by placing it on a spindle of known but adjustable stress weight connected to a displacement transducer. The shear stress is reduced after a pre-selected amount of movement. At the test temperature (from -10 °C to 30 °C) the weight applied and the selected sensitivity are such that a 50 % chart deflection is obtained within a timescale permitting manual removal of the weight on the spindle.

C. Elastic recovery using the ARRB elastometer (Oliver et al., 1988). In this test, the annular sample is first sheared between an outer, fixed cylinder and an inner moving cylinder. The level of strain is determined by the annular gap (sample film thickness) and the displacement of the inner cylinder. The rate of strain is determined by the annular gap and the speed of the inner cylinder. The moving cylinder is pulled upwards by a motor drive assembly at a selected speed. The force required to deform the sample is measured by a load cell. When the selected displacement has been reached, the drive disengages so that there are no external forces acting on the sample. The proportion of the strain that is recovered (% elastic recovery) is measured using a displacement transducer. The "time under stress" and the "recovery phase" can be obtained from the elastomer chart output.

D. Elastic recovery using the controlled stress rheometer (Jøgensen, 1988). The Controlled Stress Rheometer is an effective tool for studying creep and elastic properties of PMBs. With the cone and plate sensor geometry, the volume of samples is less than 0.5 ml. The electronical temperature control (5 °C - 90 °C) allows quick temperature setting. The temperature range can be expanded (-100 °C to 300 °C) with optional units. In the creep test, a constant force is applied to the sample and the resulting strain is monitored. The elastic recovery is the ratio of recovered strain to maximum strain. The rheometer makes it possible to measure elastic recovery at different temperatures, stresses, strain levels and times under stress.

E. Elastic recovery using Höppler consistometer (Svetel, 1985). A specimen, height 1 cm, is loaded at 0 °C using a pressure of 0.1 MPa, for 60 s. After unloading, the part of the deformation recovered after 60 s is measured. Testing can also be performed at other temperatures (-40 °C - +25 °C) and pressures (0.025 MPa - 1.5 MPa).

F. Torsional recovery test (Thompson and Hagman, 1958). A bolt and disc assembly is supported in an a shallow container. The melted binder is poured into the container until it is flush with the surface of the disc. The assembly is conditioned at room temperature for a minimum of 2 h, after which the disc is rotated at a steady rate through an arc of 180° by means of a wrench and locknuts on the bolt. The disc is then released immediately and the angle of the recovery measured after 30 seconds and 30 minutes. The percentage values for torsional recovery are recorded.

1.2.3 TENSILE TESTS

A. Toughness and tenacity test (Benson, 1955). 36g bitumen heated to 190 °C is poured into a metal cup. All air bubbles are removed from the sample. The tension head, which consists of a polished metal hemisphere having a radius of 11.1 mm (7/16 inches) and a threaded stem for attachment to the testing machine head, is embedded in hot molten sample to exactly 11.1 mm and centred by means of a suitable spider mounted on the cup. After cooling, the head is pulled from the bitumen at a rate of 508 mm/min (20 inches/min) and the load deformation (stress - strain) curve is plotted. The entire area under the stress-strain curve is considered as the toughness of the bitumen, and the area under the long-pull portion of the curve is considered as the tenacity.

B. Toughness and tenacity test - modified method (Thompson and Hagman, 1958). An assembly is prepared in which a 3/4 inch diameter steel hemispherical head on a threaded stem is completely embedded in the surface of the sample. After conditioning at 25 °C for at least 12 h, the head is pulled from the sample at a rate of 305 mm/min (12 inches/min). The force - elongation curve is recorded and values for toughness and tenacity are measured.

C. Toughness and tenacity test acc. to Boussad et al. (1988). 50 g of binder (160 °C) is poured into a steel container, the holder and the cover equipped with a chromium-plated ball are fitted to the container, and the system is allowed to cool. The filled system should be preheated at 160 °C before use in order to ensure good adhesion to every part and uniform cooling with no stresses. The sphere is then pulled out of the specimen at a speed of 500 mm/min, and the force is recorded against time/elongation. Based on recorded curves, the maximum strength and the toughness (the total energy required to dislodge the sphere) are obtained. The tenacity is calculated as follows:

$$\text{Tenacity} = \text{Toughness} - 2 \times \text{Energy at Maximum Strength}$$

D. Extraction test (Woodside and Lynch, 1989). Binder is poured onto a metal plate (20 x 20 cm) and allowed to form a film with a thickness of 1.5 mm. Circular limestone discs (40 mm in diameter, 6 mm in thickness) are pressed evenly into the hot binder about 25 mm apart and the sample is cured in an air blown oven at 60 °C for 24 hours. The sample is then readjusted to room temperature and clamped to the base of an Instron Load Testing device. Hooks are centred and secured to the elevating crosshead by chain and extraction begins at 100 mm/min under constant load. The maximum load to failure and nature of bond during failure are recorded on a chart. For each extraction, the adhesion character is visually observed.

E. Force ductility test (D-4 P 226). A brass mould with both straight sides is filled with sample and allowed to reach room temperature for 30 min, after which it is placed in a water bath at 4 °C for 30 min. After trimming excess sample, the specimen is kept in the 4 °C water bath for another 90 min. The briquet is then removed from the brass plate and the side pieces detached. The ring at each end of the clips is immediately attached to the pins in the testing machine and the force measuring apparatus, respectively. The two clips are pulled apart at a speed of 5 cm/min until the briquet is elongated to rupture or the force remains constant at zero. The force-elongation curve is recorded and the maximum force in pounds and the elongation in centimetres at which the maximum force occurred are measured. The tensile stress may be calculated by dividing the force required to elongate the briquet by the initial cross-sectional area of the sample. The area under the force-elongation curve represents the work required to elongate the briquet. Variants of this test procedure have been described (Chaverot, 1989; Verburg and Molenaar, 1991), the main difference being the type of specimen mould used.

F. Direct tension test acc. to Anderson and Dongre (1993). In this test method, a dog-bone specimen geometry is used (6 mm thick and 100 mm in overall length, including the end inserts). A special gripping technique and elongation measuring system are employed. The gripping method utilises plastic end inserts and specially designed loading pins. The elongation is measured using a laser-based noncontact extensometer that provides accurate and repeatable measurements. From the stress-strain curve, the stress or strain at failure as well as the energy to failure (the area under the stress-strain curve) may be determined. This test equipment has been developed in the Strategic Highway Research Program (SHRP).

1.2.4   STATIC MECHANICAL TESTS

A. Stiffness modulus using the sliding plate rheometer (Tosh et al., 1992). Two aluminium plates are correctly positioned in a holder using a removable 3 mm spacer bar (compare test method 2.1 B above "Apparent viscosity"). Silicone release paper is placed around the gap in the plates to prevent the hot binder leaking out. The set-up is preheated in a 160 °C oven. The binder is then heated, with stirring, to 160 °C - 170 °C and the sandwich filled to excess. After cooling, the excess is trimmed off and the release paper removed. The sample is clamped into the apparatus and conditioned at the test temperature in a water bath for 30 minutes. The deformation of the sample under shear is measured and the stiffness modulus at a certain loading time t and temperature is calculated as the ratio of stress to strain.

B. Stiffness modulus using the bending beam rheometer (SHRP Method B-002, 1993). About 30 g of sample is used to prepare a beam 125 mm long, 12.5 mm wide and 6.25 mm thick. The beam is conditioned at the test temperature for 60 min and then placed on the supports and centred visually. The beam together with the supports and the lower part of the test frame is submerged in a constant-temperature bath, which controls the test temperature. A loading shaft is allowed to contact the beam. After a seating step, a constant load is applied to the beam at its midpoint. Creep stiffness and the slope of the log stiffness against the log time curve are measured at several loading times ranging from 8 to 240 s. The device is useful for measuring moduli ranging from 30 MPa to 3 GPa. These moduli typically occur in the temperature range from -40°C to 25°C. This test equipment has been developed in the Strategic Highway Research Program (SHRP).

1.2.5 DYNAMIC MECHANICAL TEST

In this type of test, the sample is subjected to an oscillatory strain or stress. Viscous and elastic parameters such as storage modulus (G_1), loss modulus (G_2), complex modulus (G^*), complex dynamic viscosity (η^*) and phase angle can be measured simultaneously. The test may be performed using frequency or temperature sweep. Different test geometries, e.g. parallel plate, cone and plate, couette, cup and plate and torsion rectangular are employed.

A. Test procedure acc. to Jovanovic et al. (1993). Two different geometries are used depending on the test temperature. For tests in the range -80 °C to 20 °C, the specimen is prepared by pouring bitumen into a metal mould 63.5 x 12.7 x 2 mm, after which it is subjected to dynamic torsion. The test temperature is varied in 10 °C increments. At each temperature, the sample is tested at frequencies ranging from 0.01 rad/sec to 100 rad/sec. Over the temperature range 30 °C to 120 °C, measurements are performed using dynamic shear mode between parallel plates (25 mm diameter, 1 mm gap). Master curves of complex dynamic viscosity or storage modulus versus reduced frequency are presented.

B. Test procedure acc. to De Ferrariis et al. (1993). Tests are run in the temperature range 20°C to 70°C using the Rheometric Asphalt Analyser. Strains are kept small at lower temperatures ($< 0.5\%$) and increased at higher temperatures, but always in the linear viscoelastic region. Frequencies from 0.1 rad/sec (0.016 Hz) to 10 rad/sec (1.6 Hz) are swept for each test temperature in five equal steps per decade. Results are analysed by investigating loss tangents as a function of temperature.

C. Test procedure acc. to Cavaliere et al. (1993). Tests are operated in torsion using a parallel plate geometry (8 mm diameter) in the temperature range -30 °C to 80 °C and frequency range 0.01 Hz to 10 Hz. The gap height varies between 5 and 6 mm. Test samples are glued to prevent slippage. A series of master curves of storage modulus against frequency are obtained on the basis of time-temperature superposition principle.

D. Test method using dynamic mechanical thermal analyser (Khalid and Davies, 1993). The Dynamic Mechanical Thermal Analyser (DMTA), which is capable of applying a sinusoidal stress to a sample in bending, shear or tensile modes, is used to determine the modulus and phase angle as functions of temperature and/or frequency. The test temperature range is -40 °C to 50 °C, and the frequency from 0.3 Hz to 200 Hz. The phase angle has been shown as a useful factor in describing the temperature susceptibility of binders.

E. Test using controlled stress rheometer (Jørgensen, 1993). This equipment, which is described above ("Elastic recovery"), can also be used for dynamic mechanical tests. The sample is subjected to a sinusoidal shear stress and different visco-elastic parameters are measured from the resulting strain.

F. Test method based on the balance rheometer (Kolb, 1985). The construction of the Balance Rheometer is based on the parallel viscometer where the parallel plates are replaced by a hemispherical jug and a hemispherical ball. The jug rotates with a given frequency around the axis. The viscoelastic parameters are measured at temperatures ranging from 5 °C to 175 °C and at circular frequencies from 78.5 sec^{-1} to 0.0196 sec^{-1}. The sample is submitted to a sinusoidal strain and the momentums resulting from shear stresses are measured using a balance.

G. Test method using dynamic shear rheometer (SHRP Method B-003, 1993). In this test, a strain or stress controlled dynamic shear rheometer is used to measure the linear viscoelastic moduli of binders in the sinusoidal mode. Measurements may be performed at different temperatures, strain and stress levels, and frequencies using various geometries (mentioned above). For specification (cf. Section 5), testing is performed at temperatures ranging from approximately 45 to 75°C for unaged and TFOT aged binders and 5 to 35°C for the Pressure Ageing Vessel (PAV) residue. Actual test temperatures and stress/strain levels used depend on the stiffness of binders. Complex modulus (G^*) and phase angle (δ) are determined at frequency of 10 rad/sec. The values of $G^*/\sin\delta$ or $G^*\sin\delta$ are presented. This test procedure has been developed in the Strategic Highway Research Program (SHRP).

1.3   Adhesion/cohesion tests

1.3.1   VIALIT TEST (Kin, Muncy and Prudhomme, 1986)

1 kg/m² of the binder to be tested is spread on a 200x200 mm plate. The binder is covered with 100 chips of aggregate which are rolled six times with a rubber cylinder, three times in each perpendicular direction. The gravelled plates are maintained at

room temperature and 100 per cent humidity (in water). The tests are carried out 20 minutes after rolling three plates in series. The plates are then inverted onto a tripod and a steel ball weighing 500g is dropped from a height of 50cm. The value of the adhesion is expressed as

$$100 - (a + d) = b + c - d$$

where

a = the numbers of chips which fall and are not coated at all with binder;
b = the number of chips which fall and are partially coated with the binder;
c = the number of chips which adhere;
d = the sum of b and c.

1.3.2 BRITTLENESS TEMPERATURE (Boussad, Muller and Touzard, 1988)

Test plates are prepared as described above for Vialit Test and stored for 14 days in a 50 °C oven. The plates are then conditioned overnight before the test at a temperature that should allow a retention of at least 90 % aggregates, usually 10°C. Plates are subjected 3 times in a row within 10 seconds to the shock of a falling steel ball (500g), and then immediately reloaded into the oven. Losses are recorded. The test is repeated on the same plate decreasing stepwise in temperature by 5 °C. 90 minutes are allowed to reach temperature equilibrium. When losses are over 90 % the test is completed. The brittleness temperature is defined as the temperature at which 50 % of the chips are lost.

1.3.3 DROPPING TEMPERATURE (Breuer, 1988a)

This is also a modified Vialit Test. The binder film thickness is 1.4 mm. Before testing, the binder is stored on the plates for 24 h at 25 °C and another 24 h at 40 °C. The dropping temperature is defined as the temperature at which 10 % of the chips are lost after three blows of an iron ball (compare Vialit Test above). Two different dropping temperatures are defined, the Cold Dropping Temperature (KST) and Heat Dropping Temperature (WST), respectively. The two dropping temperatures, which are determined by increasing (KST) or decreasing (WST) the temperature in steps of 5 °C, can be used to estimate the plasticity, cohesion and adhesion of bituminous binders.

1.3.4 CONTRACTION TEST (Molenaar, 1991)

Stone tablets measuring 5 x 5 cm are polished and covered with a thin film of bitumen (0.5 mm). The tablets are placed in 98 °C water. The contraction of the film is observed visually. The results show that the polymer modified bitumens contract less in comparison with conventional bitumens. However, the contraction of a coarsely dispersed two-phase polymer modified bitumen is similar to that of conventional bitumen.

1.3.5 COHESION TEST USING THE VIALIT PENDULUM RAM (Bononi, 1988)

The principle of this test is to measure the energy absorbed by the fracturing of a binder film under a given impact by a VIALIT pendulum ram. The binder is poured onto a steel cube equipped with spacers such that the thickness of the film is 1 mm and the area 1 cm². The surface of the cube is grooved in order to obtain good mechanical adhesion between binder and steel. The cube with sample is pressed against a grooved surface of another steel cube, put into the test rig and allowed to reach the testing temperature. The cube with the spaces is impacted using a pendulum ram, causing rupture in the binder film. The maximum swing of the pendulum, which is a measure of the cohesion of the binder, is recorded. The test is performed at different temperatures.

1.3.6 BOILING WATER STRIPPING TEST (Belgian Road Research Centre, 1991)

Coated aggregate is subjected to stripping in boiling water under specified conditions of duration and temperature using a simple device in which no local overheating can occur. The proportion of exposed aggregate surface is then evaluated through chemical attack. The stripping ratio is determined with reference to a calibration curve, and decreases as binder-aggregate adhesion increases. The reagent used in this test is hydrochloric acid for calcareous aggregates and hydrofluoric acid for silicocalcareous or siliceous aggregates.

1.4 Ageing tests

1.4.1 PRESSURE AGEING VESSEL TEST (SHRP Method B-005, 1993)

The Pressure Ageing Vessel (PAV) is used to simulate long-term ageing in the field. The test accounts for temperature effects but is not intended to account for mix variables such as air voids, aggregate type and aggregate adsorption. The procedure is performed on bituminous binders after TFOT (ASTM D 1754) or RTFOT (ASTM D 2872). If the TFOT is used, the bitumen pans used in this test are directly transferred to the PAV. If the RTFOT procedure is used, the bitumen samples from at least 2 bottles are combined and mixed to produce one homogeneous sample, 50 g of which is poured in a PAV ageing pan (the same pan as used in TFOT). The vessel is heated to the test temperature in an oven. The pans containing samples are placed in a holder and put inside the hot vessel. An air pressure of 2.07 MPa is applied to the vessel. A test temperature is chosen according to the climate in which the tested bitumen will be used. The pressure and temperature are kept constant for 20 hours. The vessel is removed from the oven and the bitumen pans are heated in a 135 °C oven for 30 min to allow air bubbles to escape. The samples are poured into storage containers for other tests. This test equipment has been developed in the Strategic Highway Research Program (SHRP).

1.4.2 AGEING TEST ACC. TO CRR (Choquet and Verhasselt, 1993)

An accelerated ageing apparatus has been developed for this test. The device is placed in an oven at 85°C. In the test, 500g sample is poured into a stainless steel cylinder 300 mm in length and 130 mm in diameter, enclosed between two welded discs. A stainless steel roller 296 mm in length and 34 mm in diameter is inserted into the cylinder and rotated to distribute the binder into a 2 mm thick film on the inner wall of the cylinder. The cylinder is rotated at 1 r/min for 144 h with a constant oxygen flow of 4.5 l/h. Aged binder is tested by penetration, ring & ball temperature and generic composition.

1.5   Methods of chemical analysis

1.5.1 SPECTROSCOPIC METHODS

A. Infrared spectroscopy (Fifield et al., 1990). The principle of this method is that absorption of electromagnetic radiation in the infrared region results in changes in the vibration energy of molecules. Fourier transform spectrometer or double-beam spectrophotometer incorporating prism or grating monochromator, thermal or photo detector and alkali halide cells are employed in the test. Infrared spectra are recorded as the transmittance (%) against the wavelength (μm) and/or wavenumber (cm^{-1}). The method is widely used in the identification and structural analysis of organic materials. In this case, infrared spectra are used in an empirical manner by comparison of samples with known materials and by reference to charts of group frequencies. IR is also applied in quantitative analysis, which is based on infrared absorption obeying the Beer-Lambert Law, as expressed in the following form:

$$\log (I_0/I) = A = \varepsilon Cl,$$

where $\log (I_0/I)$ is defined as the absorbance A, I_0 and I are the incident and transmitted intensities respectively, l and C are the thickness and concentration of the absorbing medium respectively, and ε is a constant known as the molar absorptivity.

The quantitative measurements are preceded by the development of a calibration curve. For this purpose, the same cell should be used for samples and standards. The applications of the IR method to polymer modified bitumens include determination of polymer content and the functional groups (Choquet et al., 1991, 1992; Fifield et al., 1990; He et al., 1991; Little1987; Öster et al., 1989), prediction of the PMB compatibility (SHRP A-004, 1990) and investigation of the dissolution of polymer in bitumens (Jovanovic, et al., 1991).

B. Nuclear magnetic resonance spectroscopy (Fifield et al., 1990). Nuclear Magnetic Resonance Spectroscopy (NMR) measurement is based on the fact that absorption of electromagnetic radiation in the radio-frequency region results in changes in the orientation of spinning nuclei in a magnetic field. The nuclei which can give rise to an NMR spectrum should have non-zero *spin quantum numbers such* as 1 and 1/2. These nuclei include ^1H, ^{13}C, ^{31}P, ^{15}N and ^{19}F etc. An NMR spectrometer consists of a powerful and highly homogeneous electromagnet, a radio-frequency signal generator and detector circuit, an electronic integrator and glass sample tubes. The NMR spectra may be obtained by scanning the magnetic field at a fixed frequency (usually 60 to 100 MHz), or by scanning the operating frequency of the transmitter at a fixed magnetic field. The technique can be used in the identification and structural analysis of organic materials and in the study of kinetic effects. It is also useful for quantitative analysis but not widely applied. Since ^1H has the highest relative sensitivity and ^{12}C and ^{16}O are inactive (having *spin quantum numbers* of zero), ^1H NMR or Proton Magnetic Resonance Spectroscopy (PMR), has become one of the most useful techniques in the qualitative analysis of organic compounds. Interpretation of PMR spectra is accomplished by comparison with reference spectra and reference to chemical shift tables. Concerning its applications to bitumen, NMR method has been reported to estimate saturate/aromatic carbon ratios, measure the content of carboxylic acids, phenol and benzylic protons (Giavarini et al., 1989; Goodrich et al., 1986), and determine chemical generic compositions (Santagata et al., 1993).

1.5.2 CHROMATOGRAPHIC METHODS

A. Gas chromatography (Fifield et al., 1990). In Gas Chromatography (GC) samples are introduced into a gas flow (mobile phase) via an injection port located at the top of a column containing a stationary phase. The continuous flow of gas elutes the components from the column in order of increasing distribution ratio. The eluted solutes are monitored by a detector (e.g. ionization detector) and chromatograms are obtained using a recording system. The mobile phase is usually chosen as helium, nitrogen, or hydrogen. The stationary phase may be an active solid material (e.g. silica, synthetic zeolites) or a high boiling point liquid (e.g. squalane, polyethylene) which has been coated or immobilised on an inert supporting material. Identification of the component peaks of a chromatogram can be achieved in two different ways: comparison of retention volumes and trapping the eluted components for further analysis, respectively. In the latter case other analytical techniques such as infrared and mass spectrometry are used. The integrated area of a peak is directly proportional to the amount of solute eluted. Quantitative analysis is performed by internal normalisation, internal standardisation or standard addition. In GC, the samples to be analysed must be volatile and thermally stable at the operating temperatures. Gas Chromatography is relatively rapid and simple, and very useful for the analysis of organic mixtures. It is also suited for characterising the volatile components in bituminous binders (Ruud, 1989).

B. High pressure liquid chromatography (Fifield, 1990). In High Pressure Liquid Chromatography (HPLC) the sample is injected into a pressurised flow of liquid mobile phase either by syringe or valve injector. The mobile phase with sample passes through the column containing a stationary solid. Components migrate through the column at different rates due to differences in solubility, adsorption, size or charge. The eluted components are monitored by a detector (e.g. fluorimetric detector) and chromatograms are obtained using a recording system. Methods for qualitative and quantitative analysis are similar to those used in gas chromatography. In HPLC, unmodified or chemically modified, microparticulate silicas are normally used as the stationary phase. The chemical composition of the mobile phase should be selected appropriately. Optimum retention and resolution are often achieved by using a mixture of two solvents. If sample components vary widely in polarity, gradient elution can be employed. In normal phase separations (polar stationary phase/ non-polar mobile phase), pentane or hexane with dichloromethane, chloroform or an alcohol are frequently used. In reverse phase separations, the most widely-used mobile phases are mixtures of aqueous buffers with methanol or water with acetonitrile. HPLC has been used commonly for the separation of non-volatile substance including ionic and polymeric samples. The content of SBS in the polymer modified bitumen can be determined using HPLC (Neubauer, 1988).

C. Thin layer chromatography with flame ionization detector (Sherma et al., 1991). In Thin Layer Chromatography using Flame Ionazation Detector (TLC-FID) the sample to be analysed is dissolved in a solvent and spotted at one end of Chromarods (quartz rods coated with a thin layer of sintered silica or alumina). The rods are then developed with suitable solvent(s) as in conventional TLC, after which the solvent is removed by heating. The rods are scanned at a chosen speed through a hydrogen flame of the FID. In this process, the fractions separated on the rod are successively vaporized/pyrolyzed. The ionizable carbon is converted to ions which are detected using a collector electrode. The FID signals from each fraction are amplified and recorded as separate peaks. Quantitative calculation is carried out by area normalisation, internal standards or empirical calibration method. In this test proper choice of the operating variables, both in chromatography and scanning, is crucial for satisfactory sensitivity and reproducibility in quantitative analysis. The TLC-FID technique has found numerous applications for a wide variety of substances. It is suitable for the determination of the generic composition of bitumens. For polymer modified bitumens, the influence of bitumen composition on polymer swelling, the generic composition of the phase of the produced binder and polymer content may be investigated by means of TLC-FID (Brule et al., 1988; Torres et al., 1993).

D. Gel permeation chromatography (Fifield et al., 1990). Molecules that differ in size can be separated by passing the sample through a stationary phase consisting of porous cross-linked polymeric gel. The pores of the gel exclude molecules greater than a certain critical size while smaller molecules can permeate the gel structure by diffusion. The process is described as Gel Permeation Chromatography (GPC). Gels used for the stationary phase can be hydrophilic for separations in aqueous and other polar solvents or hydrophobic for use in non-polar or weakly-polar solvents. In GPC separation, the smaller molecules are eluted at rates dependent upon their degree of permeation into the gel. The sample components are therefore eluted in order of decreasing size or molecular weight. The components are detected using refractive index or UV absorption or by collecting and analysing separate fractions. The obtained chromatogram represents the relative amount of eluted components appearing at a given elution time. The method is applied in the separation of high-molecular weight material and also in the determination of molecular weight. The molecular mass distribution of components, polymer content in PMBs, and the degradation of polymer in the process of PMB ageing can be studied by GPC (Johansson et al., 1991; Molenaar, 1991).

1.6   References

See chapter 1.

Appendix 2

Inventory of mix design methods

M Luminari and A Fidato

Bituminous Binders and Mixes, edited by L. Francken. RILEM Report 17. Published in 1998 by E & FN Spon, 11 New Fetter Lane, London EC4P 4EE, UK. ISBN 0 419 22870 5

2.1   Australia

2.1.1   INTRODUCTION [AUS-1,2,3,6,7,10,11]

Since the late 1980s one of the main aims of Australian pavement research has been to move away from the past empirical approach and to develop practical and affordable equipment in order to achieve a better understanding of those fundamental properties of bituminous mixtures which can be useful in developing a performance-related mix design method.

The mix design method currently recommended by the Australian Road Research Board ARRB Transport Research, (ARRB TR), AUSTROADS and the Australian Asphalt Pavement Association (AAPA) is the result of an R&D program begun in 1988 and created for the purpose of improving knowledge of the in-situ properties of mixes. A particular aim was to improve rutting resistance and become more performance-related so that the properties utilised in the mix design process were those which determine the on-road performance. This design method uses a test device manufactured in Australia called the MATerials Testing Apparatus (MATTA), similar to the British NAT for mechanical characterisation of mixes, and conFig.d in such a way as to be able to determine two basic properties of the mix: the modulus (measured using the repeated loading indirect tensile test) and the rutting resistance (using the dynamic creep test). A third basic property - the fatigue strength is measured with a repeated load beam test using a special beam tester based on the MATTA. The cylindrical test specimens utilised for the first two tests are compacted using a Gyratory Shear Compactor (GYROPAC) which is first used for the volumetric analysis of the mix, which is considered the basic step in mix design and, for lower traffic mixes, the only one required. Mixes used on roads with a high heavy traffic volume, where greater rutting resistance is required, are also subjected to the wheel tracking test and refusal density.

This new Australian mix design procedure, still being perfected, and the tests associated with it, was developed by the National Asphalt Research Committee (NARC) Mix Design working party.

2.1.2   THE NEW AUSTRALIAN MIX DESIGN PROCEDURE FOR DENSE GRADED MIXES [AUS-6,8,11,13,14,15]

As in the case of the SHRP Superpave - USA and the ASTO - Finland methods the new Australian mix design method is broken down into three levels of increasing complexity: for light, medium and heavy traffic (Fig. 2.1).

2.1.2.1 Level 1 mix design
In level 1 mix design, which we can define as volumetric, the aim is to achieve a composition with correct volumetric proportions starting from a fixed grading curve provided by the Australian Standard 2150 or starting from formulas similar to those of Fuller and Thompson, and preparing three mixes with bitumen contents b, b+0.5% and b-0.5%. The trial bitumen content (b) can be selected either on the basis of experience with mixes of similar composition which have shown good performance,

or it can be calculated from a mathematical formula in such a way that a minimum bitumen film thickness is achieved.

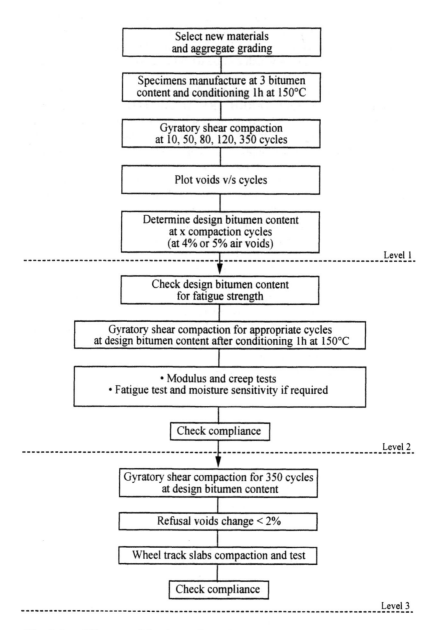

Fig. 2.1　Diagram of the Australian mix design procedure.

This calculated minimum bitumen content which assures an adequate mixture cohesiveness and fatigue resistance, is increased by about 0.5% to determine the trial bitumen content (b) with which to start the design study.

In the case of each of the three bitumen contents, five specimens are compacted in the gyratory shear compactor, each with a different number of gyrations (10, 50, 80, 120, and 350, the latter corresponding to refusal density), and their densities and other volumetric parameters are calculated.

The test parameters recommended by the Australian Standard include a gyratory angle of 2 degrees for the specimens of 100 mm diameter and 3 degrees for the specimens of 150 mm diameter, a pressure of 240 kPa and a speed of 60 rotations per minute. These differ from those specified for the SHRP gyratory shear compactor (1.25 , 600 kPa and 30 rpm). A new generation of Australian gyratory compactors called Servopac, however, permit one to control the parameters so as to be able to adapt them in simple manner to the SHRP parameters or to any other specification (France AFNOR, etc).

Before compaction, the mix is heated in an oven at 150°C for an hour, in order to simulate the bitumen hardening phenomenon which occurs during transport of the mix prior to its being laid on site and in the first two years of service.

Consideration is then given to the results yielded by the specimens of the other mixes (b+0.5, b-0.5) compacted at the same level of compaction required for said mix. The required compaction level varies depending on the type of mix and the class of traffic envisaged (light 50, medium 80, heavy and very heavy 120). The results are reported on four charts with the bitumen percentage on the abscissa and the voids percentage, the density, the VMA and the VFB (voids filled by bitumen) reported respectively on the ordinate. The design bitumen content is the bitumen percentage which corresponds to a compacted mix with a voids percentage value equivalent to the design value depending on the traffic level (for wearing courses equivalent to 4% for the light, medium and heavy traffic classes, 5% for the very heavy class). For dense mixes used for roads of light traffic, the mix design procedure concludes at this point.

2.1.2.2 Level 2 mix design
For those mixes which will have to be used on roads of medium and channelled traffic or on roads of intense traffic, level 2 of the Australian procedure provides that the so-called "trial" mix, resulting from the level 1 mix design be characterised from the mechanical standpoint.

A series of three specimens produced with a bitumen percentage equal to the design bitumen content of the level 1 mix design are compacted with the number of gyrations and voids percentage required for the traffic class envisaged. The mix must first be subjected to the same conditioning process specified for level 1. Once the three specimens are prepared, the resilient modulus is calculated at a temperature of 25°C using the indirect tensile test in accordance with the Australian Standard (AS) 2891.13.1. The same three specimens are then subjected to the dynamic creep test (without lateral confinement) at a temperature of 50°C. If both the mean modulus value and the minimum creep slope of the permanent compressive strain versus loading cycles curve fail to meet the minimum values required by the standards (for the creep test these depend on the climate and the level of traffic envisaged), it is necessary to modify the mix composition and repeat the entire mix design procedure.

As mentioned earlier, the test device preferred in Australia for determining the resilient modulus and the rutting resistance is the MATerials Testing Apparatus, even though the AS also permit the use of other test devices. The MATTA consists essentially of a load frame which permits application of loads of up to 4.5 kN on cylindrical specimens of 100 mm and 150 mm diameters. It should be noted that the Poisson coefficient assumed in Australia is 0.4 for all types of mixes.

An Accelerated Loading Facility (ALF) trial conducted in Queensland on full scale pavements showed that the unconfined dynamic creep test had a great number of deficiencies; in particular, the creep test classified the mixes of different aggregate grading in a different order to ALF. This would seem to be due to the excessive sensitivity of the rutting to the voids percentage in the creep test sample, as opposed to what happens in reality on the road. As is happening also in the British method, studies are underway to modify the creep test to include lateral confinement of the specimen.

In level 2, in certain specific cases, provision is made for the Moisture Sensitivity Test to determine to what extent the mix is prone to stripping. This is performed on a series of six specimens prepared with the design bitumen content and compacted with the gyratory compactor to a voids percentage of 8±1%, since the Australian experience has shown that this is a critical condition for the onset of said phenomenon. In the absence of a national standard (which is currently in process of being written), the moisture sensitivity is determined on the basis of the Asphalt Stripping Test AASHTO T283, developed by SHRP.

2.1.2.3 Level 3 mix design

For mixes used on the roads of very heavy and channelled traffic, where the mix required must be highly reliable with respect to rutting strength performance, the new Australian mix design procedure provides that mixtures are designed according to level 3. This provides for two distinct checks on the level 2 mix design: a check of the voids percentage of the specimen at refusal density obtainable with the gyratory compactor, and verification of the rut resistance using the wheel tracking test.

The first step consists of compacting the mix in the gyratory compactor for 350 cycles (which corresponds to refusal density), and checking that the voids percentage has not dropped more than 2% below the design voids percentage, otherwise it is necessary to change the composition and repeat the mix design procedure from the beginning (the 2% reduction results in a minimum voids at refusal density of 2.5%). This test is conducted because it is considered that the refusal compaction is a good indicator of how the mix will perform in situ if conditions worse than those envisaged occur, whether in terms of traffic or temperature. It is important that the check on the refusal density be conducted during level 1 mix design, so as to avoid having to reject the designed mix once it is reached level 3 and one has conducted all the mechanical tests to no avail.

The wheel tracking test is performed at a temperature of 60°C on three slabs of 300 cmq surface area and 75 mm thick and consists of the repeated passage of a wheel 50 mm in width subjected to a load of 700 kPa. Also in this case, if the mean result of the test on the two slabs does not meet the specifications, the mix design procedure must be repeated from the beginning.

The fatigue test (conducted at 20°C) is not obligatory, but may be required for the level 2 and level 3 mixes. The fatigue test procedure proposed by the NARC guide is modelled on the SHRP M009 based in turn on the loss of stiffness of the mix used in specimens subjected to a continuous sinusoidal load of 10 Hz. The fatigue life is defined as that corresponding to a modulus value equivalent to one-half of the original. In Australia, it is considered that the requirement for a minimum thickness of the bitumen film around the aggregate, as explained for level 2, ought to ensure a reasonable fatigue strength.

2.1.3 REFERENCES

(AUS-1) Austroads: "Pavement Design: A Guide to the Structural Design of Road Pavements", Chapter 6, Pavement Materials, AUSTROADS, Sydney, New South Wales, 1992;

(AUS-2) Sharp K.G. and Aderson A.J. : "ARRB's Contribution to the Australian Asphalt Industry's Research and Development Program, 1989 - 1992", Australian Road Research Board, Working Document, AD R192/008, 1992;

(AUS-3) Tritt B. : "Australian Development of Affordable Equipment for Determination of the Engineering Properties of Asphalt Materials", Proceedings of the AAPA Workshop, 16th Conf ARRB, Perth, 1992;

(AUS-4) Alderson A.J. : "Draft Test Methods for the Determination of the Stiffness and Deformation Properties of Asphalt", Proceedings of the Workshop on the Use of the MATTA for Laboratory Characterisation of Asphalt Mixes, ARRB, Vermont South, Vic., 1992;

(AUS-5) VicRoads Technical Bulletin 37 - Pavement Materials, Chapter 6.4, Asphalt, September 1993;

(AUS-6) Oliver J. : "Progress with the Development of an Australian Mix Design Method", Proceedings of the AAPA Members Conference, Melbourne, August 1993;

(AUS-7) Kadar P., Tritt B. : "Australian Experience with Affordable Test Equipment for the Determination of the Engineering Properties of Asphalt Mixes", Proceedings of the Conference on Strategic Highway Research Programs and Traffic Safety on Two Continents, The Hague, The Netherlands, September 1993;

(AUS-8) Tritt B. : "Asphalt Testing - Some Current Developments and Issues", Proceedings of the Workshop for Technology Transfer of Strategic Highway Research Findings, RTA / Austroads, Parramatta, May 1994;

(AUS-9) Oliver J., Alderson A.J., Tredrea P., Karim M.: "Results of the Laboratory Program Associated with the ALF Deformation Trial", APRG Report No. 12, ARRB Transport Research, Vermont South, 1995;

(AUS-10) NARC Mix Design Manual Working Party: "Selection and Design of Asphalt Mixes: Australian Provisional Guide", Complied by the ARRB Transport Research, Final Draft, October 1996;

(AUS-11) Butcher M.: "Australian Gyratory Compactors", International Workshop on the Use of the Gyratory Shear Compactor, Nantes, France, December 1996;

(AUS-12) STANDARDS AUSTRALIA. Hot Mix Asphalt, AS 2150 - 1995;

(AUS-13) STANDARDS AUSTRALIA. Method 2.2: Sample Preparation-Compaction of Asphalt Test Specimens Using a Gyratory Compactor.AS 2891.2.2 1995;

(AUS-14) STANDARDS AUSTRALIA. Method 13.1: Determination of the Resilient Modulus of Asphalt - Indirect Tensile Method. AS 2891.13.1 - 1995;

(AUS-15) STANDARDS AUSTRALIA. Method 12.1: Determination of the Permanent Compressive Strain Characteristics of Asphalt - Dynamic Creep Test. AS 2891.12.1 - 1995.

2.2   Belgium

2.2.1   INTRODUCTION [B-1, 2, 3, 4, 10, 11]

One of the major lessons drawn from the experience acquired over the past few years on the Belgian road network is that the composition of a bituminous mixture has to be determined in terms of volume.

The mix design method proposed by the Belgian Road Research Centre and developed together with certain Road Administrations was published in 1987 under the form of recommendation (CRR-R 61/87 "code de bonne pratique pour la formulation des enrobés bitumineux denses"). The method was originally limited to dense, continuously graded bituminous mixtures (void percentage < 7%), belonging to the sand skeleton grading family, but over the last few years, thanks to a change in the method of calculating volumes, it was possible to extend it also to stone skeleton mixtures rich in mastic, such as SMA (Splitmasticasphalt), as well as to those poor in mastic, such as gap graded porous asphalts. With reference to other lesser known mixtures it is necessary to proceed prudently and to resort to fundamental mechanical tests. The Belgian C.R.R. is now drawing up a new version of the "code de bonne pratique" which is the result of studies carried out and the experience acquired over the last few years though no change is made in the principal bases of the analytical method. The software, named **PRADO** (Program for Road Asphalt Design and Optimisation), developed by the B.R.R.C. to promote the practical use of the mix design method, has been fully included in the new version of the "code" thus becoming part and parcel of the design method and a vital tool for the VMA calculation. The Belgian mix design method is based on an analytical mix design method and consists of three phases (Fig. 2.2):

1) choice of materials (coarse aggregate, sand, filler and binder) and knowledge of their basic features according to the standard tender specifications and market availability;
2) analytic determination of the basic composition (the principles are explained in more detail further in this text);
3) verification of the base composition by means of tests. In the Belgian mix design method, the balance among the contrasting influences of certain composition parameters (especially the % of bitumen and the % of voids) on the resistance to permanent deformation and fatigue, is found by ensuring the two conditions of mastic filling ("remplissage") and mastic composition. In this approach, the mastic constitutes the effective binder of the asphalt mixture.

2.2.2 THE ANALYTICAL MIX DESIGN [B-3, 4, 5, 6, 7, 9, 10, 11]

In the Belgian mix design method, which takes the changes of the new code version into account, the volumetric analysis of the composition entails the following steps:
1) choice of the type of mix (depending on function of the layer and on traffic and temperature conditions);
2) choice of the aggregate constituents classes enabling one to obtain a grading curve corresponding to the desired curve in most cases already envisaged in the tender specifications;
3) determination of the voids in the mineral aggregates (VMA);
4) determination of the maximum volume available for mastic;
5) determination of the mastic composition;
6) determination of the base composition.

* VMA (Voids in Mineral Aggregates) determination
  The calculation of the percentage of VMA is based on a generalisation of the law on binary mixes of aggregates and takes into account both mechanisms of granular arrangement (filling and replacement). From the value of VMA the so-called v_Q - value (the voids in the skeleton of stones and sand) can be obtained: $v_Q = VMA + f$ (with f volume of filler). The VMA calculation is performed by the means of the PRADO calculation programme. This calculation entails the mixes to be compacted according to the Marshall method.
* Determining of the maximum volume available for the mastic, $l_D = v_Q - y$ min
  The y min is a minimum percentage of voids in the total mix so as to leave a minimum volume free without filling it with bitumen thus avoiding instability phenomena. This value is generally specified in the standards tender specifications (recommended values are also given in the "code" of good practice) and depends on the type of mix and on the category of traffic. Once l_D is determined, it is necessary to check whether this value falls within the limits of the values proposed by the "code" and, if not, decide on possible adjustments (choice of different materials, change in the grading curve).
* Determining the mastic composition.
  The mastic composition must have a thermal consistence and susceptibility so as to provide mixes with a cohesion and stability adequate to the course used in the pavement as well as to the expected road traffic conditions. Its properties are affected by the bitumen characteristics, the filler stiffness power that can be reached by determining the variation in ring and ball temperature obtained because of an addition of various quantities of filler, and finally by the k ratio of filler/bitumen volumes.
* Determination of the base composition.
  The binder content (b) that must now be determined should at the same time satisfy that the ratio k=f/b to falls within certain limits, in order to have a suitable mastic stiffness, and the upper limit of the bitumen content (bmax = l_D - f). At the same time it is necessary to check whether the corresponding void content of the mix falls within the limits of the recommendations of the "code" of good practise

and/or is in agreement with the standard tender specifications. If this is not the case, modifications are necessary, e.g. the choice of another filler, leading to a different value of f/b for the mastic and the change in the grading curve, by changing the proportions of the different constituents and/or by using other materials.

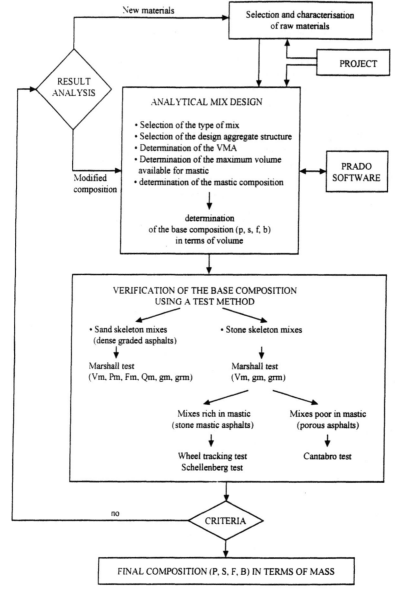

Fig. 2.2   Flow Chart of the Belgian Mix Design Method.

2.2.3 THE VERIFICATION OF THE BASE COMPOSITION BY MEANS OF TESTS [B-3,4,7,8,10,11]

Before manufacture and laying the mix, the next indispensable step in the Belgian mix design method is to verify the base composition by means of tests. This check ensures:

a) that at the moment of preparing the specimens to be submitted to testing, the mix shows adequate workability for practical purposes;
b) that once it is compacted, the mix complies with the composition criteria recommended and/or required by the tender specifications for the work to be executed (% of voids, etc.);
c) judging the mechanical performance/result of the mix, comparing the characteristics supplied by the mechanical test with the criteria associated with the test method for the use conditions envisaged (type of road, function of road, etc.).

In cases where one of the above-listed conditions is not met, the base composition must be modified. Interpretation of the results of the mechanical tests enables one to decide if it is necessary to return to step 1, to modify the choice of one or more constituents of the mix, or to step 2, to modify the original mix composition by, for example, changing the ratio between the course aggregate and the sand, or the composition (or possibly the volume) of the mastic. The basic composition thus modified must then be subjected again to the mechanical tests to verify that the required characteristics have been attained by means of the modifications effected.

In the 1987 version of the "Code de Bonne Pratique" the basic composition is verified by applying the Marshall mechanical test, whereas according to the latest version of the code of good practise, the choice of the test to be used for the verification of the base composition is based on the mineral aggregate skeleton of the mix to be design.

With reference to *sand skeleton mixes*, defined as Type I, II, III, and IV, as envisaged by the Belgian recommendation published in 1987, the choice fell on the *Marshall test* and on the *wheel tracking test*. The Marshall criteria recommended for the design of the asphalt mixes are listed according to the type of traffic and mix. Six Marshall parameters are taken into consideration in order to describe the mix's mechanical behaviour:

a) the percentage of voids in the specimen v_M;
b) the stability of the mix (mix strength at rupture point) P_m;
c) the flow of the mix (mix deformation at P_m, with the correction of the initial deformation) F_m;
d) the Marshall quotient $Q_m = P_m/F_m$;
e) percentage of aggregate voids, including filler, during compaction reached by the specimen (the percentage of lithic skeleton voids i.e. stone and sand are excluded) g_m which gives an indication of the volume available to the bitumen;
f) percentage of aggregate voids, occupied by the bitumen (bitumen filling rate) g_{Rm} (this value should not be confused with the filling rate of the mastic linked to the

lithic stone + sand skeleton). Wheel tracking tests are recommended to check the resistance to permanent deformation of the proposed mix.

As far as *stone skeleton mixes* are concerned, the *Marshall method* is recommended only to determine the % of V_M voids as well as the features of the g_m and g_{Rm} combined composition (for which the criteria are to be fully complied with). Wheel tracking, Cantabro and Schellenberg tests are added to the Marshall method according to the kind of studied mix. The *wheel tracking test*, is particularly used for mixes rich in mastic such as SMA. The "code" proposes a number of criteria to be complied with per e_{10} and $\alpha_2 = (e_{10}\text{-}e_7)/3$ (in which e_{10} and e_7 are equal to the depth of the rut in mm accounting respectively for 100.000 and 70.000 cycles), according to the type of mix and the nominal thickness of the course. The *Cantabro test* is used for mixes with a very small amount of mastic such as porous asphalts to assess their cohesion. The *Schellenberg test* is recommended to study the effect of anti-stripping agents in mixes particularly rich in bitumen ($t = 170°$, $t = 60$ minutes).

The use of an analytical mix design method is no guarantee that a bituminous mixture has a good mechanical behaviour. In order to obtain this guarantee one must verify the mix formula by using one or more laboratory tests which can be empirical or even better performance-related. However such test methods do not quickly and directly lead to the optimal mix formula and very often require preliminary studies. The analytical design approach is of paramount importance since when applied, besides avoiding serious design mistakes, it enables the number of preliminary studies as well as the number of mechanical tests to be reduced, i.e. by estimation of the volume available for the mastic. Moreover, when requirements for the test are not met, the analytical method allows one to identify the cause more quickly. Prado software, described below, is one of the most essential tools needed to fulfil this aim.

2.2.4   PRADO [B-4, 7, 10, 11]

The software named PRADO (Program for Road Asphalt Design and Optimisation) developed by the B.R.R.C., has been fully included in the new version of the "code", and now is an essential tool of the Belgian mix design method. This software is made up of the following 8 programs:

1) INTRO enables the introduction of data concerning materials (grading, mass density, aggregate angularity, characteristics of binders);
2) BINDER enables the binder characteristics to be compared to the values imposed in the standard tender specifications;
3) SAND enables the sand mix in question to be analysed by selecting several individual sands. It also enables this sand mix to be identified in the Richardson diagram;
4) MATERIALS for the choice of the basic constituents of the aggregate mix: the stone, filler and sand mix;
5) GRADING enables the grading curve of the mix to be build up;
6) MIX DESIGN for the determination of the binder content and the void content of the mix;

7) PROPERTIES which enables the mechanical properties of a bituminous mixture to be estimated, on the basis of its composition and bitumen characteristics. The estimated properties are as follows: the complex modulus and the master curve, the fatigue law, the permanent deformation law and the thermal expansion coefficient;
8) REPORT for the storage and print-out of a complete mix design study.

2.2.5 REFERENCES

(B-1) "Code de bonne pratique pour les bétons hydrocarbonés, bétons asphaltique et bétons de goudron", Recommendation of Belgian Road Research Centre CRR - 23/61;

(B-2) Verstraeten J. - Francken L.: "Sur le compromis entre la stabilité et la durabilité des mélanges bitumineux", La Technique Routiere, 4, 1979;

(B-3) "Code de bonne pratique pour la formulation des enrobés bitumineux denses", Recommendation of Belgian Road Research Centre R 61/87;

(B-4) Francken L. : "The Belgian Mix Design Method and its Implementation", Proceedings of the 4th Eurobitume Symposium, 1989 Madrid;

(B-5) Heleven L. : "A Geometrical Approach to the Mix-Design of Bituminous Pavements", Proceedings of the 3rd Eurobitume Symposium, The Hague, the Netherlands, 1985;

(B-6) Francken L. - Moraux C. : " Influence des fillers sur les caractéristiques de la consistance des mastics bitumineux", Bituminfo 49, décembre 1985;

Rapport présenté au 3éme Symposium Eurobitume de La Hague, Septembre 1985;

(B-7) Verstraeten J. : "Enrobés bitumineux à haute resistance a l'ornierage par fluage", Comité Technique AIPCR des Routes Souples (CT8), XXe Congrès Mondial de la Route, 1995 Montréal;

(B-8) Francken L. : "Etude en laboratoire du phénomène de déformation permanente". Bituminfo 37, 1979.

(B-9) Francken L. : "Granulométrie et formulation", Bituminfo N°60/1991;

(B-10) Francken L. - Vanelstraete A. : "New Developments in Analytical Asphalt Mix Design", Proceedings of the 5th Eurobitume Symposium, Stockholm 1993;

(B-11) "Code de bonne pratique pour la formulation des enrobés bitumineux denses", Recommendation of Belgian Road Research Centre - latest draft, 1996.

2.3   FINLAND

2.3.1   INTRODUCTION [FIN-1,2,3,4,5,6]

Between 1987 and 1992 in Finland an intense research program on asphalt pavements was conducted called ASTO (Finnish initials for the Asphalt Pavement Research Program), which dealt with ten special fields of research, including "Mixtures" and "Hot-Mix Surfaces". The main purpose was to increase the working life of the pavement and at the same time cut the maintenance surfacing costs.

One of the main results of the Finnish ASTO research program was the development of a mix design system linked to the properties of the mix in situ, and which embraces the entire mix design process, from the choice of raw materials down to the final testing to determine the performance properties of the mix in situ.

The ASTO program produced three different guides dealing respectively with asphalt pavement design, production and quality assurance. The ASTO guide "Asphalt Pavement Design", in particular, contains detailed information on how the pavement performance is influenced by the properties of the raw materials and of the mix, and is of particular help to the pavement designer.

In 1995, the Finnish Asphalt Specifications were revised, amended and updated so as to incorporate the results of the intense ASTO research effort; they were prepared with particular consideration to the Finnish traffic and weather conditions.

With the new Finnish specifications, covering not just the traditional mixes but also the recycled ones, the designer is much more free to try to achieve an optimal mix thanks to the introduction of the performance-related design method developed during the ASTO program, without being constrained to respect a series of prescriptions and recipes, which nevertheless still retain their importance. In fact, the recipe method (in the specifications called empirical mix design system), which up until recently was the only tool for formulating mix composition in Finland (apart from the Marshall mix design method which is still used by the Finnish Aviation Administration), has also been maintained under the new specifications.

2.3.2   ASTO MIX DESIGN SYSTEM [FIN-1,2,3,4,5]

The ASTO mix design system consists of three levels, based on the properties which may be required of the mix in situ (Fig. 2.3).

The first level, which can be defined as volumetric (and is used when the component materials are well known), consists in determining the optimal mix based on measurements of the volumetric parameters, voids in the mineral aggregate (VMA), voids filled with bitumen (VFB) and void content (VC) of the mix, and comparison of these with the design volumetric criteria depending on the type of mix. The volumetric and compaction properties are determined using a Finnish gyratory shear compactor, the Intensive Compactor Tester (ICT) which has lately been further upgraded to meet the actual European standard draft and the SHRP Superpave mix design requirements. The technical specifications include compacted specimen diameter of 150 mm, average compaction pressure of 80-650 kPa, angle of gyration 0-2,5° and rotation speed of 15-60 rev/min. The suitable binder content is considered to be that which is equivalent to 85% of the VMA, when the VMA falls to its minimum value. The optimal binder percentage can be determined for 1, 2 or 3 different grading curves, and one can then choose that which performs best.

In the Nordic countries, the need to repair the wearing courses is due, in 80% of the cases, to the damage caused by studded tires. As a result, the wear resistance is considered to be the most important characteristic of the mix in situ, and is determined by means of a special test which is conducted in level 2 of the ASTO mix design system. According to this latter, the wear resistance of the mix selected on the basis of volumetric criteria is determined using the SRK test; this consists of a device in which three rotating studded tires wear the sides of a cylindrical specimen. The results are given in terms of the "wear value", which is equivalent to the volume of material in cm^3 worn over a period of 2 hours. The in situ rutting depth due to wear can be predicted in the laboratory thanks to this test (rutting graph mm/year-SRK cm^3). The parameter which most greatly affects the wear resistance is the quality of the aggregate, which can be measured using the point loading method and the ball mill method.

The other performance properties required at level 2 for a mix in situ and which play a primary role in the mix design process for final selection of the optimal mix are the rutting resistance and the water sensitivity.

The first is normally measured using the repeated load creep test. If a more reliable measure of the rutting resistance is required, the mix is subjected to the wheel tracking test, with which it is also possible to predict the rutting performance in relation to the volumes of heavy freight traffic. The ASTO studies showed that the softening point of the bitumen has a strong influence on rutting.

The water sensitivity of a mix depends essentially on that of the aggregate (measurable by means of the water absorption test) and can be measured using the indirect tensile test.

Level 3 of mix design entails the determination of the mix resistance to low temperature cracking, another property of great importance in the Nordic countries, where the pavement temperatures may range from -40°C to +50°C. The properties of a mix at low temperatures can be measured by means of the indirect tensile test at -2°C, but given that they depend mostly on the properties of the bitumen, they can also be measured using the Fraass test or the penetration test performed at a temperature of +5°C. In particular, it is also possible to estimate the temperature of the low temperature cracking of a mix in situ using one of these test methods (graph of low temperature cracking °C - Fraass breaking point). For the typical Finnish conditions, it is considered that it is possible to limit sufficiently the phenomenon of low temperature cracking by using a bitumen with a Fraass breaking point below -20°C.

The other two properties measured in level 3 are the stiffness and the fatigue resistance, but only for purposes of determining the pavement design parameters. To increase the working life of a pavement subjected to high loads, it is deemed sufficient in Finland to utilise the SMA instead of a continuous graded mix (asphalt concrete) and/or recourse to modified bitumen.

After having determined the properties of the mix, depending on the level of mix design selected, and having checked that they meet the design criteria and assessed the annual cost, the mix is accepted. If the properties of the mix do not meet the minimum values required for the purpose the mix is to serve in situ, one then proceeds to change the proportions of the component materials or the type of one or more of the raw materials.

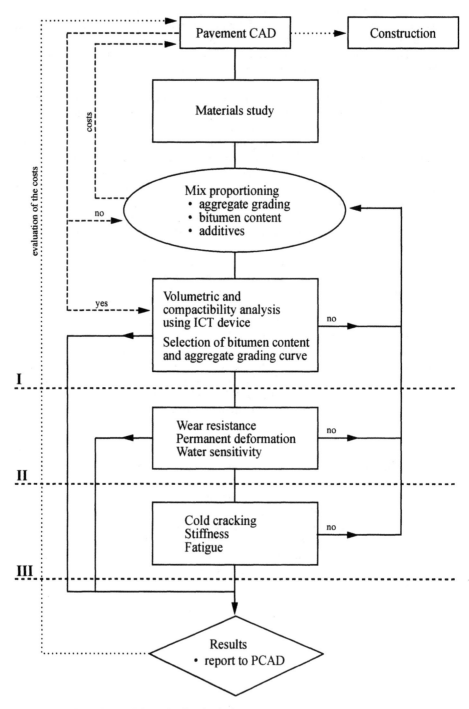

Fig. 2.3 Flow chart of the ASTO mix design system.

2.3.3 FINNISH ASPHALT SPECIFICATIONS 1995 [FIN-1,2,3,6,7]

The design of bituminous mixtures under the former Finnish specifications was based on the recipe method. In the new specifications, most of the mixes can be formulated either in this way (the so called Finnish empirical mix design system) or also on the basis of performance requirements (the so called Finnish performance-related mix design system). The choice of one or the other of these methods will depend on the type of mix, the traffic envisaged and the degree of reliability required and the characteristics of the area to be paved, and will be selected with the aid of the recommendations presented in Table 2.1.

The specifications differentiate among 4 classes of mix design: A, B, C and D, depending on the traffic volumes expressed in terms of vehicles/day (roads and streets are divided into separate groups because of their different speed limits). The characteristics of the aggregate used and the performance requirements classes are selected on the basis of the mix design class so as to take into consideration also the final cost of the pavement. The suitable type of mix and its raw materials will be selected on the basis of the required pavement performance characteristics, traffic volumes and the position of the layer within the pavement.

Table 2.1   Selection of the recommended Finnish mix design method

MIX DESIGN CLASS			REQUIREMENT FOR ASPHALT PAVEMENT		REQUIREMENT FOR GRADATION		BINDER CONTENT	
ADT(cars/day)			Empirical mix design system	Performance - based mix design system	Empirical	Performance volumetric criteria	Empirical	Volumetric criteria (gyratory compactor)
	Roads	Streets						
A	>5000	>10000		•		•		•
B	>2500	>5000	•[1]	•	•[2]			•
C	>1500	>2500	•		•		•	
D	<1500	<2500	•		•		•	
GA, PA, SIP, SOP			•		•		•	

(1 - The wear of permanent deformation resistance is tested and the water sensitivity verified.

(2 - Small changes in the grading curve are possible if the volumetric criteria require it.

For each type of mix, the specifications indicate the grading, the type of bitumen and the bitumen percentage interval, the thickness of the layer of mix to be laid and, if necessary, the percentage of additive. They also provide data regarding the minimum percentage of crushed aggregate and the percentage interval for residual voids. These latter two parameters depend on the mix design category. This information is based on experience with mixes of similar composition which have performed well in situ over time, and constitute the basis of the so called Finnish empirical mix design system.

The performance-related mix design system in the Finnish specifications is similar to the above-described ASTO system, and is based on determination of the performance characteristics required of the mix in situ, and on the compositions defined for the different mixes on the basis of the aforesaid experience.

The test mixes are compacted with the gyratory compactor (PANK 4115), after which the specimens are subjected to different tests to determine their performance properties including: wear resistance, using the Pavement Wear Resistance (SRK) test in wet conditions and at the temperature of +5°C (PANK 4209), permanent deformation resistance, measured first of all using the wheel tracking test and then by means of the repeated loading creep test (PANK 4208 - the permanent deformation resistance of gussasphalt is measured using the indentation test PANK 4401), and water sensitivity with the indirect tensile test on both conditioned and unconditioned specimens at ±10°C (PANK 4301). For the AC, GAC and SMA mixes, these properties must always be measured, and based on the specific function which the layer is to perform also other performance characteristics may be examined.

Depending on the results of the tests conducted to determine the performance parameters, the compacted mixes are divided into different classes. Among the alternative mixes studied, that which best respects the design performance requirements is selected as the optimal mix. The annual cost of the pavement remains, however, the deciding factor in choosing the optimal mix.

2.3.4 REFERENCES

(FIN-1) Saarela A.: "Asphalt Pavements, Volume 1, Design", Espoo 1993;

(FIN-2) Kettunen U., Karlsson H., Laitinen V., Nevala E., Pellinen T., Pohjola P., Wastimo E., Weckstrom L. : "Asphalt Pavements, Volume 2, Production", Espoo 1993;

(FIN-3) Kettunen U., Kollanen T., Pohjola P.: "Asphalt Pavements, Volume 3, Quality Assurance", Espoo 1993;

(FIN-4) Saarela A.: "The Finnish Asphalt Pavement Research Programme ASTO, with Emphasis on Nordic Conditions", Proceedings of the Conference Strategic Highway Research Program and Traffic Safety on Two Continents, The Hague, The Netherlands, September 1993;

(FIN-5) Kollanen T: "How to Design a Good Asphalt Mixture", Nordic Road & Transport Research No. 1, 1994;

(FIN-6) FINNISH ASPHALT SPECIFICATIONS 1995, Finnish Pavement Technology Advisory Council - PANK, Helsinki 1995;

(FIN-7) PANK Methods, 1995;

(FIN-8) Paakinen A. : "Gyratory Compaction: Precision of the Gyratory Movement", International Workshop on the Use of the Gyratory Shear Compactor, Nantes, France, December 1996.

2.4   France

2.4.1   INTRODUCTION [F-2, 5-10]

The bituminous mix design methods used in France prior to the end of the 1970s consisted of a combination of "recipes" (i.e. sets of requirements regarding the degree of bitumen penetration and thickness of bitumen chosen in function of the traffic and climatic conditions, etc.) and indications based on laboratory tests.

A new laboratory compacting method had to be developed in the 1950s, because the Duriez method, besides causing a noticeable alteration in the grading curve of the mineral aggregates due to the high stresses to which these latter were subjected during preparation of the specimens, differed too greatly from the compacting of the rollers *in situ*. The Giratory Shear Compactor ("Presse à Cisaillement Giratoire" PCG), inspired by an idea originating in Texas (the kneading gyratory compactor), is used not only to assess the behaviour of the mix under compaction, but represents an essential tool in the French procedure of bituminous mix design.

The development of numerous laboratory tests ever more adapted to fundamental research (intrinsic parameters of the material) and functional approach of the research (mix properties in relation to its use), and the need to avoid cumbersome procedures, led to the definition of the mix design method published in 1979 by J.P. Grimaux. The method currently used in France, developed by the LCPC (Laboratoire Central des Ponts et Chaussées) and largely inspired by this document, is related to the performance of the mix, rather than to descriptive prescriptions regarding material composition.

2.4.2   FRENCH APPROACH TO MIX DESIGN [F- 2,7,8,10,19]

The actual French mix design method makes distinctions among the following three situations (Table 2.2):

1. The study of a completely new formula hypothesised for use of materials which are either non-traditional or of unknown performance characteristics; in such cases all the laboratory tests mentioned above must be carried out. In particular, the PCG workability and compacting test must be performed on the type of mix selected, varying certain parameters regarding the materials and the grading; the formula selected must then be subjected to the Duriez test and the rutting test (if risk of rutting is envisaged), fatigue and modulus if structural layer, and if a structural design must be done.
2. The adaptation of a given formula with known basic characteristics where one or more components have been varied (e.g.: the nature of the aggregates, the nature of the binder and /or its dosage, the presence or absence of modifiers). Firstly the new components are identified and the density of the mix calculated, then the PCG test is performed on several variants, selecting the optimal formula, and again performing the Duriez and the rutting test (where risk of rutting is envisaged). In such cases it is then necessary to check that the properties of the new formula correspond to those of the base formula, with particular attention being paid to the

incidence of change(s) in the constituents of the chosen formula, as far as its rutting resistance is regarded.

3. Verification over time of a formula which provides assurance that the mix always maintains the same characteristics and that, with the passage of time, there has not been any variation in the properties of the constituents, which cannot be detected using current characterisation tests. In this latter case, the only tests to be performed is the PCG workability and compacting test.

The test common to all three cases is the "Presse à Cisaillement Giratoire" (PCG) which, as mentioned previously, is the basis of the French mix design method.

2.4.3   TYPES OF TESTS USED IN MIX DESIGN

2.4.3.1   Duriez test [F- 5,6,8,9,11,15,18,21]

The Duriez test is a simple compression test with free lateral expansion. It is no longer used in mix design, even if it represents an incomplete approach to determining the mechanical characteristics of the mix (compression strength according to void content). It is used to measure the sensitivity of mixes to water. The test is performed on cylindrical specimens after having immersed the latter in water. The ratio between compression strength test with immersion and without immersion, currently represents the sensitivity to water of the mix.

2.4.3.2 The Gyratory Shear Compactor (Presse à Cisaillement Giratoire) [F - 1-12, 14-17, 22]

The Gyratory Shear Compactor called "Presse à Cisaillement Giratoire" (PCG) is used to predict the evolution during compaction of the percentage of residual voids in a bituminous mixture (Dmax ≤ 20 mm) by measuring the height variations of the specimens (diameter 160 mm) taken from the mix and subjected to isothermal compacting. This test can be performed both on material produced in the laboratory or on material taken from the plant, and applied to both hot and cold bituminous mixtures.

The compacting action is produced through the combination of a gyratory shear stress and an axial force, which together are very similar to the action produced by the job site pneumatic-tired compactor. The kneading action is caused by the movement of the axis of the specimen, which generates a conical surface of revolution, with 0 vertex and vertex angle of 2φ ($\varphi=1°\pm1'$), whereas the extremities of the specimen, one of which is fixed while the other performs a circle, are maintained perpendicular at all times to the axis of the conical surface. Both the resultants according to the axis of the cone of the forces acting on the specimen and the φ angle must be kept constant during the entire duration of the test, as opposed to the height of the specimen, which drops progressively.

Table 2.2 The French Approach to Mix Design

Verification of a Mix Formula already designed and applied	Adaptation of a Mix Formula where one or more constituents have changed	New Mix Formula
PCG test (NF P 98 - 252) ↓ Duriez test at 18°C (NF P 98 - 251 - 1)	PCG test on several variants ↓ One formula selected ↓ Duriez test at 18°C ↓ Rutting test * (NF P 98 - 253 - 1)	PCG test on several variants ↓ Few formulas selected ↓ Duriez test at 18°C ↓ Rutting test * ↓ One formula selected ↓ Characterisation tests of mix mechanical performance **

* Test to be carried out in the case of heavy traffic and /or special uses, for example:
- slow channelled traffic
- large thickness
- use of rounded sand, etc.

** Test to be carried out where the mechanical properties of the mix are used for pavement structural design.
The tests used will be:
- either the test for determining the modulus or loss of linearity (NF P 98-260-1)
- the complex modulus test (NF P 98-260-2)
- the fatigue test (NF P 98-261-1)

During the test the following parameters are measured:

1) the loss of height (to a minimum height of 150 mm), which permits the evolution of the residual voids content of the mix to be determined, in function of the number of revolutions (ng max = 200 and vg = 6-32 revs/min)

$$v \% = 100 \left[\frac{h(ng) - h_{min}}{h(ng)} \right]$$

2) complementary, the evolution of the inclination of the axial force F (F = 12000 N±250 N), which represents the load needed to maintain φ constantly equal to 1°.

Studies conducted to correlate the test with real scale behaviour showed relationships between the residual voids percentage calculated after compacting of the specimens in the laboratory with the PCG and the percentage of residual voids measured on specimens taken *in situ* after compaction by the roller compactor or directly in the field by gamma rays.

The PCG thus permits laboratory simulation of in situ compaction process. In fact the number of revolutions of the PCG (n_g) corresponding to the number of passes performed with the compactor (n_p) is given by the formula $n_g = k \; e \; n_p$, where k is the factor depending on the type of compactor and compaction energy, e is the thickness (in mm).

The indications obtained by using this test are important in mix design. The course of the voids percentage in function of the compacting energy (expressed by the number of revolutions n_g of the PCG represented on a logarithmic scale) provides very interesting information regarding the mix characteristics (marked by the slope of k, which is the characteristic parameter of the mix adopted, of the right line V% = V_1 + k l_n n_g and by the V% values at n_g = 1 and n_g = 10) and regarding the steps to be taken to achieve the required characteristics.

Interpretation of the PCG test can also be accompanied by a qualitative analysis of the axial force F exercised during specimen compacting. The curves obtained are generally of two types: a) the first type in which the F force increases at the onset of compacting and stabilises at the end of this operation; this curve is generally obtained with mixes of low workability; b) the second type, in which, beyond a certain level of compacting, a sharp drop occurs in the force, reflecting a change in the bituminous mixture, owing to the reduction of internal friction; this curve is encountered in bituminous mixtures with gap graded aggregate and a high mastic content. To minimise this phenomenon, the mix is modified by reducing the percentage of sand or of fines.

The original PCG test equipment, already modified in 1985 (PCG II model) was further upgraded in 1995-96 (PCG III model). This latter model differs from the preceding versions with respect to the diameter of the test specimens (150 and 100 mm), the system for adjusting the angle of rotation (which varies from 0 to 2° with a sensitivity of 5'), the loads applied (10550 N for the 150 mm diameter and 4700 N for the 100 mm diameter) and gyration speed (30 rev/min).

It is important to note that in the French mix design method the gyratory compactor is not used in the same way as it is used in S.H.R.P. Superpave level 1.

In this last method the gyratory compactor is used to control the void content after compaction by traffic, at the end of the layer life (the void content value of 4 % must be obtained at a given number of gyrations which increase with traffic volumes and site temperature). In Superpave level 1 no mechanical tests are run to control resistance to permanent deformation, and the gyratory compactor is both a compaction and traffic simulator, while in the French method the PCG is only a compaction simulator and is supplemented by a wheel tracking test to simulate the effect of traffic (to assess rutting resistance).

2.4.3.3 Laboratory Plate Compactor [F-5,6,8,20,27]

This equipment permits laboratory compacting of parallelepiped plates (600 x 400 or 500 x 180 mm, with thickness ranging between 50 and 150 mm), these are subsequently cut into cylindrical and trapezoidal specimens and subjected to the tests required by the mix design method (rutting, modulus, fatigue).

The bituminous mixture is put in a parallelepiped mould and subjected to the action of single or twin wheels (7 bars inflation pressure). The tire wheels are moved in three directions : vertically to load them (with application of an imposed variable load of up to 520 daN), horizontally to displace them and obtain compacting action (at constant speed with passage frequency of 6 sec) and transversely to ensure that the total surface of the plate is covered. It is possible with this device to simulate the compaction in situ and find on the sample the same mechanical characteristics as in place

2.4.3.4 Wheel - Tracking Rutting Test (Orniéreur) [F- 5,6,8,10,14,15,23, 28]

The rutting simulator named "Orniéreur" is a test device which permits verification of the rutting resistance of bituminous mixtures, which have a satisfactory aptitude for compacting. The rutting test is performed if the mix is to be employed under high traffic and channelling conditions on ramps, under particularly high temperatures. For research purposes, the rutting resistance is determined by using the repeated compression triaxial test.

With the rutting simulator, one can determine the rut depth created under isothermal conditions of 60°C on a prismatic specimen (base 500 x 180 mm and thickness 100 or 50 mm) by the repeated passage of a pneumatic tire (smooth tread with pressure of 6 x 10 Pa and subjected to a load of 5000 N) in alternating movement (every 30' with 205 mm amplitude) to simulate loads in transit (loading time equal to 1 s). The rut depth is defined by the relative percentage of reduction in the thickness of the plate on the wheel path. The measurements are taken along 5 transversal profiles, interrupting the test after 30, 100, 300, 1000, 3000, 10000, 30000 and in rare cases 100000 cycles.

The rutting test is an indispensable complement to the PCG test, but must be considered only as a means of verification, given that it is not possible to establish analytic correlation between the rut depth obtained in laboratory and the rut depth that would be obtained *in situ*.

In the light of experience thus far with the simulator, we can confirm that when the specifications for the rut test are respected, rut phenomena generally do not appear at the work site. The results of many experimental studies show, respectively, the influence of the nature of the binder and of the percentage of crushed sand, as well as the void content of the mix on the rutting depth.

2.4.4 THE FRENCH STANDARDS FOR BITUMINOUS MIXTURES [F-8,10,11,13,15,19,24,25,26]

In the French standards, most of which were published in 1991, the mixes are defined by a combination of specifications on the raw materials, some limits on their proportions and on the voids in the compacted mix, and finally by specifications on some mechanical properties.

The mechanical properties specified in the French standards regarding bituminous mixtures are believed to influence the performance of the mix *in situ*, such as: a) resistance to permanent deformations, specified as the maximum rut depth resulting from the rutting test; b) the stiffness modulus, specified by the complex modulus or by a secant modulus measured in a direct tensile test; c) the fatigue strength, specified by a minimum deformation value for 10^6 load applications in a constant deformation bending test.

The test which verifies rutting is only obligatory for roads with high freight traffic volumes, whereas the tests for the mechanical characterisation of the mix behaviour, because they are very demanding in terms of cost and time, are not obligatory for all situations. The test to determine the stiffness modulus and the fatigue strength must be performed only for mixes which play a structural role in the pavement and in cases of pavements differing from those presented in the French Pavement Catalogue (for particular thickness) or to verify that the minimum values of the prescribed mechanical characteristics have been attained.

We must stress the basic fact that the French specifications for bituminous mixtures are not based exclusively on the mechanical properties, but also introduce references to the raw materials. The minimum bitumen content to assure sufficient mix durability and thermal cracking resistance is calculated analytically, in function of the grading curve of the mix, starting from the "module de richesse" K (supplied in function of the type of mix), by the specific conventional surface Σ and by an α factor, to correct the density.

Each material's standards define the laboratory tests to be performed for the mix design, the number and type of which vary depending on which of the three afore-mentioned studies is under examination. For the Duriez test and the PCG test, the minimum and maximum required values for the residual voids percentage are indicated. Other specified mix requirements include not just the mechanical performance values (Duriez compression strength, rut depth at n cycles, and possibly the values for the complex modulus, the secant modulus and loss of linearity, fatigue deformation at 10^6 cycles), but also of course the laying temperature of the mix and the residual percentage of voids to be obtained *in situ* and some time the binder content of tack coat and the minimum of macrotexture.

It is important to note that for each of these standard types there are different aggregate quality classes, and often different performance classes for the mix. The French standards do not define where and when to use the various types of bituminous mixtures, i.e. location, traffic levels, etc.; these problems are left to the design engineer, who has to take into account the suggestions of the Administration.

2.4.5 REFERENCES

(F-1) Moutier F. : "La presse a cisaillement giratoire", Bulletin de Liaison de L.P.C., 68, 1973;

(F-2) Grimaux J. P.: "Vers une nouvelle méthodologie d'étude des enrobés bitumineux", Bulletin de Liaison de L.P.C., 104, 1979;

(F-3) LCPC: "Essai de compactage à la presse à cisaillement giratoire", Avant-Projet de Mode Opératoire, December 1981;

(F-4) Moutier F. : "Prévision de la compactabilité des enrobés bitumineux à l'aide de la P.C.G.", Bulletin de Liaison de L.P.C., 121, 1982;

(F-5) Linder R. et A.A.: "Essais mécaniques pratiques de formulation et de controle des enrobés bitumineux", Rapport National Français, RILEM, 1983. Bulletin Liaison L.P.C. 132, 1984;

(F-6) Bonnot J.: "Asphalt Aggregate Mixtures", Transportation Research Record 1096, TRB 1986;

(F-7) Ballie M., Delorme J.L., Hiernaux R., Moutier F. : "Formulation des enrobées-Bilan des essais à la presse à cisaillement giratoire (PCG)", Bulletin de Liaison de L.P.C., 170, 1990;

(F-8) Delorme J.L.: "Méthode française de formulation des enrobés", RGRA, 1991;

(F-9) AASHTO - FHWA - NAPA - SHRP - TAI - TRB: "European Asphalt Study Tour", June 1991;

(F-10) L.C.P.C.-Nantes: "Essai Interlaboratoire R.I.L.E.M.: Partie 1- Formulation", June 1992;

(F-11) Bonnot J. : "French Specifications For Asphalt Road Construction Work", ACMA Seminar, 1992;

(F-12) Moutier F. : "Utilisation de la presse à cisaillement giratoire et de l'ornieureur dans la methode française de formulation des enrobés bitumineux", Proceedings of the 5th Eurobitume Symposium, Stockholm 1993;

(F-13) Bonnot J. : "French Experience of Performance-Related and Performance-based Specifications for Asphalt Concrete", Proceedings of the Conference, Strategic Highway Research Program and Traffic Safety on Two Continents, The Hague, The Netherlands, September 1993;

(F-14) Delorme J.L. : "Tests used in the French Method for the Type Testing of Bituminous Mixes: Specification and Results", Proceedings of the Conference on Strategic Highway Research Program and Traffic Safety on Two Continents, The Hague, The Netherlands, September 1993;

(F-15) Bonnot J. : "The performance based specifications in the french standards for bituminous mixtures", Technical Workshop on new specifications for bituminous products, Spanish Road Association, Barcelona, Spain, 1995;

(F-16) Gallier S., Gaschet J., Vialletel H., Moutier F.: "PCGIII", Proceedings of the International Workshop on the Use of the Gyratory Shear Compactor, Nantes, France, 12-13 december 1996;

(F-17) Bonnot J. : "Preparation of the European GSC testing Standard: a progress report", Proceedings of the International Workshop on the Use of the Gyratory Shear Compactor, 12-13 december 1996, Nantes, France;

(F-18) Hadrizinski F.: "Comparison between GSC and Duriez voids content", Proceedings of the International Workshop on the Use of the Gyratory Shear Compactor, Nantes, France, 12-13 december 1996;

(F-19) NF P 98-130: "Couches de roulement et couches de liaison: bétons bitumineux semi-grenus", Décembre 1991;

(F-20) NF P 98-250-2: "Confection de plaques au compacteur de laboratoire";

(F-21) NF P 98-251-1/4: "Essai Duriez sur mélanges à chaud et à froid (à l'emulsion de bitume);

(F-22) NF P 98-252: "Détermination du comportement au compactage des mélanges hydrocarbonés: essai de compactage à la presse à cisaillement giratoire (PCG)", Décembre 1993;

(F-23) NF P 98-253-1: "Déformation permanente des mélanges hydrocarbonés - Partie 1: essai d'ornierage";

(F-24) NF P 98-260-1: "Détermination du module et de la perte de linéarité en traction directe";

(F-25) NF P 98-260-2: "Détermination du module dynamique en flexion sinusoidale";

(F-26) NF P 98-261-1: "Essai de fatigue par flexion à l'amplitude de flèche constante".

(F-27) Brosseaud Y. et de La Roche C.: "Influence du mode d'élaboration des enrobés bitumineux sur leurs propriétés mécaniques mesurées en laboratoire", MBTM 5th RILEM Symposium, Lyon, may 1997;

(F-28) Brosseaud Y. et Hiernaux R. : "Etude de sensibilité aux déformations permanentes de bétons bitumineux européens et japonais par l'orniéreur LPC", MBTM 5th RILEM Symposium, Lyon, may 1997.

2.5   Germany

2.5.1   GERMAN MIX DESIGN [D-1-6]

The German regulations and specifications on bituminous mixtures, which are mostly recipe-based, include detailed descriptions of the tests to be performed according to the kind of course and mix types (Table 2.3). The following kinds of bituminous mixtures are used in Germany: Asphaltbeton (asphalt concrete) for wearing course, basecourse and roadbase, Splittmastixasphalt (stone mastic asphalt) for wearing course, Gussasphalt for wearing course and OffenporigerAsphalt (porous asphalt) for wearing course.

There is no real design method which is applied to asphalt concrete in Germany. Mixes which are to be used for wearing course, basecourse and roadbase are selected, as above mentioned, from an array of standard mixes defined as "recipes". These have been developed through years of experience and have been applied to the various levels of traffic volume by using the Marshall test to analyse the voids in the mix and to select the percentage of bitumen. This kind of mix design requires a careful selection of raw materials, a rigid check during the production of the mix of all variables together as they must be in strict compliance with all the specifications.

Table 2.3 Tests to be performed according to the kind of course and mix.

TEST BITUMINOUS MIXTURES	MARSHALL				INDENTATION
	Residual Voids	Density	Stability	Flow	
WEARING COURSE					
Asphalt Concrete					
Splittmastixasphalt	●	●	-	-	-
Gussasphalt	●	●	-	-	-
	-	-	-	-	●
BASECOURSE					
Asphalt Concrete	●	●	-	-	-
ROADBASE					
Asphalt Concrete					
• Simple	●	●	●	●	-
• Dual - Function (Wearing Course and Roadbase)	●	●	●	●	-

REF. STANDARDS: ZTV bit - StB 94 (wearing course and basecourse); ZTVT - StB 95 (roadbase)

In Germany the Marshall test is regulated by the DIN standard established in 1996. It is divided into two parts: part 4 (specimens preparation) and part 11 (testing of Marshall stability and flow).

The technical requirements and guidelines needed for mix design used in wearing course and basecourse (ZTV Asphalt-StB 94), envisage the specifications both on the percentage of the Marshall specimens residual voids and on the density needed to work out the degree of compaction. The rules concerning mix design used in roadbase (ZTVT- StB 95) on the other hand, envisage specifications both on the percentage of residual voids and on the Marshall stability and flow.

Since mixes enable a certain freedom in SMA, in particular with reference to the grading curve, the bitumen and additive percentages, the analysis of voids in the Marshall design method is sometimes applied to limit the "job mix formula" specification. In fact, thanks to a grading curve some Marshall specimens are prepared with a 135 ± 5°C temperature with different bitumen percentages and are compacted with 50 blows from each side. The bitumen percentage corresponding to the mix with a 3 percent of the residual voids is chosen for the "job mix formula". A bitumen drainage test of the job mix is executed. The Marshall stability, "creep compliance" function and the resilient module don't seem to be necessary and are misleading when selecting the SMA final "job mix formula".

In Germany the static and dynamic creep tests, as well as the wheel tracking test and the gyratory shear compactor, are deemed complementary tests and are used only for special design studies.

At the present time, studies are underway which might permit the inclusion of the gyratory shear compaction method alongside the Marshall test in the German testing procedure. This will be possible, however, only after a clear relationship has been established with the Marshall test, which in Germany is considered to be the only reliable test for a good mix design, and only when this will have been done for all the types of mixes utilised. For this purpose a comparative study is being conducted wherein both Marshall and gyratory specimens are subjected to performance - related tests. In particular, the dynamic creep test showed that the gyratory specimens are less subject to deformation. This may be due to a more rapid and constant orientation of the aggregate due to the kneading action occurring in the gyratory shear compactor. In any case, it has been shown that the compaction curve of the gyratory specimens is similar to that of the Marshall ones.

2.5.2 REFERENCES

(D-1) DIN 1996, part 4
(D-2) DIN 1996, part 11
(D-3) ZTV Asphalt - StB 94
(D-4) ZTVT-StB 95
(D-5) AASHTO - FHWA - NAPA - SHRP - TAI - TRB: "European Asphalt Study Tour": Germany, June 1991;
(D-6) Wallner B.: "Comparison of the Compactibility of Asphalt Based on the properties of Marshall and Gyrator Specimens", International workshop on the use of Gyratory Shear Compactor, France, Nantes, December 1996.

2.6   Italy

2.6.1   INTRODUCTION [I-12-13, 21]

In Italy in recent years there has been growing recognition of the need for specialisation of bituminous mixtures, especially with regard to mixtures used in motorway maintenance. Achievement of such specialisation is based on the properties which define the utilization envisaged and the specific problem to be resolved, taking into account different utilisation requirements relating to the route, terrain, climate, traffic, as well as diversification of mix structural function within the pavement layers.

The studies for the design of bituminous mixes used in roadbases, basecourses and wearing courses of Italian road pavements, which were traditionally based on "recipe" or "empirical" type methods, make increasing reference to the "performance-related" approach, particularly for motorways.

There is consequent increasing integration of test procedures for assessing the performance characteristics of a mix, the results of which are interpreted in combination with the volumetric properties. This integration, especially in view of the need for laboratory forecasts to be verified performance-wise at full scale, is currently undergoing study and definition on the part of both road management agencies and the universities.

Despite the aforesaid growing need for mix specialisation and consequent performance-related mix design methods, there are as yet no standards or guidelines regarding the design of bituminous mixtures for road pavements.

Only recently, in the official documents of National Research Council (C.N.R.) No. 125/1988 ("Instructions for Road Maintenance Planning") and No. 178/1995 ("Catalogue of the Road Pavements"), is reference made to the criteria and acceptance requirements for certain types of mixtures used in road pavements, including bituminous mixtures for wearing courses, basecourses and roadbases. This lack of national standards regarding bituminous mix design is, nevertheless, made up for by technical tender specifications prepared by the main administrations, such as ANAS, for the state highways, expressways and normal roads and the various other road management agencies and motorway concessionaires. Among these latter is Autostrade S.p.A., whose technical standards apply to the initial construction and subsequent maintenance of the network entrusted to it under concession (some 3000 km, equal to half of the entire Italian motorway system).

2.6.2 MIX DESIGN IN ITALY [I-1-10, 11, 14-21]

Given a certain type of stone aggregate, normally divided into several grading fractions, as well as a given type of bitumen, also in Italy reference is had to the traditional requirements regarding grading, binder content and volumetric characteristics of the compacted mix. These requirements, specified in function of the proven suitability of the bituminous mixture in situ in terms of established physical and mechanical mix performance characteristics, are reported in the aforesaid tender technical specifications.

The mixture of aggregates, or the grading curve, is determined, possibly also with the aid of some dry compaction tests, in function of the final utilization of the mix and the voids to be achieved. In defining the binder percentage to be used, recourse is had either to theoretical procedures (using the specific surface area of the aggregate) or to experimental procedures, using conventional methods such as the Marshall test, which measure the stability of the mix. The optimal binder dosage is thus determined on the basis of the Marshall parameters of stability (S), flow (s), quotient (S/s) and voids percentage after compacting with 75 bowles per side (%v). The volumetric assessment of the binder dosage and residual voids parameters, obtained in this fashion, provides an indirect measure of the mix performance characteristics. In the tender specifications, this is translated into prescriptions regarding the minimum values for residual voids and the maximum voids occupied by binder (resistance to permanent deformations), the volume occupied by the binder and the voids in the aggregate mixture (fatigue resistance), as well as the minimum binder content and minimum residual voids (resistance to ageing).

The methods of mix characterization (especially mechanical type) in function of the short and long term performance of the pavement came to be included, initially for understandably precautionary reasons, in the design, first of modified and, later, of recycled bituminous mixtures introduced in the 1980's, due to the need to cut costs and employ more economical raw materials in maintenance interventions.

Besides the traditional mechanical and physical parameters derived from the Marshall results, therefore, consideration is also given to those parameters determined by rheological, resistance and deformability tests, both cyclical and non-cyclical, performed on both the binders and the mixtures, sometimes considered as mix design, and sometimes as mix verification.

The method used by the Autostrade Company (Fig. 2.4) provides for the following main steps:

— characterization of the raw materials using up-to-date methods, especially with respect to the binder rheology;
— a grading analysis which, following the preliminary verification with a short Marshall study of the mechanical performance a several alternative aggregate gradings, permits determination of the design aggregate grading;
— a Marshall test, in conjunction with other mechanical tests, to optimize the complete Marshall study results with the other mechanical strength and deformability characteristics to define the optimal bitumen content.

These latter parameters, termed additional with respect to the traditional ones, are obtained from tests measuring static creep parameters (permanent deformation), complex modulus, indirect tensile strength and coefficient (thermal cracking and binder-aggregate adhesion), conducted at various temperatures in accordance with italian standards (C.N.R.), also on cylindrical specimens cored from slabs prepared in the laboratory using the roller compactor (with dual rubber wheels).

The un-confined compression static creep test (cyclical and non) provides the J1 parameters (further indication of low temperature cracking potential), the alfa coefficient and the Jp, taken respectively from the first two of the strain data in the first 10 s of load and the last after a specimen recovering time amounting to three times the interval under load, with tests conducted at temperatures of 10, 25 and 40°C.

The un-confined compression static creep test permits the calculation of the complex modulus values at a given frequency also obtained from the aforesaid creep parameters. The indirect tensile test provides both the strength as well as the strain at the ultimate load at temperatures of 10, 25 and 40°C.

It should be noted that in the case of the more recently developed types of bituminous mixtures, the use of modified binders permits thicker layers in relation to the aggregate grading, with larger residual voids to provide better surface texture and to respond to drainage, sound-absorption and skid resistance needs.

It should also be recalled that the indirect tensile strength test, along with the Marshall test, is also used to evaluate one of the durability aspects of these types of mixture, verifying that the parameters which define the mechanical behaviour after saturation (brought about by immersion in water for 15 days) do not fall below 75% of the original values. The post-compaction is also determined by measuring the density of Marshall specimens compacted with differentiated and increasing (from 75 to 150) number of blows per face.

A procedure which provides for integration of the experimental values relative to the aforesaid parameters as performance indicators (Fig.2.4) was also used by the laboratory of the Autostrade Company involved in the mix design part of the interlaboratory test program organised by the RILEM 101 BAT Technical Committee in 1991-1993.

For motorway applications, the fatigue test is conducted by applying the tensile-compression or four-point bending test to cylindrical or prismatic specimens obtained also from slabs produced also using the roller compactor. Given the time required to perform it, this test is not currently used in mix design studies, but is still employed in the type characterization of mixtures to determine the pavement design stage the service life of a structure. A study is currently underway at university level to assess the reliability of applying cyclical indirect tensile stress to determine fatigue resistance, based on results recently reported in Great Britain using a procedure developed by the University of Nottingham. This study is based not just on an analysis of repeatability, but also on a university interlaboratory comparison with results obtained using the three-point bending test.

This latter type of stress was applied during the above-mentioned RILEM inter-laboratory test program in the part two devoted to the dynamic characterization of bituminous mixtures by means of modulus and fatigue tests, by the Italian laboratory involved belonging to the Institute of Road Construction of the University of Palermo. The three-point bending stress was applied to prismatic specimens taken from slabs derived from the pavement constructed for the FORCE-OECD project at the LCPC circular test-track at Nantes.

Finally, mention should be made of several studies currently underway in Italy and which may result in procedures for the design of bituminous mixtures. In the motorway field, the above-described methodology is being updated through harmonisation of the volumetric part with the performance-related part, thanks to the relations existing between the volumetric composition and the mechanical properties and the verification of the compactibility of a mixture via the parameters relating to said behaviour obtained using the gyratory shear compactor, as well as the inclusion in the design procedure of the rheological and mechanical characterization of bituminous binders as such, modified and aged with the new rheometer and test apparatus derived from the SHRP experience.

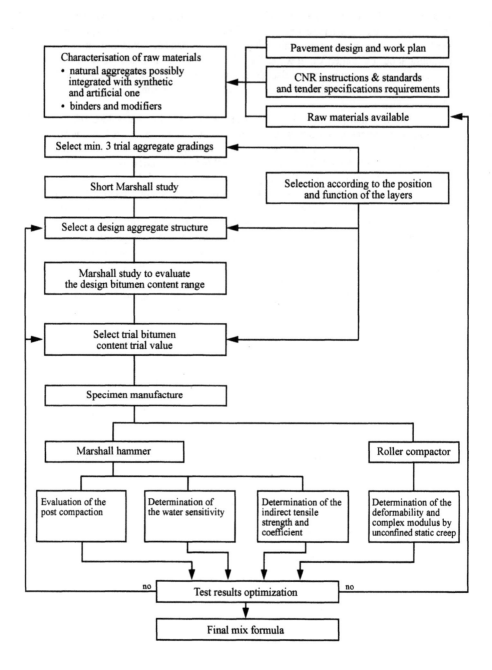

Fig. 2.4 Flow Chart of the Italian Mix Design Method for Motorways.

2.6.3 REFERENCES

(I-1) Giannini F. , Cassinis C. : "Experimental Control of the Rheological Behaviour of a Bituminous Mixture by Means of Indirect Tensile Tests", XVI PIARC National Road Congress, Salerno, 1970;

(I-2) Giannattasio P., Pignataro P. : "Characterisation of Bituminous Mixtures: Input Data for the Rational Design of Pavements", Naples, 1983;

(I-3) Battiato G.F., Peroni G. : "Recycling of Bituminous Mixtures for Road Pavements - New Design Methods", Autostrade Magazine, Vol. XXV, No. 9, 1983;

(I-4) Domenichini L. : "Evaluation of Bituminous Mixtures with the Diametral Compression Test", Autostrade Magazine, Vol. XXVI, No. 1, 1984;

(I-5) Celauro B. - Giuffre O. : "Criteria for Design of Bituminous Mixtures for Road Pavements", Binder No.9, Università degli Studi di Palermo, 1987;

(I-6) Celauro B. : "Use of Dense Mixtures with Modified Bitumens". P.I.A.R.C. Technical Committee on Flexible Pavements, 1990;

(I-7) Peroni G., M. Luminari : "20 Years of Using Modified Bitumens", Autostrade Magazine, Vol. XXXIII, No. 4, October-December 1991;

(I-8) Celauro B. : "Contribution à l'experience interlaboratoire du RILEM CT 101 BAT-partie 2 sur les essais dinamiques de module complexe et de resistance à la fatigue", Institut de Costruction Routières de l'Université des Etudes de Palerme, juillet 1992;

(I-9) AUTOSTRADE S.p.A.: "RILEM Inter-Laboratory Circuit of Tests, Part 1a, Mix Design - Feasibility Study for a Type B Wearing Course Mix Using the Prescriptions Reported by N.T.A.", Autostrade, 1992 Edition.

(I-10) Celauro B., :"Dense Bituminous Mixtures",P.I.A.R.C., T.C. on Flexible Pavem.,'94;

(I-11) Fidato A.: "Bituminous Mixture Design Methods", Thesis, Università degli Studi di Roma "La Sapienza", Academic Year 1993-94;

(I-12) Azienda Nazionale Autonoma delle Strade: "Special Tender Specification, Part Two, Technical Specifications", ANAS, Rome, 1993;

(I-13) AUTOSTRADE S.p.A.: "Pavement Maintenance - Interventions on Pavements - Tender Technical Specifications", 1995 edition;

(I-14) Santagata E., Bassani M., De Palma C. : "A Rational Framework for the Design of Bituminous Mixtures", Eurobitume & Eurbituminous Congress, May, 1996;

(I-15) Santagata E. : "Structural Mix Design of Bituminous Mixtures", Acts of the SIIV Congress on "Road Pavement Materials", Ancona, 1996;

(I-16) C.N.R. B.U. No. 30/1973: "Determination of the Stability and Creep of Bituminous Mixtures and Stone Aggregates Using the Marshall Apparatus";

(I-17) C.N.R. No. 106/1985: "Determination of Unconfined Static Compressive Creep of Bituminous Mixtures and Calculation of the Complex Modulus";

(I-18) C.N.R. No. 134/1991: "Determination of the Indirect Tensile Strength and Breaking Strain of Mixtures of Stone Aggregate and Bitumen";

(I-19) C.N.R. No. 138/1992: "Standards on Aggregates: Criteria and Acceptance Requirements for Aggregates Used in Road Pavements";

(I-20) C.N.R. No. 149/1992: Assessment of the Effect of Immersion in Water on the Properties of a Mixture";

(I-21) C.N.R. No. 178/1995: "Catalogue of Road Pavements".

2.7 Switzerland

2.7.1 INTRODUCTION [CH-1, 2]

When considering the mix design of bituminous mixtures adopted in Switzerland reference must be made to the SN 640 431a Swiss standard on asphalt concretes. It was published in 1988 and includes detailed descriptions pertinent to the choice of raw materials and mix composition. The standard is the result of considerable experience acquired by the Swiss road experts during the 70's.

It should be noted that the above mentioned regulation, which may also be applied to mixes with a percentage of recycled asphalt concrete, doesn't provide ready "solutions" but merely fixes the general framework. It induces the operator to accurately interpret such "frame-conditions" recommending that he resort to the Marshall test in view of the mix optimisation.

2.7.2 THE SWISS STANDARD SN 640 431A [CH-1]

The Swiss standard SN 640 431a published in 1988, which includes detailed recommendations and specifications concerning the materials to be used and the limits to be fixed for some mix parameters, distinguishes, between the most used bituminous mixtures in Swiss, three asphalt concretes depending on the climate and the traffic condition (expressed in TF, daily average measurable time): L (light), N (normal), S (heavy). It doesn't take the basecourse mixes into consideration but only that of wearing course (AB), whose main function is to resist wear, roadbase mixes (HMT), whose key function is to distribute loads and finally dual function coarse mixes. These mixes are subdivided according to the nominal maximum size of the aggregate and the acceptable thickness per course depending on the type and subtype of mixes is also included.

Depending on the type of asphalt concrete the standards lists the specifications concerning the proportions of the crushed aggregates (sand, fine and coarse aggregate), the binder dosage, the type of bitumen and the percentage of Marshall voids (Table 2.4). With reference to N asphalt concretes, the L ones allow for a higher rounded aggregate percentage.

Furthermore they employ softer bitumen, a higher binder dosage and a lower percentage of Marshall voids; whereas the S asphalt concretes require a higher crushed aggregates percentage, they use stronger bitumen, a lower binder dosage and also allow for a higher percentage of Marshall voids.

In the standard only lower and higher limits are fixed for the Marshall stability and flow criteria: the gap limits are not provided. For the mix design, firstly a bitumen dosage should be fixed between the interval envisaged by the standard for the binder dosage and percentage of residual voids in the Marshall specimens, successively the optimal grading, which enable the expected Marshall residual voids to be obtained. This should come about starting from the grading envelopes prescribed according to the type of mix used. In particular starting from the Talbot function, high fine aggregate content mixes should be employed in the case of asphalt concretes defined as S type, thus generating open mixes ($x = 0,5 - 0,6$). High sand and filler content mixes on the other hand should be used with asphalt concretes defined as L type thus generating more compact mixes with lower void volume ($x = 0.4 - 0,45$).

2.7.3 REFERENCES

(CH-1) Schweizer Norm SN 640 431 a - 1988;
(CH-2) MATHIAS BLUMER: "La construction routière et la maintenance des chaussées à l'aide de béton bitumineux", Schweizerische Mischgut - Industrie SMI, 1991.

Table 2.4 Specifications on bituminous mixtures according to the SN 640 431a

ASPHALT CONCRETE TYPE	Wearing Course (AB)			Roadbase (HMT)		
TRAFFIC & CLIMATE TYPE	L (Light)	N (Normal)	S (Heavy)	L (Light)	N (Normal)	S (Heavy)
RESIDUAL VOIDS % in vol.	2,5 ÷ 4,0	3,0 ÷ 4,5	3,5 ÷ 5,0	3,0 ÷ 5,0	3,0 ÷ 5,0	3,5 ÷ 5,5
VOIDS FILLED WITH BITUMEN % in vol.						
AB 3*	79 ÷ 89	-	-	-	-	-
AB 6*	78 ÷ 88	-	-	-	-	-
AB 11*, HMT 11*	76 ÷ 86	73 ÷ 83	70 ÷ 80	62 ÷ 82	67 ÷ 81	64 ÷ 78
AB 16*, HMT 16*	75 ÷ 85	72 ÷ 82	68 ÷ 78	67 ÷ 81	66 ÷ 80	63 ÷ 77
HMT 22*	-	-	-	66 ÷ 80	65 ÷ 79	62 ÷ 76
HMT 32*	-	-	-	-	63 ÷ 77	59 ÷ 73
MARSHALL STABILITY min (kN)	6	8	10	5	7	10
MARSHALL FLOW max (mm)	4,5	4,0	3,5	4,0	3,5	3,5

*Types of mixes in function of maximum aggregate size for each asphalt concrete type

2.8   The Netherlands

2.8.1   INTRODUCTION [NL-2-5]

The bituminous mix design method which has been used in The Netherlands since 1960 is based on the Marshall test. In fact, the specifications for bituminous mixtures, belonging to the "recipe" types, meet not only the requirements regarding aggregates, filler, bitumen type and aggregate grading, but also the prescriptions regarding stability, flow, Marshall quotient (this latter expressed as the stability/flow ratio), void percentage and void filled with bitumen, so as to permit the formulator to determine the optimum mix composition. It is worth noting that these prescriptions differ with the level of traffic considered (five levels), and not with climatic conditions.

For every type of bituminous mixture, a complete mix design study is conducted once a year (particularly if the origin of the material changes).

During the mix production season, the formulator checks the properties of the mix by means of an internal quality control system which includes the initial control of the delivered material and the verification of the density of the mix produced.

In 1989, as part of a review of the existing Dutch mix design method, a special work-group called "Asphalt Technology" was set up at CROW (Centre for Research and Contract Standardisation in Civil and Traffic Engineering). The activity of this group was divided into two phases. The first was dedicated to the re-examination of the existing Dutch mix design method based on the Marshall test, and the second to devising a modern, performance-based mix design method. The group ended its activity in 1994.

Mechanical characterisation of bituminous mixtures in the Netherlands is essentially based on certain tests, such as the 4-point bending test, the triaxial test and the low temperature cracking test. These types of tests which, due to their complexity, cost and execution time are not used in mix design, were developed for the determination of fundamental characteristics of mixes, and were used to measure the properties of new mixes (especially with recycled asphalt) in order to compare traditional mixes and recycled mixes on a fundamental basis. These tests, together with full-scale road tests, are comprised in a performance-based assessment system, by means of which one can classify new types of mixes and establish sensitivity to changes in composition. In order to define a new routine mix design system, the results of these investigation must be inserted within a more easily implemented framework, and will include recipe specifications with volumetric parameter and possibly some simple tests. This new method must therefore be valid both for daily routine use, and for new mix research and development using more sophisticated performance-related tests (Fig. 2.5).

2.8.2   BASIC STRUCTURE OF THE CROW MIX DESIGN METHOD [NL-1-9]

The proposed method (Fig. 2.6) is applied to bituminous mixtures made with locally available materials, such as quarry aggregates or by-product materials. It first requires ascertainment that the mineral aggregate and the binder are compatible. The pavement design must be based on a knowledge of certain characteristic properties of the mix, particularly so as to ensure that the layers are of such dimensions as to support the high levels of frequency and loading of traffic, associated with the severe weather conditions.

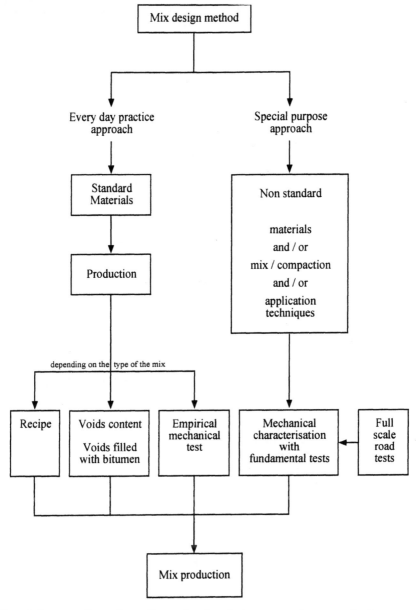

Fig. 2.5   General outline of mix design according to CROW w.g.

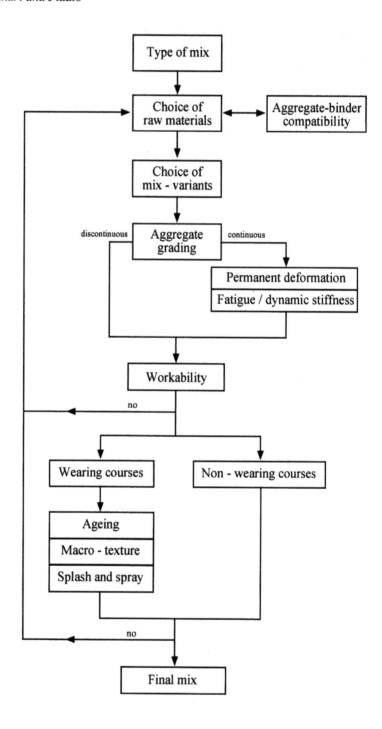

Fig. 2.6 Basic format of the CROW w.g. mix design procedure.

The mix design method must moreover ensure, that the requested properties and the mix characteristics are effectively reached. Given the conditions normally present in the Netherlands, the mixes must not only be resistant to fatigue and permanent deformations, but also possess acceptable values for dynamic modulus and workability (this latter takes into account both compaction and desegregation requirements) and, in the case of materials used for wearing courses, also resistance to ageing, high skid resistance and reduction of "splash and spray" effect.

If all these seven requirements are satisfied, the mix is produced; otherwise, modifications have to be made with respect to the choice of raw materials or its composition. Some of the aforementioned requirements may be overlooked, depending on the type of mix, the position of the layer in the pavement (if used in the wearing course or not, for example), the same apply to certain mix requirements regarding raw materials and composition. The approach can then be applied to any type of mix, depending on the traffic class of the road to be designed. In the proposal of the Dutch mix design method, the dynamic creep test with lateral confinement, conducted on specimens compacted with the kneading gyratory compactor, is used to determine the resistance of the mix to permanent deformations

2.8.3 REFERENCES

(NL-1) Hopman P.C., Kunst P.A.J.C., Pronk A.C. : "A renewed interpretation method for fatigue measurements: verification of Miner's rule", Proceedings of Fourth Eurobitume Symposium, 1989, Madrid.

(NL-2) AA.VV. "RAW Standard Conditions of Contract for Works in Civil Engineering", Centre for Research and Contract Standardisation in Civil and Traffic Engineering, 1990, Ede, The Netherlands.

(NL-3) Hopman P.C. - Kunst P.A.J.C. - Pronk A.C. - Molenaar J.M.M. - Molenaar A.A. : "Asphalt Research in The Netherlands", Proceedings of the Conference Strategic Highway Research Program and Traffic Safety on Two Continents, Gothenburg, September 1991;

(NL-4) Hopman P.C. - Valkering C.P. - Van Der Heide J.P.J. : "Mixes and Design Procedures: Search for a Performance-Related Mix Design Proceed", Proceedings of annual meeting of the A.A.P.T.,1992;

(NL-5) Hopman P.C. - Dijink J.H.: "A Modern Performance Based Mix Design Method", Proceedings of the conference Strategic Highway Research Program and Traffic Safety on Two Continents, The Hague, The Netherlands, September 1993;

(NL-6) Hopman P.C.: "Fundamental Research on Asphalt Mixes; practical design and application", Proceedings of the Conference Strategic Highway Research Program and Traffic Safety on Two Continents, The Hague, The Netherlands, September 1993;

(NL-7) Molenaar J.M.M. - Westera G.E. : "A Survey into Performance Related Specification of Asphalt Concrete", Proceedings of the 5th Eurobitume Symposium, Stockholm 1993;

(NL-8) AA.VV. "On the route to a more functional mix design method for asphalt mixtures", final report of the CROW-Working group "Asphalt Technology", November 1994;

(NL-9) J.M.M. MOLENAAR, H.A. VERBURG, G.E. WESTERA: "Characterisation of permanent deformation behaviour of asphalt mixtures", Proceedings of the Conference Strategic Highway Research Program and Road Safety in Europe, Prague, September 1995.

2.9 The United Kingdom

2.9.1 INTRODUCTION [UK-7, 8, 13, 16]

With the exception of the use of a version of the Marshall method of the Asphalt Institute as an alternative to the recipe method of design for the determination of the binder content to be used in a hot rolled asphalt wearing course mixture, in the United Kingdom bituminous mixtures are formulated trusting in the use of so-called "recipe" specifications or "recipe" methods.

One should recall that already back in 1945 the essential requirements for a road pavement in the United Kingdom were that it be durable and economical. At that time the majority of county roads and many trunk roads were surfaced in macadam, in accordance with British Standards (BS) 1621. With the increase in traffic, this material proved progressively less satisfactory, and it was determined that increased durability could be obtained using more impermeable dense mixtures with less than 10% residual voids. This led to the increasing use of rolled asphalt (BS 594 of 1950, 1958 and 1961), and the development of dense bitumen macadam, which appeared for the first time in the 1961 edition of BS 1621.

During the mid 1960's there was a further notable increase in traffic, in terms of both volume and axle loads. At the beginning of the 1970's, it was noted that the bituminous pavement mixtures designed using "recipe" methods, which up until that time had provided satisfactory in service, were deforming under this heavy freight traffic. This led, in 1973, to a partial reconsideration of the general philosophy of mix design up to that time, and the acceptance of the Marshall method for determining the binder content of hot rolled asphalt used in wearing courses. This method, suitably modified, introduced into the BS 594 in 1973, was later superseded by BS 598:1985, and was more recently updated with BS 598:1990.

In the mid-1980's, the Department of Civil Engineering of the University of Nottingham started to develop a method for the bituminous mix design geared more toward performance, based on the characterisation of the mechanical behaviour of the mix using three different tests, i.e. repeated-load indirect tensile stiffness, uniaxial creep and repeated load axial creep. These tests, already recognised individually and commonly used in numerous laboratories throughout the world, can be carried out using a single, versatile machine designed by the University of Nottingham and known as the NAT (Nottingham Asphalt Tester).

2.9.2   RECIPE METHOD [UK-7, 8, 13]

With the exception of the mix design method for rolled asphalt wearing course mixtures, in almost all the highway authorities in the UK bituminous mixtures are formulated by using the "recipe method". The recipe formulations are derived from the accumulated experience of road pavement constructions and from the considerable amount of information gathered in the past few years on the behaviour of numerous types of mixtures under different traffic and weather conditions. In short, it is based on the experience of mixtures of known composition which performed successfully in situ over long periods of time. A "recipe specification" defines a bituminous mixture in terms of its grading curve, mixture composition, binder grade, binder content (which is a function of the type of aggregate used), layer thickness and mixture characteristics through the manufacture, laying and compaction stages.

2.9.3   BRITISH MARSHALL MIX DESIGN METHOD [UK-7, 8, 13, 19]

The mix design method, defined in the standard BS 598:1985 and 1990, evaluates the behaviour of the mixture as a whole, that is, including the coarse aggregate fraction retaining to the BS 2.36 mm test sieve. This standard broke with the traditional approach of using "recipe" specifications, as well as with the preceding edition of the standards dealing with this matter (BS 594:1973), which, using the Marshall test, tended first to determine the optimal bitumen content of the sand-filler mixture, and subsequently calculate, on the basis of a mathematical formula, the optimal bitumen content of the mixture with the addition of the coarse aggregate.

This mix design method consists in a testing procedure which starts with three specimens for each bitumen content, examining a minimum of 9 bitumen contents at 0.5% increments. Depending on the percentage in mass of the coarse aggregate in the mixture, the standard indicates the mean bitumen percentages for the intervals to be subjected to testing. From the test results, four graphs are plotted which show the variation of Marshall stability, flow, mix bulk density (Sm, which is the ratio mass in air/volume) and the compacted aggregate density [Sa = Smx(100-Wb)/100Sa] with the bitumen content (Wb in %).

The design bitumen content is calculated as the mean value of the contents determined at the maximum Marshall stability, maximum mix density and maximum aggregate density plus an additional binder content depending on the coarse aggregate content (retained on the BS 2.36 mm test sieve). This latter is accepted only if the Marshall specimens, prepared with the said design bitumen content, respect the Marshall stability and flow criteria required by the BS 594 for the design type HRA mixtures.

It should be added that the British Marshall method, beside not adopting any volumetric criteria in the determination of the design bitumen content, introduces an apparently arbitrary corrective factor depending on the coarse aggregate content to determine the design bitumen content. Moreover, in the UK the Marshall method is applied to gap graded mixtures, as opposed to the practice in the USA where it has been introduced for continuously graded mixtures.

2.9.4 PERFORMANCE-RELATED MIX DESIGN METHOD OF THE UNIVERSITY OF NOTTINGHAM [UK-1-18, 20, 21]

The performance-related mix design method developed at the Civil Engineering Department of the University of Nottingham, which is illustrated by the flow diagram of Fig. 2.7 , was carried out for the continuously-graded road base and base courses dense macadam mixtures (similar to the mixes of the Asphalt Institute, MS-2, 1984) and is based on the determination of three basic parameters which greatly influence both the mechanical properties of bituminous mixtures and, consequently, the performance behaviour of the road pavement. These parameters are the aggregate grading curve, bitumen content and degree of compaction of the mixture. The operational procedure of this performance-related mix design method consists essentially of two main steps described below.

The first step entails the testing of the stone aggregates to determine the physical properties (such as the shape, real and apparent volumetric mass, absorption, hardness, etc.) and the optimal grading of the aggregates mixture. To calculate the latter in terms of the percentage aggregate sieve, a modified version of Fuller's maximum density curve is used; the modifications have been introduced to allow the grading and the fines content to be varied so as to maintain the filler content at a prefixed value.

The procedure for the first step entails for each of the three aggregate grading curves corresponding to the values of $n = 0.5$, 0.6 and 0.7, the manufacture of three mixtures corresponding to the bitumen percentages $Mb = 3.5\%$, 4.1% and 4.7% and the subsequent compaction of the resulting nine mixtures to three energy levels, thus obtaining 27 mixture combinations.

Compaction is determined using a particular methodology originally developed at Nottingham University called "Percentage Refusal Density" (PRD) similar to the equipment which is normally used to determine the degree of compaction of samples cored from the pavement. Using this method 150 mm diameter by approximately 70 mm height specimens are prepared, handled and compacted using a vibratory hammer. By varying the time of compaction and the temperature of the material, three compaction levels of 93, 97 and 100 percentage refusal density are achieved.

The volumetric composition of the 27 specimens are determined and, through certain limits placed on the percentage of voids, the void mineral aggregate (VMA) and the percentage volume of the bitumen, an initial selection of mixes is obtained.

The second step involves the determination of the mechanical properties of the selected mixtures, such as the indirect tensile stiffness modulus and resistance to permanent deformation. These are determined by a series of low cost practical tests carried out in the Nottingham Asphalt Tester (NAT), which is comprised of a versatile test apparatus consisting of an actuator powered by air and controlled by a conventional personal computer which, by means of specific user friendly software, sets up, controls, acquires, processes and stores the experimental data. The NAT is conFig.d in such a way as to be able to perform four types of test including the Indirect Tensile Stiffness Modulus (ITSM) test with application of the load through steel strips to measure the stiffness modulus, the Indirect Tensile Fatigue Test (ITFT) to measure fatigue cracking resistance, as well as the Uniaxial Creep (UC) and Repeated Load Axial (RLA) tests to measure resistance to permanent deformations..

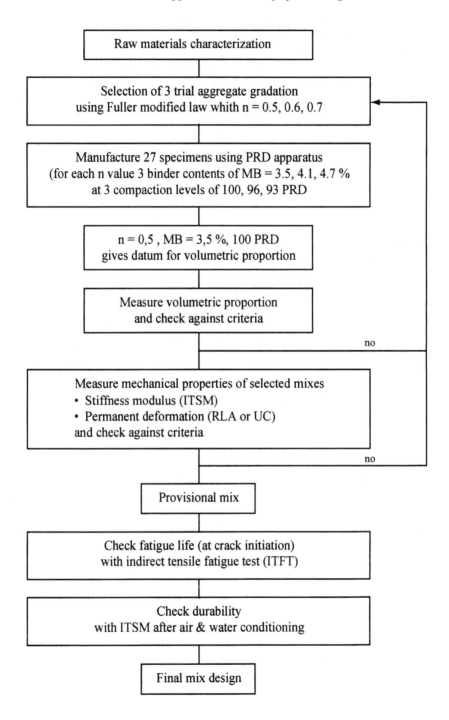

Fig. 2.7 Flow diagram of the Nottingham mix design procedure.

The RLA test is preferred to the UC in the case of mix design because it allows one to quantify the effects of variations in mix composition on the resistance to permanent deformations. The RLA test, which is used without lateral confinement in the normal configuration, is only considered to be applicable for dense mixtures. For this reason work is in progress to develop an RLA test with the lateral confinement in particular for the assessment of open-graded mixtures such as porous asphalts.

Preliminary results from tests carried out indicate that a confining pressure of 70 KPa is appropriate for all mixtures types

Preliminary suggested values for dense macadams are pointed out for stiffness modulus (> 2500 MPa at T=20°C with load pulse rise time of 120 ms) and permanent deformation (< 1% axial strain at T=40°Cwith axial stress of 100 kPa and number of load applications of 3600).

If the mechanical property values of the selected mix do not meet the minimum acceptance levels, or prove highly susceptible to variations in the PRD compaction, the grading curve of the mix has to be modified with repetition of the two step procedure described above. If several mixtures satisfy these minimum acceptance values the optimal mixture is determined on the basis of certain additional criteria which take into account durability, the workability of the mixture, the availability of the component materials and finally, certain considerations of an economic nature.

In the context of recent studies on the performance-related mix design the fatigue performance of a bituminous mixture may be obtained from the indirect tensile test, performed with the same NAT, based upon the life to crack initiation. In this Indirect Tensile Fatigue Test (ITFT) the specimen (generally of dimensions of 100 mm in diameter and 40 mm thick) is subjected to repeated load pulses of 1Hz until failure, which is defined by a specified amount of vertical diametrical deformation.

The results obtained indicate that the ITFT is able to characterise the fatigue working life of a bituminous mixture by subjecting a limited number of specimens. It is interesting to note the similar pattern of the fatigue curves obtained from the results of the ITFT and from the two-point bending test performed on trapezoidal specimens which better reproduces the in situ behaviour of the mixture.

This performance-related mix design method may be combined with the assessment of the mix durability, which is the resistance offered by the material to environmental damage caused by air and water. The protocol to assess durability uses the Indirect Tensile Stiffness Modulus (ITSM) to determine the stiffness modulus before, during and after air conditioning (85°C for 5 days) and the partial saturation in water with cycles (6 hours at 60°C and 15 hours at 5°C) so as to check for possible variations (which should be lower than 30%).

One of the principal objectives of the research and development project on the design and testing of bituminous mixtures, known as "Bitutest", and which involved the University of Nottingham, the UK Department of Transport, nine companies and 14 county highway authorities, is to assess whether this performance-related method would be applicable in practical situations. The aim is to re-structure the method so it could be extended to cover a wider range of mixture types including "non -standard" or innovative materials and would apply equally to roadbases, basecourses and wearing courses.

2.9.5 REFERENCES

(UK-1) Brown S.F. et alii: "The analytical design of bituminous pavements", Proceedings of Institute of Civil Engineering, part 2, 1985, 79.

(UK-2) Brown S.F. et alii: "Improved road bases for longer pavement life", Proceeding of the 3rd Eurobitume symposium, vol.1. The Hague, 1985.

(UK-3) Rowe G.M., Cooper K.E. : "A practical approach to the evaluation of bituminous mix properties for the structural design of asphalt pavements", Asphalt Paving Technology, Vol. 57, February 1988.

(UK-4) Bell C.A., Cooper K.E., Preston J.N., Brown S.F. : "Development of a New Procedure for Bituminous Mix Design", Proceedings of the 4th Eurobitume Symposium, Madrid 1989.

(UK-5) Cooper K.E., Brown S.F.: "Development of Apparatus for Repeated Load Testing in Creep and Indirect Tension", Proceedings of the 4th Eurobitume Symposium, Madrid 1989.

(UK-6) Brown S.F., Preston J.N., Cooper K.E.: "Application of New Concepts in Asphalt Mix Design", Annual Meeting, Association of Asphalt Technologists, Seattle, Washington, March 1991.

(UK-7) AASHTO - FHWA - NAPA - SHRP - TAI - TRB: "European Asphalt Study Tour", June 1991.

(UK-8) Whiteoak D. : "The Shell Bitumen Handbook", Shell Bitumen UK, September 1991.

(UK-9) Cooper K.E., Brown S.F., Preston J.N., Akeroyd F.M.L.: "Development of a Practical Method for the Design of Hot Mix Asphalt", Paper Presented to the Annual Meeting, Transport Research Board, Washington D.C. 1991.

(UK-10) Gibb J.M., Brown S.F. "A repeated load compression test for assessing the resistance of bituminous mixtures to permanent deformation", Proceedings of the 5th Eurobitume symposium, Vol.1B, Stockholm, June 1993.

(UK-11) Cooper K.E., Brown S.F.: "Assessment of the Mechanical Properties of Asphalt Mixes on a Routine Basis Using Simple Equipment", Proc.s of the 5th Eurobitume Symposium, Stockholm, June 1993.

(UK-12) Brown S.F., Cooper K.E.: "Simplified Methods for Determination of Fundamental Material Properties of Asphalt Mixes", Proceedings of the Conference Strategic Highway Research Program and Traffic Safety on Two Continents, The Hague, The Netherlands, September 1993.

(UK-13) O'Flaherty C.A. : "Highway Engineering", Vol. 2, Edward Arnold, 1993;

(UK-14) AA.VV.:"Bituminous pavements: materials, design and evaluation", Lecture notes of a residential course of the University of Nottingham, Department of Civil Engineering, 11-15 April 1994.

(UK-15) Brown S.F., Cooper K.E.: "Simplified methods for determination of fundamental properties of asphalt mixes, Proceedings of the conference strategic highway research program (SHRP) and traffic safety on two continents, The Hague, Vol.1, September 1994.

(UK-16) Brown S.F., Cooper K.E., Gibb J.M., Read J.M., Scholz T.V.: "Practical Tests for Mechanical Properties of Hot Mix Asphalt", Proceedings of the 6th Conference on Asphalt Pavements for Southern Africa, 1994.

(UK-17) Brown S.F.: "Practical test procedures for mechanical properties of bituminous materials", Proceedings of Instn. Civ. Engrs. Transp., 1995, 111, November.

(UK-18) Read J.M. & S.F. Brown : "Practical evaluation of fatigue strength for bituminous paving mixtures", Proceedings of Eurasphalt&Eurobitume Congress '96.

(UK-19) Brown S.F., GIBB M., Read J.M. & Scholz: "Laboratory protocols for the design and evaluation of bituminous mixtures", Proceedings of the Eurasphalt & Eurobitume Congress 1996.

(UK-20) British Standard Institution: "Sampling and examination of bituminous mixtures for roads and other paved areas, BS 598:Part 107, London, 1990.

(UK-21) British Standard Institution: "Method for determination of the indirect tensile stiffness modulus of bituminous materials, Draft for Development 213, London, 1993.

(UK-22) British Standard Institution: "Methods for assessment of resistance to permanent deformation of bitumen aggregate mixtures subject to unconfined uniaxial loading", Draft for Development 185, London, 1994.

2.10   USA

2.10.1 MARSHALL MIX DESIGN METHOD [USA-1-7,13]

In the United States, the bituminous mix design method used by 38 state highway agencies is that developed by Bruce Marshall, an engineer for the Mississippi State Highway Department, and subsequently improved, after numerous experiments, by the US Army Corps of Engineers.

This method is based on the Marshall test, the procedures for which have been standardised by the American Society for Testing and Materials in the "Resistance to Plastic Flow of Bituminous Mixtures Using the Marshall Apparatus" (ASTM D 1559).

The Asphalt Institute provides details of the traditional mix design method in MS-2 and MS-22.

The method is applicable to asphalt concretes and other types of hot bituminous mixtures used for wearing courses and base layers with a maximum mineral size aggregate of 25 mm (1in.).

The method entails the preparation of at least 3 specimens, 63 mm (2.5in) in height, and about 102 mm (4in.) in diameter, for each of at least 5 bitumen contents, varying by 0.5% in mass, for each grading combination considered. The mix is compacted with a hammer, undergoing 35, 50 or 75 blows, depending on the design traffic class (light, medium, heavy) and several volumetric parameters are measured (i.e. VMA, Gmb, etc.). In the case of the bitumen, it is also necessary to identify the temperatures of equiviscosity during the mixing (170 ± 20 cST) and compaction stages.

After thermosetting the specimens at a temperature of $60 \pm 1°C$, a displacement is imposed to the specimen, using a special loading test having a forward progress speed of 51 mm (2in.) per minute, until their rupture. The results of the test are the stability, equal to the maximum load resulting at the rupture of the specimen, and the flow, equal to the maximum deformation at the rupture of the specimen.

The mix with the optimal bitumen is selected, after having reported on a diagram the mean values of all the parameters for each bitumen content and calculating the mean of the bitumen percentages which correspond to the maximum value of apparent bulk density, the maximum stability value and the design value for the residual voids percentage. Different criteria must then be checked to determine whether the mix corresponding to the optimal bitumen percentage is acceptable. In particular, the values for flow, VFA and VMA corresponding to the optimal bitumen percentage must fall within the specified limits. Changes in the criteria of the MS-2, were introduced at the beginning of 1990.

2.10.2 SUPERPAVE MIX DESIGN METHOD [USA-8,9,11,15-19]

The Superpave mix design method, which was the end result of the "Asphalt" research effort of the Strategic Highway Research Program (S.H.R.P.), which operated from 1987 to 1992 in the USA, represents the integration of more than 25 research areas in a single system for the characterisation and design of asphalt mixes.

The procedure entailed in this new method includes the specifications of the raw materials, the test methods with the specially developed equipment, the strict mix design method itself, and the related software system ("core").

Superpave is supposed to substitute the specifications on materials and the bituminous mix design methods currently utilised in the 50 states of the USA, bringing them into a single, performance-based system which can provide results tailored to the different climatic and traffic conditions present for the different classes of roads in the United States and Canada.

This method is applicable to mixes such as Hot Mix Asphalt -HMA, whether virgin or recycled, of closed grading, with or without modified bitumen, as well as a variety of special mixes such as Stone Matrix Asphalt (SMA). It can also be applied to newly constructed wearing courses, connecting and base layers and for restoring of deteriorated pavement surface layers, with the aim of selecting appropriate materials, reducing and controlling permanent deformations, and cracking, whether due to fatigue or low temperature. The flexibility of the system, in any case, permits mix design taking into consideration, both separately and in combination, the three main distress factors, and predicting, moreover, the influence of ageing and sensitivity to water at the onset of the aforesaid types of deterioration.

The Superpave mix design system has five main distinguishing characteristics:

1. the criterion for selecting a bituminous mix design, which refers to its various properties, whether these be performance-based, which directly govern the pavement response to loading, or performance-related, which are indirectly linked to and influence these latter, but do not control them;
2. performance-based binder specifications;
3. performance-related mineral aggregate specifications;

4. specifications for bituminous mix design based on pavement performance over a pre-determined pavement working life, expressed in terms of rut depth, areas of fatigue cracking and areas of low temperature cracking;
5. interaction between the bituminous mixture and the pavement structure, traffic and weather conditions in predicting performance. The traditional mix design methods (for example, Marshall and Hveem) are neither performance-based nor performance-related; these methods are concerned, instead, with obtaining an economic mix of aggregate and binder which will be characterised by sufficient workability to permit efficient laying. Both methods try to predict performance by measuring certain empirical properties, but neither method can ensure that the designed mix conforms to specific pavement performance criteria.

The Superpave mix design system contains 3 distinct formulation levels (Fig. 2.8). This permits the selection of the mix design method best suited to the traffic (in terms of either loading or number of passages) envisaged for the pavement to the designed (expressed as the total number of single axles of a load equivalent of 80 kN - ESALS = equivalent single axle loads) during the working life of the pavement. In the case of weather conditions, all three mix design levels take into account their effect on pavement performance.

The complexity of the mix design process increases notably between level 1 to 3, requiring a larger number of tests and specimens. This naturally entails a long design time, but the probability that the mix in situ provides satisfactory levels of service (reliability) under the expected weather and traffic conditions, also increases accordingly.

Level 1 also provides reasonable guarantees of adequate performance if all the specified criteria are respected. The element common to all three levels of mix design is the volumetric analysis. Level 1 of the mix design method is based on the gyratory compaction, as opposed to levels 2 and 3, in which not only is gyratory compactor used but also the SHRP Shear Test Device is used. Level 3, moreover, entails a series of optional surveys which are recommended to confirm the results of the performance-based tests in cases where the reliability of the mix design must be absolute.

In the Superpave mix design system, control of ageing is achieved though the combined application of the TRFOT (Rolling Thin Film Oven Test) and the PAV (Pressure Aging Vessel), which enables one to determine the bitumen ageing tendency over the short and long term. Seeing that the laboratory mixes are prepared with bitumen which has not undergone an ageing process, the mix ageing procedure must take into account both in-plant and *in situ* ageing. The short term ageing procedure for bituminous mixtures involves exposure of the loose mix, immediately after his production, to four hours at 135°C in a forced draft oven, and subsequent compaction at the appropriate temperature.

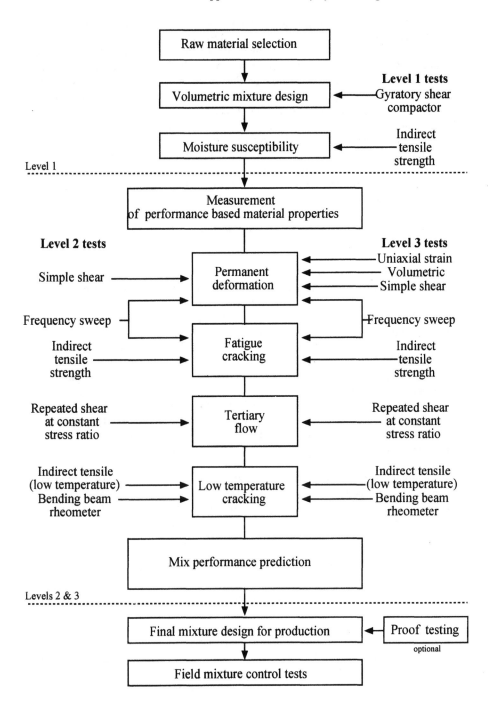

Fig. 2.8 Flow Chart of the Superpave Mix Design System.

This procedure simulates the ageing process which occurs during production in plant and in the process of laying the mix. In the long-term asphalt mix ageing procedure the specimens are made with mix which has undergone short-term ageing and subsequent exposure to 85°C temperature in a forced draft oven. The time which it is kept in the oven will vary according to the duration of the working life of the pavement which is being simulated. The recommended time is two days, which correspond to roughly ten years of pavement service.

To evaluate the moisture sensitivity of trial mixes the Superpave mix design system uses either the AASHTO T 283 test, "Resistance of Compacted Bituminous Mixture to Moisture Induced Damage", or the S.H.R.P. test method M-006, "Determining the Moisture Susceptibility of Modified and Unmodified Hot Mix Asphalt with the Environmental Conditioning System". The Environmental Conditioning System used in S.H.R.P. M-006, is a triaxial test unit, suitably modified, in which the dynamic resilient modulus of the specimen is measured as moisture is forced inside. Moisture susceptibility is characterised by the resilient modulus ratio of specimens before and after conditioning.

2.10.2.1 Superpave Software

The Superpave software consists of a program called "core", which contains:

1. the SHRP performance-based asphalt binder specifications and the algorithms necessary to chose the appropriate binder performance
2. a database of environmental conditions based on $30 \div 80$ years of observations performed at 700 stations scattered throughout the United States and Canada (the possibility of inserting other data permits software to be applied to other particular regions of the world);
3. a database of the results of the tests performed on the bitumen and the aggregates;
4. the algorithms needed to perform a volumetric mix design (level 1); the control needed to transfer the data obtained from the characterisation tests on the raw materials and the mixes to the performance prediction models.

The Superpave software, which integrates all the instruments necessary for the mix design of asphalt mixes, provides not only the "core" program , but also other so-called "included" or "associated" programs of which are generally independent and which have precise purposes and were created to interact with the Superpave structure.

The "included" programs type from the Superpave software, and to which one obtains access automatically through the control routines, provide:

1. the algorithms which determine the fundamental properties of the materials used in the performance prediction models, in function of the results of the load related and the non-load related performance tests;
2. the performance prediction models which, based on data regarding the mix materials properties and the weather, structural and traffic conditions, supply realistic estimates of the development, over the working life of the pavement, of permanent deformations, fatigue cracking and low temperature cracking;

3. a modified version of the Federal Highway Administration model of the effects of weather conditions, which generates files on pavement temperatures, used in the performance prediction models.

The "associated" programs type form of the Superpave software, which are not a part of this latter, permit preliminary analysis of the test results linked to associated or non-associated performance and maintain all the data in the files directly accessible by the Superpave software. Using the Superpave performance prediction model it is possible to obtain a mix design minimising a specific distress or a combination of different distresses.

The model consists of four parts relating to the properties of the materials, the effects due to weather conditions, the pavement response and the pavement distress (Fig. 2.9). The properties of the materials are used by the "pavement response" program to predict the behaviour of a mix subjected to specific weather and traffic conditions. Several of these material properties are also used in the "pavement distress" program. The "pavement response" program is a finite elements program, with which it is possible to calculate the stresses and strains of a bituminous mixture subjected to particular loads and weather conditions. The "pavement distress" program calculates the permanent deformations on the basis of the relevant properties of the mix and the stresses generated within it.

2.10.3 SUPERPAVE MIX DESIGN LEVEL 1(LIGHT TRAFFIC) - VOLUMETRIC [USA-16,17,18]

Besides the performance-based binder specifications and the performance-related aggregate specifications, Level 1 mix design uses only the volumetric formulation approach.

This is for the purpose of obtaining a mix which behaves *in situ* in satisfactory fashion for light traffic levels ($ESAL_{80KN} \leq 10^6$) without the need to conduct performance-based tests.

Level 1 of the Superpave mix design system consists of 3 main stages:

1. selection of raw materials (mineral aggregate and bitumen);
2. determination of the design grading composition (expressed in terms of grading curves);
3. determination of the design bitumen percentage.

Fig. 2.9 shows the level 1 mix design flow chart.

The mineral aggregates are characterised by the angularity of the coarse, and by the fine aggregates, the clay content, the elongated particles and fines, and their gradation.

The grading curve of a mixture of mineral aggregates for dense bituminous mixtures is selected using the Superpave mix design system and the grading charts of the Federal Highway Administration (representing on the ordinate the percentage of passing, and on the abscissa the sieve opening for n = 0.45 maximum density) in which 8 control and a "restricted" zone are indicated in function of the nominal size of the aggregate.

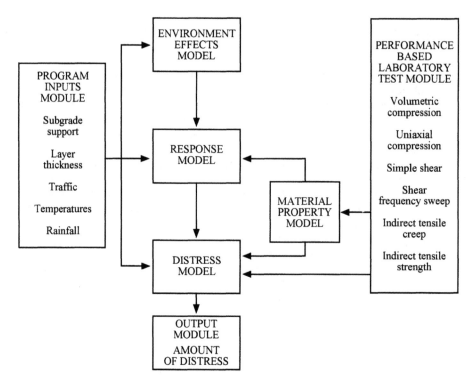

Fig. 2.9 Flow Diagram of the Superpave Performance Prediction Model

The "restricted" area, through which gradation curves are not allowed to pass, permits the inclusion of a large percentage of natural sand to be limited and avoids adoption of the grading curves which fall along the line of maximum density, and which often have inadequate levels of VMA.

In the Superpave mix design system, the design bitumen percentage is determined after having identified the level of performance (PG Performance Grade) of the bitumen (modified or not). This is obtained by comparing the results of the performance-based tests with limits contained in the specifications on the basis of both the minimum design temperature of the pavement and the mean of the maximum daily temperatures for the seven hottest consecutive days of the year.

The Superpave performance-based specifications for bitumen have been designed for the purpose of quantifying and maximising the capacity of the bitumen to resist permanent deformations and fatigue cracking induced both by repeated load applications and low temperatures.

The appropriate choice of the design performance grade of the bitumen permits low temperature cracking to be avoided, but does not eliminate the risk of either permanent deformations, which depend strongly on the properties of the aggregate and the volumetric properties of the mix, or fatigue cracking, which depends greatly on the pavement structure.

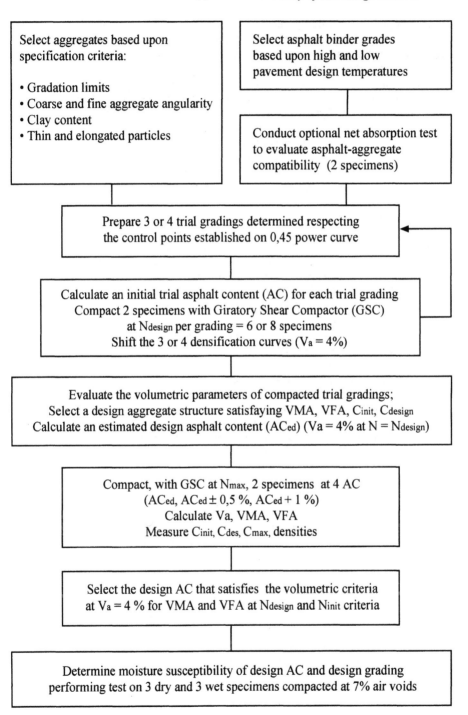

Fig. 2.10 Superpave Mix Design Level 1 - Volumetric.

Level 1 Superpave mix design requires that the specimens be compacted using the SHRP Gyratory Compactor. This device simulates the *in situ* compaction process, and ensures that the engineering properties of the specimens compacted in laboratory are equivalent to those of a mix *in situ*.

The Gyratory Compactor permits cylindrical specimens to be manufacture by means of the combination of a vertical pressure and a dynamic shear force when meshes the aggregates. While the Gyratory Shear Compactor has the capability to produce 150 mm high specimens; for mix design, specimens are compacted to 115 mm +/- 5 mm.

The Gyratory Compactor enables not only the bulk density and the percentage of residual voids to be determined in real time and during the compaction (in function of the number of gyrations), but also that the compatibility of the mix be assessed (including a prediction of the final residual voids percentage under forecast traffic conditions).

The Superpave gyratory compactor has the following characteristics:
gyration angle = $1.25 \pm 0.02°$; 30 gyrations per minute; vertical load of 600 kP$_a$ during the gyrations; capacity to produce cylindrical specimens d = h = 150 mm; the compaction is performed at equiviscous temperature determined with the SHRP M 002 test method, or at an appropriate temperature chosen by the designer.

The density is determined in correspondence to three points on the Densification Curve, corresponding to 3 compaction pressures expressed in numbers of gyrations:
N_{init}, N_{design}, N_{max} → C_{init}, C_{design}, C_{max}.
N_{init} and N_{max} are utilised to determine the compatibility of the mix, whereas N_{design} is used to determine the bitumen percentage.

The design grading structure of the aggregate and design bitumen percentage are chosen (in level 1) in such a way as to produce a density curve which passes for 96% of the maximum theoretical value of the apparent bulk density in correspondence to the number of design gyrations N_{design}.

The design bitumen percentage is determined for V_a = 4% in correspondence with N_{design}. The value of this latter is determined in function of the level of traffic and the mean of the maximum daily temperatures for the seven hottest consecutive days of the year.

The selection of the design grading distribution of the mineral aggregate is such that:

- the VMA value is adequate to N = N_{design} and V_a = 4%.
- it respects the density requirements, respectively, for N = N_{init} and N = N_{max}.

This requires that, after having chosen the aggregates which respect the Superpave criteria, the various pieces are combined in such a way as to respect the control points, developing 3 or 4 grading curves and that the results of the mixing analysis is reported on a graduation chart with n = 0.45 power, in order to verify that each of these respect the Superpave control points.

For each test grading curve, once the specimens manufactured with an initial trial bitumen content have been compacted with the SGC, the densification curves of the mix are then determined.

So as to permit comparison between the different densification curves, these latter are translated, to obtain the shifted densification curves to assess both the VMA and the VFA of each mix, with reference to the same theoretical percentage of residual voids $V_a = 4\%$ (naturally, it is also necessary to calculate the value of the variation of the bitumen content, so as to vary the Va percentages at the value of 4%).

One can, therefore, compare the volumetric parameters with the Superpave criteria (V_a, VMA, C_{init} and C_{max}) and select the design grading curve which satisfies these.

Where the VMA criteria, and consequently those for C_{init} and C_{design} are not satisfied, one can proceed either by changing the proportions of the grading fractions (increasing the VMA as the grading curve moves away from the maximum density curve $n = 0.45$ power), or by varying the characteristics of the form and texture of the aggregates.

Once the design grading composition has been determined, one then proceeds to calculate the design bitumen percentage corresponding to it, defined as that percentage of bitumen which yields a value of $V_a = 4\%$ in correspondence to $N = N_{design}$. Operatively, one thus proceeds to the compaction of the specimens having the following percentages of binder with respect to the P_b value assessed by the shifted densification curve: $P_b - 0.5\%$, P_b, $P_b + 0.5\%$, $P_b + 1\%$.

The non-compacted mix must also be subjected to short term ageing in accordance with the SHRP M 007 procedure.

Two specimens for each of the 4 bitumen percentages at the value of Nmax are compacted, and the V_a, VMA and VFA values for this value are calculated.

For each bitumen percentage it is necessary to assess the densification curve and measure the correct densities of the specimens, C_{init}, C_{design} and C_{max} corresponding to N_{init}, N_{design} and N_{max}.

The V_a, VMA, VFA and C_{design} values are reported on 4 graphs in function of the bitumen percentage. By graphic interpolation of the results, one can determine the value of the design bitumen percentage P_b corresponding to $Va = 4\%$, and verify that the mix conforms to the Superpave requirements for the VMA and VFA values corresponding to P_b.

2.10.4 PERFORMANCE TESTING IN SUPERPAVE MIX DESIGN [USA-8-12, 15-21]

In the Superpave mixture design and analysis system, performance tests are used only in situations involving moderate to high traffic. This means that they are required only for Levels 2 and 3 mixture designs.

Performance testing utilises new equipment and procedures to ensure that Suprepave mixtures exhibit acceptable amounts of the distress types that were considered by SHRP researchers: permanent deformation, fatigue cracking, and low temperature cracking. Two performance test devices were developed: the Superpave Shear Test (SST) and the Indirect Tensile Tester (IDT).

The extent of use of performance testing for Levels 2 and 3 mix design are shown in Table 2.5 for a new two layer HMA system, which is the newest considered by Superpave.

If an overlay is being designed, Superpave does not attempt to predict fatigue cracking or low temperature cracking. Only permanent deformation is considered. Consequently, the extent of use of performance testing for asphalt mixtures used for overlays is shown in Table 2.6.

Table 2.5 Performance Tests, Levels 2 and 3 (New Construction)

Design	Performance Distress Mode		
Level	Permanent Deformation[1]	Fatigue Cracking	Low Temperature Cracking
2	Simple shear test at constant height at T_{eff}(PD). Frequency sweep at T_{eff}(PD).	Simple shear test at constant height T_{eff}(FC). Frequency sweep test at constant height T_{eff}(FC). Indirect tensile strength at T_{eff}(FC).	Indirect tensile creep compliance at -20, -10, 0°C. Indirect tensile strength at -10°C. Binder creep stiffness (s) and creep rate (m) from bending beam test.
3	Frequency sweep test at constant height at 4°, 20°, 40°C.		Indirect tensile creep at -20, -10, 0°C
	Uniaxial strain test at 4, 20, 40°C. Volumetric test at 4, 20, 40°C. Simple shear test at constant height 4, 20, 40°C.	Indirect tensile strength (50 mm/min) at -10, 4, 20°C.	Indirect tensile strength (12.5 mm/min) at -20, -10, 0°C.

[1] To check for tertiary flow, Level 2 and 3 require repeated shear test at constant stress ratio T_o

Table 2.6 Performance Tests, Levels 2 and 3 (Overlay Construction)

Design Levely	Permanent Deformation[1]
2	Simple shear test at constant height at T_{eff}(PD)
	Frequency sweep test at constant height at T_{eff}(PD)
3	Frequency sweep test at constant height at 4, 20, 40°C
	Uniaxial strain test at 4, 20, 40°C
	Volumetric test at 4, 20, 40°C
	Simple shear test at constant height 4, 20, 40°C

[1] To check for tertiary flow, Level 2 and 3 require repeated shear test at constant stress ratio T_o

2.10.4.1 Shear Test Device (STD)

This device is used to assess the capacity of a bituminous mixture to resist permanent deformations and fatigue cracking. The STD permits the simultaneous application of both vertical and horizontal loads to the specimens, thus simulating both the compression and the shear forces which are applied to pavements by wheel loads.

The STD test apparatus consists of a servo-hydraulic loading system with two perpendicular load applicators, applying respectively an axial force and a shear force to the specimens, a confinement pressure and temperature control system (environmental chamber), a computer for setting and controlling the test operations and the data acquisition, and finally of several LVDT type transducers which when, attached to the specimens to measure their response to the test loads applied, provide feedback on the entire testing system.

All the various tests which can be performed with the STD, besides determining the properties of the material, also yield data on pavement performance, thanks to the prediction models included in the Superpave method.

2.10.4.2 Indirect Tensile Test Device (IDT)

This test is used to predict the deformability and strength of a bituminous mixture to fatigue cracking induced by repeated of loads applied both by traffic and by climate. Two types of test are conducted with the IDT at low and at average temperatures:

1. the *Indirect Tensile Creep Test*, in which the specimen (150 mm in diameter), after a diametral preloading up to uniform deformation, is then stressed, at constant temperature, with a constant load for 100s, recording the vertical and horizontal deformations by means of LVDT transducers;
2. the *Indirect Tensile Strength Test*, in which the specimen (150mm in diameter) is stressed to rupture under a compression load, with a constant gradient of vertical deformation, along the diametral axis.

2.10.5 SUPERPAVE MIX DESIGN LEVEL 2 (INTERMEDIATE TRAFFIC) - PERFORMANCE BASED [USA-8-12, 15,16,18-21]

Level 2 of Superpave mix design is applied for intermediate traffic ($10^6 \leq$ ESALs $_{80Kn} \leq 10^7$), when level 1 mix design, in which the mechanical properties are not measured, is not sufficiently reliable. This level of mix design is best suited for highway type applications, even routine ones, because it permits mixes with the required performance characteristics but without excessive mix design time and costs to be obtained.

Level 2 also includes the volumetric mix design procedure used in level 1, but the fact must be kept in mind that the design traffic level lies between 10^6 and 10^7 ESALs$_{80kN}$. From the pattern of the graphs for V_a, VMA and VFA, calculated for N = N$_{design}$, in function of the percentage of bitumen, one can determine the values of the bitumen contents in correspondence to $V_a = 3\%$, $V_a = 4\%$ and $V_a = 6\%$ which are indicated as the high, design, and low values, respectively.

Using level 2 mix design, one can predict pavement behaviour with relation to permanent deformations and fatigue cracking, summarising the entire "history" of the pavement temperatures in a single effective temperature at which the properties of the constituent materials are measured and at which the said performance is predicted.

The effective temperatures of the tests on fatigue (T$_{eff}$(FC) and on the permanent deformations (including tertiary creep) [T$_{eff}$(PD)] are different. The Superpave software calculates both the effective temperatures which are used as temperatures for the performance-based tests.

2.10.5.1 Calculation of Optimum Bitumen Content

To identify the optimum bitumen % it is necessary to make up and compact 3 series of 4 specimens (each at $V_a = 7$ and at high, design and low bitumen percentages) to be subjected to testing to determine the fatigue strength associated with the loads, the permanent deformations and the indirect tensile stresses.

Each pair of specimens from each of the 3 series are subjected to frequency sweep tests at constant height, $T_{eff}(FC)$ and $T_{eff}(PD)$ temperatures and at frequencies of 10, 5, 2, 1, 0.1, 0.05, 0.02 Hz , as well as simple shear test at constant height, at $T_{eff}(PD)$ and $T_{eff}(FC)$ temperatures, to determine the values of axial deformation, tangential deformation, axial loading and tangential loading during the load cycles.

The remaining pair of specimens in each series are subjected to indirect tensile strength tests at $T_{eff}(FC)$ temperature and a load gradient of 50mm/min.

The data regarding these tests are transferred to the Superpave software, which predicts the permanent deformations for a standard load applied to the pavement, using the pavement temperature and the stiffness value of the mix obtained from the results of the frequency sweep test. The permanent deformations of the layers are calculated using the results of the simple shear test, and consequently calculating the total permanent deformation (expressed as rut depth in mm) in function of the time. The fatigue cracking is assessed considering an effective temperature for each season. The stresses within each pavement layer are calculated for the stiffness of the layer at the representative temperature, under a standard load.

The dissipated energy is calculated for each season, and the cumulative dissipated energy is used to estimate the percentage of the fatigue-cracked area during the design life of the pavement.

Following this, it is necessary to prepare and compact another three series of 3 specimens (each at $V_a = 7\%$ and at high, design and low bitumen percentage), and subject these to the indirect tensile creep test at temperatures of -20°C, -10°C and 0°C and to the indirect tensile strength test at T = -10°C, determining the load and the vertical and horizontal deformations.

The Superpave software is used to process this data so as to predict the low temperature cracking on the basis of the low temperature creep and the low temperature cracking stresses. To predict the course of the cracking on the basis of the tensile strength data, recourse is had to fracture mechanics, which permit tensile stresses to be calculated and to estimate the development of crack spacing over time.

To determine the optimum bitumen percentage, use is made of the values thus obtained for permanent deformation (rut depth), fatigue cracking (percentage area of the pavement) and low temperature cracking (crack spacing) in function of the high, design and low bitumen percentages used in preparing the specimens subjected to the tests. These values are reported on 3 graphs (Fig. 2.11), together with the performance values suggested for level 2 of the Superpave mix design for the 3 distress factors. These latter can thus be modified or completely substituted by the state highway agencies.

The bitumen percentages to the left of the vertical line traced by the intersection between the curve of the test values and the straight horizontal line starting from the suggested values satisfy the requirements with regard to permanent deformations.

On the other hand, the bitumen percentages to the right of the aforesaid vertical line, satisfy the low temperature fatigue requirements.

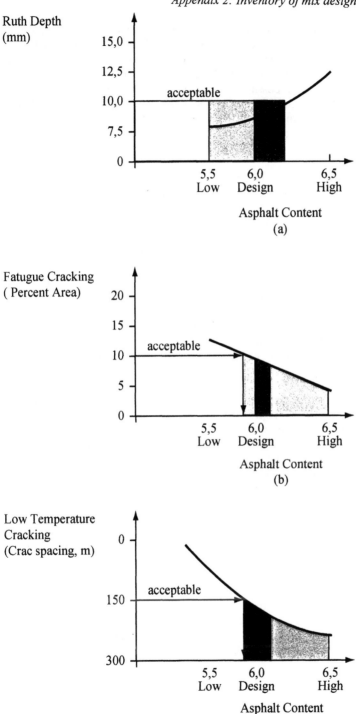

Fig. 2.11 Selection of Optimum Asphalt Content Range Satisfying Three Distress Factor Conditions.

Since the area resulting from the preceding provides the bitumen percentage range which satisfies all three requirements related to the design levels accepted by the agencies, it is necessary to select a bitumen percentage falling within this range as the optimum bitumen percentage which will be employed in the production of the mix. On the other hand, if it is not possible to determine a bitumen percentage range which satisfies all three requirements, the above-mentioned level 2 test method will have to be repeated, after having modified the proportions of the mix constituents or after having introduced a modified bitumen.

2.10.6 SUPERPAVE MIX DESIGN LEVEL 3 (HIGH TRAFFIC) - PERFORMANCE BASED [USA-8-12, 15,16,18-21]

Level 3 of the Superpave mix design differs from level 2, in so far as it characterises an bituminous mixture in terms of a more complete series of performance-based properties, a circumstance which also increases the number of specimens which have to be made up.

In level 3 not only is particular study devoted to the non-linear elastic field by means of the uniaxial strain test and the volumetric test, but the level 2 tests, such as the simple shear at constant specimen height, the repeated shear at constant stress ratio, the frequency sweep at constant specimen height, the indirect tensile creep, and the indirect tensile strength, are conducted at various temperatures.

Besides this, the data on weather conditions, such as the pavement temperatures, and on the moduli of the lower pavement layers are assessed for each of the four seasons of the year.

Also in level 3 the design study starts with the volumetric analysis of level 1, but considers a design traffic of N_{design}, used in the gyratory compactor, > of 10^{E-7}, when is selected in any case by the state administrations. The N_{init} and N_{max} values are tabulated. In function of N_{design}, the V_a, VMA and VFA are calculated and reported in function of the bitumen content. On the basis of these latter, the three bitumen percentages are then determined, defined as high, design and low, in relation to V_a = 3%, 4% and 6%. The mixes which might become unstable and demonstrate tertiary creep at the T_c temperature are discarded.

A comparison of Level 2 and 3 Superpave mix design is shown in Table 2.6.

When the mix design must be highly reliable and has to support high levels of traffic or be produced with non-traditional type raw materials with maximum aggregate size exceeding 50 mm (2 in.), recourse is had to "proof testing", independent of the Superpave mix design.

The Superpave mix design system, with the addition of "field control" to its procedures, exceeds the capabilities of existing mix design methods, in which quality control is limited exclusively to the proportions of the mix. To check in plant or *in situ* the mix design performance using the Superpave procedures, four general levels of field control are provided, which though are not currently being used in the USA at the moment.

Table 2.7   Comparison of Level 2 and Level 3 Superpave Mix Design

	Permanent Deformation/ Fatigue Cracking	**Low-Temperature Cracking**
Test Types	Level 3 considers more states of stress, and requires two additional test methods	No difference between level 2 and level 3
Test Temperatures	Level 3 considers range of temperatures from 4 to 40 °C Level 2 uses one effective temperature for fatigue cracking and one for permanent deformation	Level 3 considers three temperatures Level 2 considers tensile strength at one temperature only
Performance Prediction	Level 3 breaks the year into seasons Level 2 considers the entire year as a single season	No difference between level 2 and level 3

2.10.7 SUPERPAVE: CURRENT DEVELOPMENTS

In the last two years mistakes have been identified in the distress prediction models that were developed during the SHRP project, and consequently significant changes are expected in not only the models but the data aquisition for the mix analysis testing. For these reasons, mix design has been separated as a stand alone process in Superpave. The "mix design system", attemps to develop a bitumen-aggregate mixture with the appropriate volumetric properties. In the "mix analysis system" a designed mix is then evaluated and expected performance determined. Consequently rather a Superpave Mix Design is performed for a potential project, at any level of traffic and the terms Superpave level 1, 2 or 3, are no longer used, Superpave mix design now has only one level. It seems that in the USA optimum bitumen content, for now and the near future will be selected, at all traffic levels, using mix design, Ndesign at 4% air voids.

   If further analysis is required an enginerr has the option to perform an additional sets of tests designated as "Intermediate Mix Analysis" and "Complete Mix Analysis".

The testing data for this analysis is obtained using the Superpave Shear Tester (SST) and Indirect Tensile Tester (IDT). This data is then input into performance perdiction models.

2.10.8 PERFORMANCE BASED MIX DESIGN - SHRP A-698 [USA-12,14,20-24]

2.10.8.1 Performance Based Mix Design Concepts SHRP A-698

The objective of the Performance Based Mix Design Method proposed by SHRP A-698, is to select the optimum asphalt content in a mix that will simultaneously satisfy the rut resistance and fatigue cracking requirements for a given set of conditions (i.e. traffic and environment).

To evaluate rut and fatigue cracking resistance of the mixes the performance related tests developed during the SHRP A-003A project are used in this analysis: 1) the repetitive simple shear test at constant height (RSST-CH) and 2) the flexural bending beam test. These tests are executed in conditions most critical for the distress mechanism they are trying to evaluate. The four point flexural bending beam fatigue test should be executed at temperatures between 15° to 30°C where the conditions that cause fatigue cracking are generally most severe. The RSST-CH test is execute at the maximum average 7 days mean pavement temperature at 5 cm depth. The effects of aging and moisture on the mix are addressed as they affect performance.

This method does not take into consideration the thermal cracking because it is assumed that this problem is controlled by selecting the proper PG binder grade.

Using this performance-based approach there is no need to set explicit limits on volumetric characteristics. If a mix does not meet a particular value of VMA (voids in mineral aggregate) or VFA (voids filled with asphalt), but still has the desired performance, at the expected in-situ volumetric characteristics, it would be acceptable. The selection of asphalt content is based on the desired level of performance for each mode of distress. For instance, in this framework a change of asphalt type, even within the same binder grade, can have a significant effect on the predicted fatigue life.

In this mix design method the selection of asphalt content is based on the site specific selected test temperature. For cooler climates this permits the engineer to take advantage of the binder stiffness while for warmer and hot climates it might emphasise the need to rely on the aggregate skeleton to provide adequate resistance. This aspect is sharply different from the single test temperature inherent to the Marshall procedure. Considering that the mix should contain as much asphalt as possible to improve fatigue and ravelling behavior the highest asphalt content that would satisfy the rutting criteria should be selected.

As far as rutting is concerned the underlying assumption in this mix design approach is that permanent deformation is primarily a plastic shear flow phenonomen at constant volume, occuring near the pavement surface, and on the hottest days caused by the shear stresses occuring below the edge of the heavy truck tyres.

This phenonomen appears to be best captured by the repetitive symple shear test at constant height (RRST-CH) executed at the heighest seven-day pavement temperature at five cm depth (Tmax 5cm). It seems in fact that most of the permanent deformation due to the shear stresses developing near the edge of the tyres takes place at about 5cm. One of the advantages of this test is that it does not cause any change in volume in the specimen during testing. The RRST-CH test is normally performed on a 15cm diameter by a 5cm heigh cylindrical specimen.

The procedure for the determination of the fatigue resistance, for typical mixtures containing conventional asphalt binders, requires approximately 24-hours and is conducted by testing a minimum of 4 beam specimens in the controlled-strain mode of loading at a frequency of 10 Hz.

The concepts adopted in the mix design methodology proposed for both the performance based rutting evaluation and the fatigue evaluation follow similar approaches that can be summarized into four major steps:

1. Design Requirement definition.
2. Conversion of those design requirements into critical strain levels in the asphalt aggregate mix by means of Transfer Functions.
3. Evaluation in laboratory of the performance of the mix at the critical strain levels and environmental conditions (i.e., aging, temperature and moisture) identical to those expected in the field.
4. Conversion of the observed laboratory performance into expected field performance by means of shift factors calibrated based on previous experience.

Rutting Evaluation
A summary of the Performance Based Rutting Evaluation Procedure is presented in Fig. 2.12 . Initially in the Design Requirement stage the number of ESALs required to reach a given rut depth is selected. Through the Transfer Function the permanent shear strain is determined for the rut depth selected.

A correlation was found between the number of cycles in the RSST-CH to reach a given permanent shear strain and the number of ESALs that cause the equivalent rut depth. It was observed that the variation of the maximum shear strain in the pavement and the rut depth is linear. This relationship is given by the equation:

Rut depth (mm) = 279 x maximum permenant shear strength

RSST-CH tests are executed in the laboratory at high temperatures representative of those encountered in the top 50 mm (2 in.) of the pavement during the warmest days of the year. Tests should also be executed at the selected aging and moisture condition protocols as they affect performance. As a mixture can densify under traffic it should be designed so that when it reaches the air void content at which it exhibits its best performance it will be able to take the design traffic. This emphasises the need to compact the mixtures in the laboratoy to certain air void content levels with a compaction procedure that creates aggregate structures similar to those obtained in the field after construction and after traffic densification and not to arbitrary compaction energies or "standard compactions".

Field data have shown that dense-graded mixes tend to become unstable when the void content falls below 3%. It is therefore considered reasonable to execute the simple shear test in specimens with void contents slightly heigher than 3%.

The number of ESALs that can be carried by the mix before the desired rut depth is reached is determined using the relationship between ESALs in the field and RSST-

CH No. of cycles obtained in the laboratory: log (Cycles) = -4.36 + 1.24 log (ESAL). Under those conditions the results of the test can be converted into Equivalent Single Axle Loads (ESALs) to a required rut depth using the appropriate shift factors.

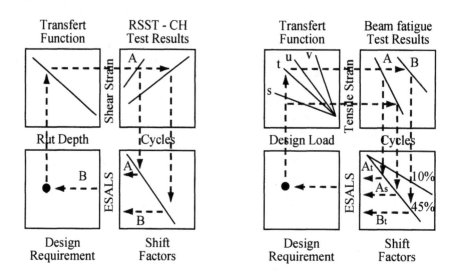

Fig. 2.12 Summary of the procedure to evaluate rutting and fatigue propensity of the mixes in a given pavement condition and environment.

2.10.9 REFERENCES

(USA-1) The Asphalt Institute: "Principles of Construction of Hot - Mix and Asphalt Pavements", Manual Series No.22 (ms-22), January 1983;

(USA-2) The Asphalt Institute: "Mix Design Methods for Asphalt Concrete", Manual Series No.2 (MS-2), May 1984;

(USA-3) Lees G.: "Asphalt Mix Design for Optimum Structural and Tyre Interaction", Proceedings of the 6th International Conference on Structural Design of Asphalt Pavements, Ann Arbor Michigan, 1987.

(USA-4) Edwards J.M.: "Design of Bituminous Mixes", Nottingham BITMADE course notes, 1988;

(USA-5) Huber G.A: "Marshall Mix Design Criteria Changes", Asphalt, The Magazine of the Asphalt Institute, Winter 1991/92;

(USA-6) Whiteoak D. : "The Shell Bitumen Handbook", Shell Bitumen U.K., September 1991;

(USA-7) O'Flaherty C.A. : "Highway Engineering", Vol. 2, Edward Arnold, 1993;

(USA-8) Huber G.A.: "Superpave: The Software System", Proceedings of the Conference Strategic Highway Research Program and Traffic Safety on Two Continents, The Hague, The Netherlands, September 1993;

(USA-9) Kennedy T.W. , Moulthrop J.S., Huber G.A.: "Better Performing Asphalt - the Superpave Steps", Proceedings of the Conference Strategic Highway Research Program and Traffic Safety on Two Continents, The Hague, The Netherlands, September 1993;

(USA-10) Monismith C.L., Deacon J.A., Sousa J.B. : "SHRP Deformation Test - the Key Component", Proceedings of the Conference Strategic Highway Research Program and Traffic Safety on Two Continents, The Hague, The Netherlands, September 1993;

(USA-11) R.J. COMINSKY, E.T. HARRIGAN: "Early Field Experience with Superpave": Proceedings of the Conference Strategic Highway Research Program and Traffic Safety on Two Continents, The Hague, The Netherlands, September 1993;

(USA-12) Sousa J.B., Tayebali A., Harvey J., Weissman S.L., Monismith C.L. : "New Developments in Fatigue and Permanent Deformation of Asphalt-Aggregate Mixes from the SHRP A-003A Team", Proceedings of the Conference Strategic Highway Research Program and Traffic Safety on Two Continents, The Hague, The Netherlands, September 1993;

(USA-13) AA.VV.:"Bituminous pavements: materials, design and evaluation", Lecture notes of a residential course of the University of Nottingham, Department of Civil Engineering, 11-15 April 1994;

(USA-14) Sousa J.B. : "Evaluation of the Effect of Modifiers on Mix Permanent Deformation", Proceedings of the Conference Road Safety in Europe and Strategic Highway Research Program (SHRP), Lille, France, September 26-28, 1994;

(USA-15) Harrigan E.T., Youtcheff J.S.: "SHRP - A - 379: The Superpave Mix Design System Manual of Specifications, Test Methods and Practices", National Research Council, Washington, DC 1994;

(USA-16) Cominsky R.J.: " SHRP - A - 407: The Superpave Mix Design Manual for New Constuction and Overlays", National Research Council, Washington, DC 1994;

(USA-17) Cominsky R.J.: "SHRP - A - 408: Level One Mix Design: Materials Selection, Compaction and Conditioning",National Research Council, Washington, DC 1994;

(USA-18) Kennedy T.W.: "SHRP - A - 410: Superior Performing Asphalt Pavements (Superpave): The Product of the SHRP Asphalt Research Program", National Research Council, Washington, DC 1994;

(USA-19) Austin Research Engineers, Inc.: "SHRP - A - 417: Accelerated Performance - Related Tests for Asphalt - Aggregate Mixes and Their Use in Mix Design and Analysis Systems", National Research Council, Washington, DC 1994;

(USA-20) Sousa J. , et al. "Permanent Deformation Response of Asphalt-Aggregate Mixes". Report no. SHRP-A-414. Strategic Highway Research Program, National Research Council, Washington, D.C., 1994.

(USA-21) Deacon J. A., Tayebali A., Coplantz J., Finn F. and Monismith C., "Part III - Mixture Design and Analysis," in A. Tayebali et al. Fatigue Response of Asphalt-Aggregate Mixes. Report no. SHRP-A-404. Strategic Highway Research Program, National Research Council, Washington, D.C., 1994.

(USA-22) Sousa J. B, Way G., Harvey J. and Hines, "Comparison of Mix Design Concepts", Transportation Research Board, Record n.1492, Washington, D.C., 1995.

(USA-23) Sousa J., Way G., Bouldin M.G. and Harvey J.T.: "Performance Based Field Quality Control for Asphalt - Aggregate Mixes", Symposium on Quality Management of Hot -Mix Asphalt, Philadelphia, 1996.

(USA-24) Leahy R.B., Hicks R.G., Monismith C.L. and Finn F.N., "Framework for Performance-Based Approach to Mix Design and Analysis " AAPT.

Appendix 3

Elements of linear viscoelasticity

The assumption of a linear viscoelastic behavior implies that the material response to solicitation made of some elementary solicitations is the sum of the responses to each of these elementary solicitations (called also Boltzmann superposition principle [Boltzmann, 1874 and 1876]).

Creep and relaxation functions
- The creep function corresponds to the evolution of the materiel strain with time when a constant stress is applied to it :

  if σ (t) = 0 for t $<$ t_0 and σ (t) = σ_0 for t $\geq$ t_0

  For a linear viscoelastic material, the strain response is then such as :

  for t $<$ t_0, ε (t) = 0
  for t $\geq$ t_0, ε(t) = σ_0 f(t, t_0) with f growing function with time

  f includes the possible strain jump linked to the instantaneous elasticity of the material at time t = t_0.
  f is called creep function of the material.

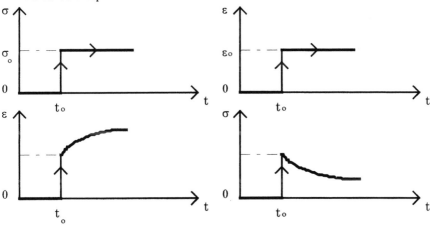

Fig. 3.1 Creep experiment at time t_0 **Fig. 3.2** Relaxation experiment at time t_0

Bituminous Binders and Mixes, edited by L. Francken. RILEM Report 17. Published in 1998 by E & FN Spon, 11 New Fetter Lane, London EC4P 4EE, UK. ISBN 0 419 22870 5

● The relaxation function corresponds to the material stress evolution with time when it is submitted to a constant strain (fig. 3.2) :

if $\varepsilon\,(t) = 0$ for $t < t_o$ and $\varepsilon\,(t) = \varepsilon_0$ for $t \geq t_o$

The stress response is then :

for $t < t_o$, $\sigma\,(t) = 0$

for $t \geq t_o$, $\sigma(t) = \varepsilon_0\,r(t, t_o)$ with r decreasing function with time. r includes the possible stress jump corresponding to the instantaneous elasticity of the material at time $t = t_0$.

r is called relaxation function of the material.

If the material is not ageing (i.e. if its properties don't vary with time, f and r are functions only of $(t - t_0)$.

These functions allow to calculate the strain response (resp. stress) of a material submitted to any time function stress (resp. strain).

Remark : Strain or stress recovery experiments are also used to characterize the viscoelastic behavior of a mix [Eurobitume, 1996].

The Boltzmann principle leads to the following formulas :

$$\varepsilon(t) = \int_0^t F(t-\tau)\,\dot{\sigma}(\tau)\,d\tau = F \otimes \dot{\sigma} \qquad (1)$$

$$\sigma(t) = \int_0^t R(t-\tau)\,\dot{\varepsilon}(\tau)\,d\tau = R \otimes \dot{\varepsilon} \qquad (2)$$

with $\otimes$ symbol of the Riemann convolution.

Use of the Laplace Carson transformation
Applying this transformation to a real function g, leads to the function $\tilde{g}$ defined as :

$$\tilde{g}(p) = \int_0^\infty p\,e^{-pt}\,g(t)\,dt$$

This transformation and the Riemann convolution are such as $\left(f \tilde{\otimes} \dot{g}\right) = \tilde{f}\,\tilde{g}$

Its use allows to obtain, from behavior equations (1) and (2) the following multiplying relations in the transformed space, which are similar to elasticity relations:

$$\tilde{\varepsilon}(p) = \tilde{f}(p)\,\tilde{\sigma}(p)$$
$$\tilde{\sigma}(p) = \tilde{r}(p)\,\tilde{\varepsilon}(p)$$

Sinusoidal solicitation - Relation between complex modulus and relaxation function.

If a sinusoidal stress with the form $\sigma(t) = \sigma_O \sin(\omega t)$, is applied to a linear viscoelastic material, then $\varepsilon(t)$ is sinusoidal in steady state, with a phase lag compared to the stress : $\varepsilon(t) = \varepsilon_O \sin(\omega t - \varphi)$

When noting :

$$\sigma^*(t) = \sigma_0\, e^{i\omega t}$$
$$\varepsilon^*(t) = \varepsilon_0\, e^{i(\omega t - \varphi)}$$

then
$$\sigma^*(t) = r^*(i\omega)\,\varepsilon^*(t)$$

$$r^*(i\omega) = \frac{\sigma^*(t)}{\varepsilon^*(t)} = E^*(\omega) \text{ (complex modulus) (3)}$$

The equation (3) is a classical relation for the linear viscoelasticity : the complex modulus is equal to the Laplace-Carson transformation of the relaxation function at point iω. This formula partly explains the interest of applying sinusoidal loading (which also allows to characterize the mix for low time) to characterize linear viscoelastic behavior.

References

Mandel J. : Cours de mécanique des milieux continus, Tomes 1 et 2. Ed. Gauthier-Villars, Paris 1988.

Salençon J. : Cours de calcul des structures anélastiques, *Viscoélasticité*. Presses de l'Ecole Nationale des Ponts et Chaussées, Paris 1983.

Appendix 4

Linear Viscoelastic Rheological models [Eurobitume, 1996]

Many rheological models made of springs and dashpots are used in order to try to describe behaviour of bituminous mixtures. The two simples models of Maxwell and Kelvin Voigt (Fig. 4.1) which are well known but not able to describe correctly bituminous mixtures behaviour are not described here.

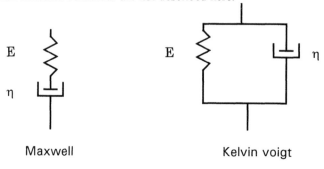

Maxwell   Kelvin voigt

Fig. 4.1 Maxwell and Kelvin-Voigt models.

Burgers Model
This rheological model is made of two Maxwell models in parallel

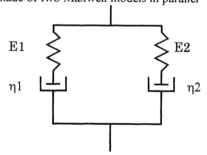

Fig. 4.2 Burgers Model.

The complex modulus associated to this model is given by [Huet, 1966] :

$$E^*(\omega) = \frac{E_1}{1 + \left(i\omega \dfrac{\eta_1}{E_1}\right)^{-1}} + \frac{E_2}{1 + \left(i\omega \dfrac{\eta_2}{E_2}\right)^{-1}}$$

Bituminous Binders and Mixes, edited by L. Francken. RILEM Report 17. Published in 1998 by E & FN Spon, 11 New Fetter Lane, London EC4P 4EE, UK. ISBN 0 419 22870 5

This is the most simple of the generalized Maxwell models (made with several Maxwell models in parallel) [Eurobitume, 1996]. According to [Huhtala, 1995], the burger's model represents globally in a satisfactory way the viscoelastic behaviour of bituminous mixtures . In fact, the fitting with experimental values is far from being satisfactory in the whole frequency range. The divergence in the fitting can be seen on the example presented Fig. 4.3:

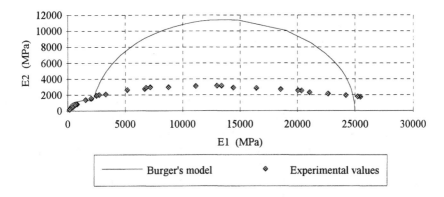

Fig. 4.3 Complex modulus in the Cole and Cole representation. Experimental values and values obtained from Burger's Model for a standard French semi-granular bituminous mixture (5.4 % binder).

Generalized Maxwell or Kelvin-Voigt models [Eurobitume, 1996], can describe in a satisfactory way the behaviour of bituminous mixtures provided that a sufficient number of parameters is used.

Huet's Model [1963]

The Huet's model is a rheological model comprising a spring (stiffness Einf) in series with two dashpots (showing parabolic creep functions with parameters h and k of the form f(t) = At^h and f(t) = Bt^k (4.4)).

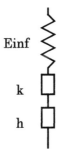

Fig. 4.4 Representation of the Huet's model [1963].

The complex modulus of the system can be expressed from the model's parameters by :

$$E^*(\omega) = \frac{E_{inf}}{1 + \delta(i\omega\tau)^{-k} + (i\omega\tau)^{-h}}$$

with : ω pulsation.

   τ parameter with time dimension, which value varies with temperature.

   h, k δ parameters of the parabolic elements of the models such as $0 < k < h$
   < 1 for   bitumens and bituminous mixtures.

   Einf,   instantaneous modulus of the model, obtained for $\omega\tau \to$ inf, i.e. for high frequencies or low temperatures.

The identification of this model has been made in the Cole and Cole representation diagram of the complex modulus, by analogy with those authors works regarding variation of the dielectric constant of a body according to the frequency.

   The fitting of the four parameters h, k, δ and Einf of the model, for a given bituminous mixture can be done graphically in such a way to obtain the best possible fitting with experimental results. In the Cole and Cole representation, the h and k parameters can be obtained by measuring the angle between the curve and the real axis both at the origin and for $E_1 = $ Einf [Huet, 1963] (Fig. 4.5).

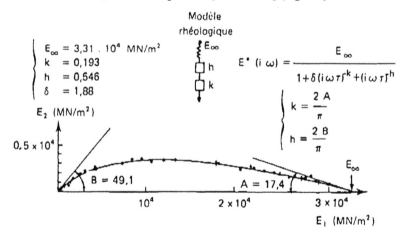

Fig. 4.5 Graphic determination of parameters h and k of Huet's model in the Cole and Cole representation.

The determination, though this model of the complex modulus evolution with temperature is done by determining τ variation with temperature. Huet proposes a law of Arrhénius type.

   If the model allows a very good fitting in the Cole and Cole representation, the accuracy of this representation is bad for low modulus. The representation in the

Black diagram ($|E^*|$ versus φ) allows to better compare experimental values and values obtained from the model for low modulus values (Fig. 4.6).

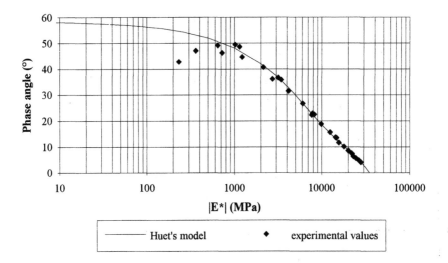

Fig. 4.6 Fitting of the complex modulus according to Huet's model in the Black diagram.

Huet-Sayegh Model [1965]
As the fitting of Huet's model was not satisfactory for low frequency, this model has been adapted by Sayegh [1965] by adding a spring with a very low stiffness compared to Einf.

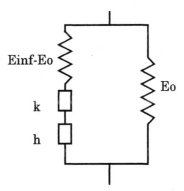

Fig. 4.7 representation of Huet-Sayegh model.

The complex modulus of the system can be expressed from the model's parameters by:

$$E^*(\omega) = Eo + \frac{E_\infty - Eo}{1 + \delta(i\omega\tau)^{-k} + (i\omega\tau)^{-h}}$$

with the same quotations as previously and Eo, the static modulus.

This model allows to represent bituminous mixtures behaviour at low frequencies (Fig. 4.8).

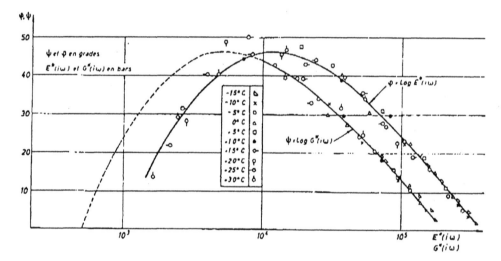

Fig. 4.8 Complex modulus in the Black diagram (after [Sayegh, 1965]).

However, it has to be noted that Huet-Sayegh model, while allowing a better fitting with experimental results requires the fitting of a fifth parameters and above all prevents to get an analytical expression of the creep function in the temporal domain.

For a modelisation in the temporal domain, Huet's model should be preferred, provided that very low modulus are not concerned.

References

Boltzmann L : Zur Theorie der Elastischen Nachwirkung Wien Berichte, 1874 ;
Pogg. Ann., Bd. 7, 1876.

Mandel J : Cours de mécanique des milieux continus, Tomes 1 et 2 Ed. Gauthier et Villars, Paris, 1966.

Salençon J. Cours de calcul des structures anélastiques Viscoélasticité. *Presses de l'Ecole Nationale des Ponts et Chaussées* Paris, 1983.

Symbols and units

a :	Slope of the fatigue law
α_T :	Shift factor
D :	Displacement (m , mm or µm)
d :	Damage
ε :	Strain (m/m)
δ :	Phase angle (of binder)
η :	Viscosity (MPa.s)
φ :	Phase angle (of mix)
µ :	Mass factor
γ :	Form factor
ν :	Poisson's ratio (dimensionles)
σ :	Stress (MPa)
ρ :	Density (t/m3)
S :	Secant modulus, stiffness (MPa)
C1 C2 :	Parameters of the WLF formula
E* :	Complex modulus (in bending compression or tension- compression)
|E*| :	Norm of the complex modulus also called stiffness modulus (MPa)
E1 :	Storage modulus (MPa) (real part of complex modulus E*)
E2 :	Loss modulus (MPa) (imaginary part of complex modulus E*)
E∝ :	Purely elastic modulus (MPa)
R* :	Reduced modulus (dimensionless = E*/ E∝)
F :	Force (N or MN)
f :	Creep function
Fr :	Frequency (cycles/s or Hz)
G* :	Complex shear modulus
|G*| :	Norm of the complex shear modulus (MPa)
G1 :	Real part of complex modulus G* (MPa)
G2 :	Imaginary part of complex modulus G* (MPa)
K :	Intercept of fatigue law at 10^6 cycles
K* :	Bulk modulus
n,N :	Number of load cycles
N_f :	Number of cycles at fatigue failure
R :	Universal gas constant
r :	Relaxation function
T :	Temperature (°C or °K)
t :	Time (s)
τ :	Time constant (relaxation or retardation time)
ω :	Angular frequency (rad/s)
ε_0 :	Initial strain

ε_6 :	Strain leading to fatigue failure after 1 million cycles
σ_6 :	Stress leading to fatigue failure after 1 million cycles
log :	Decimal logarithm (base 10)
ln :	Natural (neperian) logarithm (base e)
ΔH :	Activation energy (kJ/mole)
WLF formula :	William Landel and Ferry formula to calculate shifting factors
Va :	Volume % of aggregates
Vb :	Volume % of binder
Vv :	Volume % of voids (Va + Vb + Vv = 100%)
Vfb :	Voids filled with bitumen
VMA :	Voids in Mineral aggregates
W :	Dissipated energy (J/m^3)
W_f :	Dissipated energy cumulated over fatigue life N_f (J/m^3)

Abbreviations

ALF :	Accelerated loading facility
SHRP:	Strategic Highway Research Program
STD :	Shear test device
MATTA :	Materials testing apparatus
NAT :	Nottingham Asphalt Tester
PCG :	Presse à cisaillement giratoire
GSC :	Gyratory Shear Compactor
ASTO :	Finnish initials for Asphalt Pavement Research Program
STD :	Shear test device
IDT :	Indirect tensile test device
ITFT :	Indirect tensile fatigue test
ITSM :	Indirect tensile stiffness modulus
PAV :	Pressure Ageing Vessel
PMB :	Polymer Modified Bitumen (binder)
RTFOT:	Rolling Thin Film Oven Test
TFOT :	Thin Film Oven Test
ASTM :	American Society for Testing Materials
CEN :	European Committee for Standardization
CROW :	Centre for Research and Contract Standardization in Civil and Traffic Engineering (The Netherlands)
LCPC :	Laboratoire Central des Ponts et Chaussées (France)
NARC :	National Asphalt Research Committee (Australia)
TC :	Technical Committee
WG :	Working Group
VC :	Void Content
ICT :	Intensive Compactor Tester (Finland)
FHWA :	Federal Highway Administration (USA)
PIARC :	Permanent International Association of Road Congresses
ESAL :	Equipment Standard Axle Loads
AAPA :	Australian Asphalt Pavement Association
ESAL :	Equivalent Standard Axle Load

Index

Milton Keynes UK
Ingram Content Group UK Ltd.
UKHW021628071024
449327UK00020BA/1233